AF871266

Nanochitosan Applications for Enhanced Crop Production and Food Security

Scrivener Publishing
100 Cummings Center, Suite 541J
Beverly, MA 01915-6106

Publishers at Scrivener
Martin Scrivener (martin@scrivenerpublishing.com)
Phillip Carmical (pcarmical@scrivenerpublishing.com)

Nanochitosan Applications for Enhanced Crop Production and Food Security

Edited by

Charles Oluwaseun Adetunji

Microbiology Department, Edo State University Uzairue, Edo State, Nigeria

Maulin P. Shah

Industrial Wastewater Research Lab, Enviro Technology Ltd., Ankleshwar, Gujarat, India

Yerima Mohammed Bello

Department of Microbiology, Sokoto State University, Nigeria

Daniel Hefft

Institute of Sustainability and Food Innovation, University Centre Reaseheath, UK

Jay Singh

Department of Chemistry, Institute of Science, Banaras Hindu University, Varanasi, Uttar Pradesh, India

Shyam S. Pandey

Graduate School of Life Science and Systems Engineering, Kyushu Institute of Technology, 2-4 Hibikino, Wakamatsu, Kitakyushu, Japan

and

Ravindra Pratap Singh

Department of Biotechnology, Indira Gandhi National Tribal University, Amarkantak, Madhya Pradesh, India

Scrivener Publishing

WILEY

This edition first published 2025 by John Wiley & Sons, Inc., 111 River Street, Hoboken, NJ 07030, USA and Scrivener Publishing LLC, 100 Cummings Center, Suite 541J, Beverly, MA 01915, USA
© 2025 Scrivener Publishing LLC
For more information about Scrivener publications please visit www.scrivenerpublishing.com.

All rights reserved. No part of this publication may be reproduced, stored in a retrieval system, or transmitted, in any form or by any means, electronic, mechanical, photocopying, recording, or otherwise, except as permitted by law. Advice on how to obtain permission to reuse material from this title is available at http://www.wiley.com/go/permissions.

Wiley Global Headquarters
111 River Street, Hoboken, NJ 07030, USA

For details of our global editorial offices, customer services, and more information about Wiley products visit us at www.wiley.com.

Limit of Liability/Disclaimer of Warranty
While the publisher and authors have used their best efforts in preparing this work, they make no representations or warranties with respect to the accuracy or completeness of the contents of this work and specifically disclaim all warranties, including without limitation any implied warranties of merchantability or fitness for a particular purpose. No warranty may be created or extended by sales representatives, written sales materials, or promotional statements for this work. The fact that an organization, website, or product is referred to in this work as a citation and/or potential source of further information does not mean that the publisher and authors endorse the information or services the organization, website, or product may provide or recommendations it may make. This work is sold with the understanding that the publisher is not engaged in rendering professional services. The advice and strategies contained herein may not be suitable for your situation. You should consult with a specialist where appropriate. Neither the publisher nor authors shall be liable for any loss of profit or any other commercial damages, including but not limited to special, incidental, consequential, or other damages. Further, readers should be aware that websites listed in this work may have changed or disappeared between when this work was written and when it is read.

Library of Congress Cataloging-in-Publication Data

ISBN 978-1-394-21257-6

Front cover image courtesy of Wikimedia Commons
Cover design by Russell Richardson

Set in size of 11pt and Minion Pro by Manila Typesetting Company, Makati, Philippines

Printed in the USA

10 9 8 7 6 5 4 3 2 1

Contents

Preface

Several scientific breakthroughs and innovative advances have been documented over the years, especially in agriculture, to ensure massive food output that could help feed the ever-increasing global population. It has been projected that the population of humankind will rise to 10 billion people by the year 2050. Furthermore, several challenges have been typical impediments to sustainable agriculture, including unstable climatic changes, enhanced greenhouse gas emissions, land degradation, post-harvest losses of agricultural commodities, chronic food insecurity and malnutrition, high levels of soil salinity and alkalinity, massive deforestation, extreme temperatures, water scarcity, soil depletion, environmental pollution with toxic compounds and metals, and the prevalence of pathogens and diseases.

To achieve major transformations in agricultural systems and rural economies, there is a need to search for naturally available biological and sustainable resources that could help resolve the global problems of food insecurity, malnutrition, and securing a healthy future for all people and the planet. To this end, the application of nanomaterials derived from natural sources has been highlighted as a sustainable and innovative tool that could boost crop production, minimize losses, and augment the effectiveness of agricultural inputs, thereby ensuring sustainability.

The use of chitosan in the fabrication of nanochitosan has been identified as a sustainable, innovative, and cutting-edge technology capable of addressing many challenges affecting agricultural productivity. This can be attributed to its high biocompatibility, cost-effectiveness, biodegradability, and versatile applications, including its use as an environmentally friendly tool for enhancing crop production and protecting agricultural crops from pests and pathogens.

This book provides detailed information on the application of nanochitosan for increasing agricultural productivity. It explores its use as a biofertilizer and bioinsecticide, including applications in seed treatment and foliar spraying of agricultural crops, soil amendment, and protection against pathogens and pests. Additionally, it highlights the use of nanochitosan in manufacturing nanosensors for precision farming to monitor crop growth, soil conditions, agrochemical penetration, diseases, and environmental pollution while ensuring plant and soil health.

The modes of action through which nanochitosan performs its various biological activities are also discussed. The book provides state-of-the-art information on recent advances in nanochitosan applications, such as targeted delivery, genetic manipulation, antimicrobial uses (antibacterial, antiviral, antinematode, antiparasite, and antifungal), curing infections in plants, and delivering biologically active constituents. Moreover, it explores the application of nanochitosan in evaluating carbon dioxide concentrations and humidity in controlled greenhouse environments and its use as pressure sensors in agrichemical spraying equipment.

Information on the commercialization, mass production, derivation of various biological materials, socioeconomic perspectives, industrial applications, patents, and the roles of policymakers, institutions, governments, farmers, and agricultural ministries is also emphasized. The specific nutritional composition of plants treated with nanochitosan and its practical field applications are detailed.

This book is aimed at a diverse audience, including global leaders, industrialists, individuals in the food industry, agriculturists, agricultural microbiologists, plant pathologists, botanists, and professionals in agriculture, microbiology, biotechnology, nanotechnology, environmental microbiology, and microbial biotechnology. It also caters to investors, innovators, farmers, policymakers, extension workers, educators, researchers, and those in other interdisciplinary scientific fields. Additionally, it serves as an educational resource and project guide for undergraduate and postgraduate students and educational institutions conducting research in agriculture and nanotechnology.

This book is highly recommended for professionals, scientists, environmentalists, industrialists, researchers, students in higher education, innovators, and policymakers interested in agriculture.

I want to express my deepest appreciation to all the contributors who have dedicated their time and efforts to making this book a success. Furthermore, I sincerely thank my co-editors for their hard work and dedication during this project. Finally, I wish to gratefully acknowledge the suggestions, help, and support of Martin Scrivener and others from Scrivener Publishing.

Professor Charles Oluwaseun Adetunji
(Ph.D, AAS affiliate, FRSB (UK) FNYA; FBSN; FNSM, MNBGN)
Director of Research and Innovation, Edo State University, Uzairue, Nigeria
December 2024

1

The Role of Nanomaterials in Agriculture as Nanofertilizers

Wuna Muhammad Muhammad[1]*, Abdulqadir Bala Ibrahim[2], Job Oloruntoba Samuel[1], Ahmadu Shekwaga Khalifa[1], Oluwafemi Adebayo Oyewole[1], Charles Oluwaseun Adetunji[3,4], Eniola K. I. T.[4], Mohammed Bello Yerima[5] and John Tsado Mathew[6]

[1]Department of Microbiology, Federal University of Technology, Minna, Nigeria
[2]Perishable Crops Research Department, Nigerian Stored Products Research Institute, Kano, Nigeria
[3]Department of Microbiology, Edo State University, Uzairue, Nigeria
[4]Department of Biological Sciences, Joseph Ayo Babalola University, Ikeji Arakeji, Osun State, Nigeria
[5]Department of Microbiology, Sokoto State University, Sokoto, Sokoto State, Nigeria
[6]Department of Chemistry, Ibrahim Badamasi Babangida University, Lapai, Niger State, Nigeria

Abstract

Nanomaterials form the fundamental building blocks of both nanoscience and nanotechnology. The use of nanomaterials has revolutionized agriculture through the innovation of novel techniques and products. Multiple applications of nanomaterials exist in agriculture including its use in precision farming, nanofertilizers, nanopesticides, nanoformulations, and nanosensors to trail products and nutrients levels, to increase the productivity without the contamination of soil and water, to enhance crop productivity, and to provide protection against various biotic and abiotic stresses. The implementation of nanofertilizers has generated considerable concerns pertaining to human safety, food safety, and security. There exist apprehensions regarding the transportation, toxicity, and bioavailability of these chemicals, alongside the possibility of unknown environmental ramifications on biological systems. These problems impede the acceptance and utilization of sustainable agriculture practices. The safety issues associated with nanoparticles

**Corresponding author*: wunamuhd@gmail.com

Charles Oluwaseun Adetunji, Maulin P. Shah, Yerima Mohammed Bello, Daniel Hefft, Jay Singh, Shyam S. Pandey and Ravindra Pratap Singh (eds.) Nanochitosan Applications for Enhanced Crop Production and Food Security, (1–38) © 2025 Scrivener Publishing LLC

in agriculture arise from their reactivity and unpredictability, which might potentially affect the well-being of agricultural workers who come into contact with these xenobiotics during the manufacturing and application procedures in the field. Additional investigation is required in order to ascertain the viability and applicability of incorporating these innovative intelligent fertilizers, taking into account their projected benefits.

***Keywords*:** Nanomaterials, nanofertilizers, nanotechnology, sustainable agriculture

1.1 Introduction

The invention of the atomic force microscope (AFM) in 1986 has significantly contributed to the advancement of nanostructured materials in the field of nanotechnology. These materials hold promise for enhancing production and manufacturing processes across multiple industries [1]. The combination of nanotechnology and biotechnology has greatly expanded the use of nanomaterials in various fields such as mechanics, medicine, and the food industry [2, 3].

Nanotechnology involves the investigation of minute structures and the manipulation of individual atoms, molecules, or compounds to develop materials and technologies with distinctive characteristics [4]. Nanotechnology is an interdisciplinary field that combines various disciplines, including biology, chemical engineering, mechanical engineering, and electronics engineering. Its focus is on understanding, manipulating, and constructing devices and systems with exceptional functionalities and qualities at the atomic, molecular, and supramolecular levels [5]. Nanotechnology involves studying structures, devices, and materials that have at least one dimension between 1 nm and 100 nm. When the size of particles is reduced below this limit, the resulting material exhibits distinct biological and physicochemical properties compared to macroscale materials made from the same substances [5].

The prefix "nano" originates from the Greek term meaning "dwarf" [6]. The term "nano" is commonly used to describe things that are very small or miniature in size. This perception has led to the understanding that nanomaterials are tiny structures that act as a single unit, possessing unique characteristics and performance. These materials typically have sizes ranging from 1 nm to 100 nm, which has sparked a heightened level of interest and involvement. Nanomaterials exhibit notable mesoscopic properties, including a high surface-to-volume ratio, enhanced chemical and biological activity, catalytic behavior, strength, penetrability, enzymatic activation, and quantum characteristics, in comparison to bulk materials [7]. They are commonly used in various sectors due to their distinctive properties that enhance their ability to support diverse biological and biochemical activities [1, 5]. The exploration of nanomaterials in various aspects such as synthesis, categorization, applications, and evaluations has been stimulated [8].

Agriculture plays a crucial role in developing countries, providing support to over 60% of the population [9, 10]. The expanding human population has led to increased nutrient extraction to boost food grain production, reducing arable land, limiting water resources, degrading soil organic matter, and contributing to climate change. Consequently, the adoption of advanced agricultural technologies like nanotechnology has become imperative [11, 12]. The development of an innovative and efficient technology is crucial for improving production and reducing food waste, which is essential for maintaining sustainable and living standards and enhancing food security at a national level. Nanotechnology has the potential to improve the production of high-quality foods and enhance their nutritional bioavailability. Numerous research studies are currently investigating the broadening application of nanotechnology in agricultural production and food processing [13].

Nanotechnology is widely utilized in various aspects of agricultural product production, processing, storage, packaging, transportation, and marketing facilities [11]. The objective of employing nanomaterials in agriculture is to minimize chemical usage and distribution, decrease nutrient losses in fertilization, and enhance crop yield by controlling pests and nutrients [9, 14].

Nanotechnology has gained considerable attention in agriculture due to its potential applications, including the use of nanofertilizers and nanopesticides. These applications offer benefits such as product and nutrient tracking, increased productivity without soil and water contamination, and protection against biotic and abiotic stresses [15, 16]. Nanotechnology has the potential to serve as sensors for monitoring soil quality in agricultural fields, thereby contributing to the maintenance of crop health [17]. Nanotechnology has significantly impacted agriculture and the food industry through the implementation of innovative techniques like precision farming. Precision agriculture has several benefits, including enhanced plant nutrient absorption, improved input utilization, disease detection and control, increased resilience to environmental stresses, and efficient processing, storage, and packaging systems [18–21]. In addition, the application of nanoclays and zeolites enhances fertilizer efficiency and promotes soil fertility restoration by releasing bound nutrients [22]. Abd-Elrahman and Mostafa [11] conducted research on smart seeds that have a nanopolymer coating and are specifically engineered to germinate in favorable conditions.

1.2 Nanotechnology in Agriculture

The need for nanotechnology is emphasized by emerging issues such as the growing demand for safe and healthy food, heightened disease risk, and the potential impact of changing weather patterns on agricultural and fishery productivity [22, 23]. Nanotechnology has the potential to significantly impact the agriculture and food industries through its ability to enhance molecular disease treatment, facilitate rapid disease diagnosis, and improve plant nutrient absorption, among other applications [24]. Smart sensors and delivery systems have the potential to assist the agriculture industry in combating viral infections and other crop diseases. Joshi *et al.* [12] suggest that the use of nanostructured catalysts can enhance the effectiveness of pesticides and herbicides, thereby reducing the required dosage. Nanotechnology can contribute to environmental preservation through the utilization of renewable energy sources, filters, and catalysts, which effectively mitigate pollution and remediate existing contaminants [25].

Nanotechnology has the potential to revolutionize agriculture by aiding in the understanding of biochemical processes in crops. This can be achieved by replacing traditional methods of analyzing environmental challenges and applying nanotechnology to enhance production [13]. Nanotechnology, environmentally friendly technologies, and agricultural biotechnology have the potential to significantly impact various aspects of the agricultural-value chain, leading to synchronized public benefits, legal considerations, moral implications, and environmental effects [13]. The use of nanoscale agrochemicals in agriculture, including nanofertilizers, nanopesticides, nanosensors, and nanoformulations, has significantly transformed conventional agricultural methods, enhancing their sustainability and efficiency (Figure 1.1).

Nanotechnology has various agricultural applications, such as wastewater treatment, soil remediation, and enhancement of agricultural productivity through infection detection sensors [8, 14, 24]. Nanobiosensors are a subset of nanotools used in advanced agricultural practices. They demonstrate the practical and potential applications of nanotools in managing

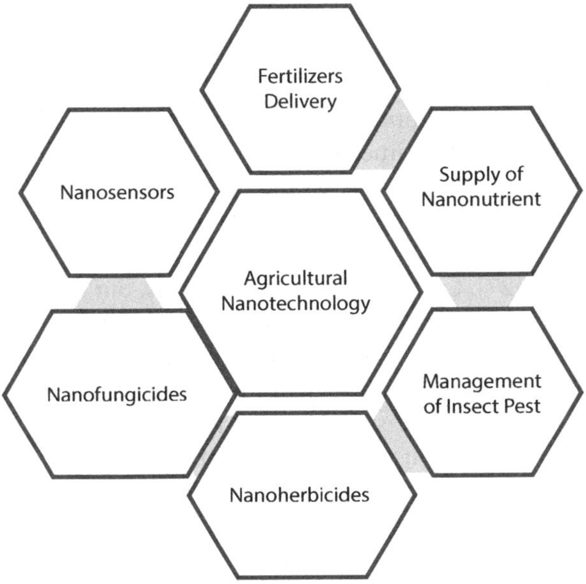

Figure 1.1 Schematic representation of nanotechnology applications of agriculture.

agricultural inputs and enhancing precision in farm management [26]. Nanotechnology in agriculture has several positive impacts. These include the utilization of nanopore carrying zeolite for improved efficacy and slow discharge of enrichers, the use of nanosensors to detect soil quality, and the implementation of efficient supply systems for herbicides [15]. Nanoparticles (NPs) commonly used for monitoring plant diseases include carbon, silica, silver, and alumino-silicates [26]. Nanomaterials are suggested for use in agriculture to replace chemical spraying by providing a consistent source of energetic molecules. Usman *et al.* [16] found that improved water and nutrient management can enhance yields and reduce nutrient waste in fertilizer application.

1.3 Nanomaterials

Nanomaterials play a crucial role in the fields of nanoscience and nanotechnology. Nanostructure science and technology is a rapidly growing interdisciplinary field that has gained significant global attention in recent years. This technology has the potential to revolutionize the production of materials and products, expanding the range and variety of available functionalities. The business influence of nanomaterials is already significant and is expected to further expand in the future [27].

Nanoscale materials are defined as substances with a dimension smaller than 100 nm. A nanometer is equivalent to one millionth of a millimeter, making it 100,000 times smaller than the width of a human hair. Nanomaterials are intriguing due to the occurrence of distinctive optical, magnetic, electrical, and other properties at the nanoscale. These emerging features possess significant potential to exert a substantial influence across diverse industries [27]. Certain NPs occur naturally, but the significance lies in manmade nanomaterials, which are purposefully engineered and widely used in various commercial products and processes [28]. Nanoparticles are commonly found in various everyday products such as sunscreens, cosmetics, sporting goods, stain-resistant clothes, tyres, and electronics [29, 30]. Additionally, they are also used in the field of medicine for purposes such as diagnosis, imaging, and drug delivery [31]. Engineered nanomaterials are molecular-level resources that leverage their small size and unique properties, which are typically absent in their larger forms [32]. The distinct characteristics of nano-sized materials can be attributed to two primary factors: the increased relative surface area and the emergence of novel quantum effects [33]. Nanomaterials exhibit a significantly greater surface area–to–volume ratio compared to conventional materials, resulting in enhanced chemical reactivity and decreased strength [34]. At the nanoscale, quantum effects have a substantial impact on material properties and features, resulting in distinct optical, electrical, and magnetic behaviors [35].

Nanomaterials are characterized by their small size, typically measuring 100 nm or less. They can exist in various dimensions, such as surface films (one dimension), strands or fibers (two dimensions), or particles (three dimensions) [36]. Microscopic particles can exist in various forms, including solitary, fused, aggregated, or agglomerated states, and can exhibit morphologies that are spherical, tubular, or irregular. Nanotubes, dendrimers, quantum dots (QDs), and fullerenes are prevalent nanomaterials. Nanomaterials possess unique physical and chemical properties that differentiate them from conventional chemicals. Examples of nanomaterials include silver NPs, carbon nanotubes, fullerenes, photocatalysts, carbon nanomaterials, and silica NPs [37, 38].

1.3.1 Nanoparticles

Gavanji [39] defines NPs as "clustered atoms" between 1 nm and 100 nm in size, which, due to their small size, may have unique properties compared to the bulk material. Their unique properties, which differ significantly from those of their coarse-grained counterparts, have garnered considerable interest in both their isolated and consolidated forms. Their high surface-to-volume ratio increases their ability to penetrate cell membranes and possibly biochemical activity. Nanoparticles are increasingly being incorporated into a wide range of commonplace items, from medical treatments to the production of food to the storage of energy in solar and oxide fuel batteries [40].

1.3.1.1 Properties of Nanoparticles

The distinct characteristics of NPs arise from differences in their size and scale. The size and shape of a particle determine its surface area–to–volume ratio. In the case of an NP, it is characterized by being very small in at least one dimension. Nanoparticles exhibit various phenomena, including the random motion of small particles, quantum tunneling, the discreteness of energy, the uncertainty of matter, and the dual nature of mass and energy for wave particles [41]. Nanomaterials exhibit altered electrical, optical, surface-related, mechanical, and magnetic properties at the nanoscale. These properties resemble those observed in NPs, as discussed more below.

1.3.1.1.1 Electrical Properties

The confinement of electrons at the nanoscale restricts their movement, leading to alterations in electrical properties. For instance, bulk conductor/semiconductor materials exhibit superconducting or conducting behavior when confined to the nanoscale. Nano gold and nano silver, which have a size smaller than 10 nm, do not possess electrical conductivity [38].

1.3.1.1.2 Optical Properties

The optical characteristics of nanomaterials are influenced by their size. At the nanoscale, electron mobility is restricted. Electrons exhibit distinct responses to light due to their confinement. Gold exhibits a golden color at the macroscale, whereas gold particles at the nanoscale display a red hue. Nanoscale zinc oxide particles do not scatter visible light, whereas larger zinc oxide particles scatter visible light and exhibit a white appearance. The optical appearance of QDs varies with decreasing particle size, leading to a range of colors [37].

1.3.1.1.3 Surface Properties

The surface area of nanomaterials significantly affects their surface-dominated properties, such as melting temperature, reaction rate, capillary action, and adhesion. Due to their high surface area, nanomaterials exhibit distinct properties compared to their bulk counterparts. Gold exhibits a melting point of 1,064°C in bulk form. However, when the particle size is reduced from 100 nm to 10 nm in diameter, the melting temperature decreases to 100°C. The melting point decreases by approximately 50% when the size is reduced to around 2 nm [42].

1.3.1.1.4 Mechanical Properties

Harish *et al.* [31] observed various alterations in the mechanical properties of materials at the nanoscale. These changes include an increase in Young's modulus, tensile strength (fourfold), reduced plastic deformation, heightened hardness, increased brittleness, deformations in grain boundaries, decreased elongation, lower density of dislocation moments, and an augmented short distance of dislocation moments.

1.3.1.1.5 Magnetic Properties

In nanomagnetic materials, every spin exhibits magnetic properties at the NP level [43]. The spin exchange interaction is the dominant interaction between adjacent spins. The majority of materials possess a J value of zero and exhibit nonmagnetic properties, either being paramagnetic or diamagnetic. After removing the external magnetic field, nano superparamagnets, similar to paramagnets, exhibit a return to zero magnetization. The occurrence is attributed to the small size of the particles rather than the inherently poor interaction between individual moments [31].

1.3.1.2 Synthesis of Nanoparticles

Metallic NPs are typically synthesized using either the "bottom-up" or "top-down" approach. Bottom-up fabrication involves constructing a material by adding atoms, molecules, or clusters one by one. In contrast, top-down fabrication involves cutting a bulk material into smaller particles, specifically in the nano-size range. The "bottom-up" technique is commonly preferred for NP synthesis. This method involves a homogeneous system where catalysts, such as reducing agents and enzymes, are used to synthesize nanostructures. The catalysts play a crucial role in controlling the formation of NPs. The "top-down" approach is used to work with materials in their bulk form. To reduce the size of the material to the nanoscale, specialized techniques such as thermal decomposition, mechanical grinding,

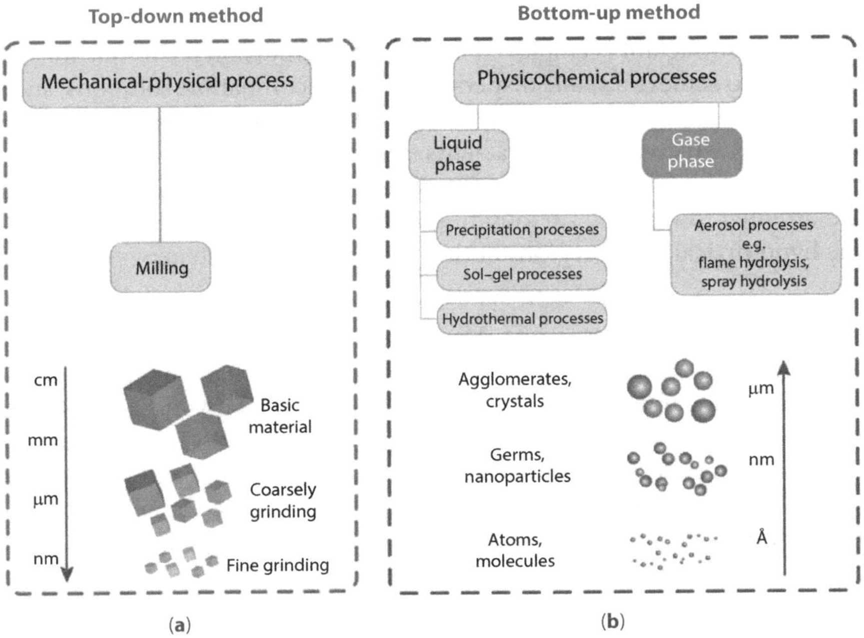

Figure 1.2 Schematic of the top-down and bottom-up approaches for synthesizing nanoparticles. Source: Neme *et al.* [13].

etching, cutting, and sputtering are employed [40]. The specific physical, optical, and chemical properties of NPs play a crucial role in optimizing their application in various contexts [44]. During synthesis, it is crucial to consider various factors such as surface property, size distribution, apparent morphology, particle composition, dissolution rate, and the types of reducing and capping agents used [45].

In spite of the methods of AgNPs synthesis described above (Figure 1.2), these nanoparticles can also be synthesized via the chemical method, physical method, and biological method.

1.3.1.2.1 Chemical Method of Nanoparticle Synthesis

Chemical methods are preferred for synthesizing NPs of different sizes and shapes due to their convenient synthesis in solution [46]. Various chemical methods, such as microemulsion/colloidal procedures, sonochemical techniques, electrochemical techniques, chemical reduction methods, and electrochemical methods, have been utilized for NP synthesis [47].

1.3.1.2.1.1 Microemulsion/Colloidal Technique

A suitable surfactant is employed in the micro-emulsion technique for NP synthesis to achieve a stable dispersion of two incompatible solutions, such as water in supercritical carbon dioxide (W/SC-CO_2) or oil in water or water in oil. Hydrophobic surfactants in nanoscale micelles and oils localize within the core of the aggregate, whereas the hydrophilic head groups are oriented toward the surrounding water solvent [48]. Figure 1.3 shows the hypothetical phase regions of the microemulsion systems.

1.3.1.2.1.2 Sonochemical Method

Sonochemical synthesis involves the application of high-intensity ultrasonic radiation (ranging from 20 kHz to 10 MHz) to induce acoustic cavitation in molecules as shown in Figure 1.4. In solo electrochemical manufacturing, NPs are produced using ultrasonic pulses and electrolytes [49]. The application process is simple, can be conducted at ambient temperature, and enables easy adjustment of NP size through modification of precursor concentrations. Silver NPs are generated using a sonochemical technique in the presence of gelatin as a stabilizer. Silver NPs have potential applications as antimicrobial agents in food packaging and coatings [49].

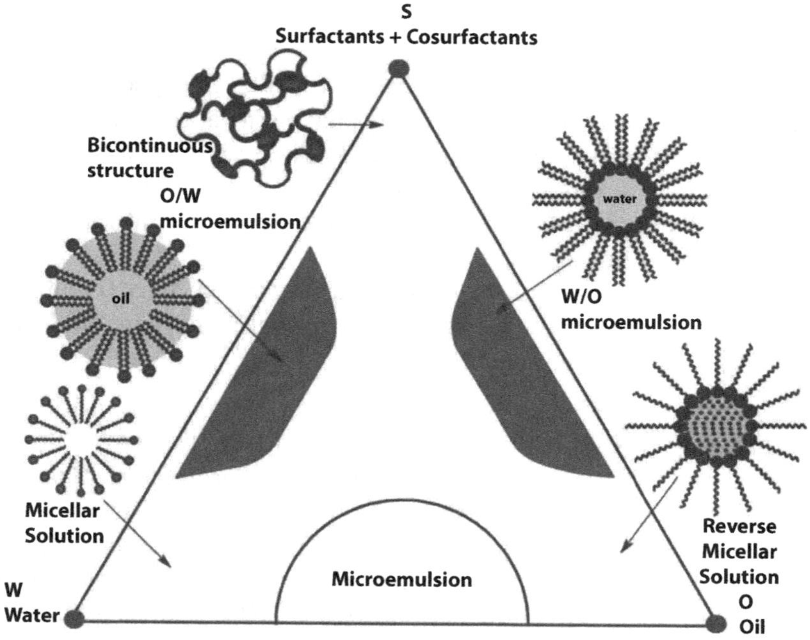

Figure 1.3 Hypothetical phase regions of microemulsion systems. Source: Singh *et al.* [2].

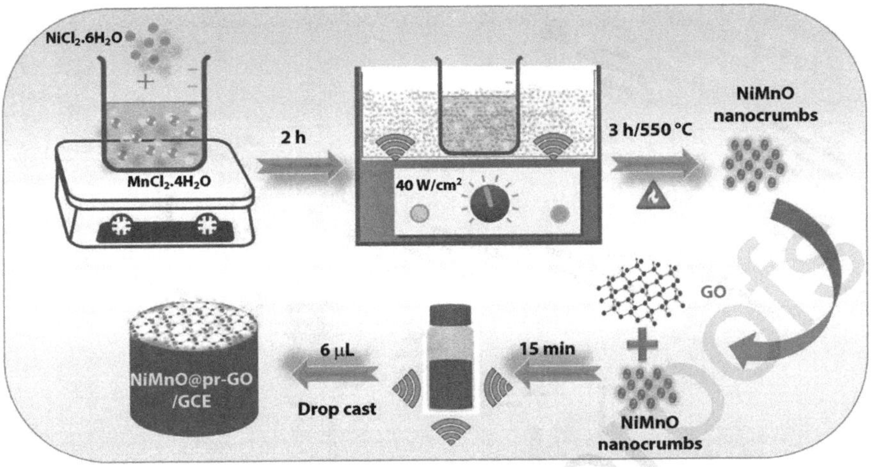

Figure 1.4 Sonochemical synthesis of nanoparticles. Source: Fadhil [1].

1.3.1.2.1.3 Electrochemical Method

In this scenario, an electrolyte is positioned between two electrodes, facilitating the flow of electric current (Figure 1.5). Electricity plays a crucial role in the development of NPs at the electrode-electrolyte interface, acting as a driving or regulating force. This technique offers several advantages, such as the absence of a vacuum system, low costs, ease of operation, high flexibility, reduced contamination leading to pure products, and an environmentally sustainable process. Abbas *et al.* [50] demonstrated that silver NPs smaller than 20 nm can be produced through an electrochemical method. This method involves using a 0.01-mm solution of silver nitrate ($AgNO_3$), a glassy carbon electrode as the working electrode, and silver metal as the counter electrode.

The primary benefit of this method is its capability to produce NPs of superior quality, with the ability to modify their size by manipulating the current density through an electrochemical technique. In addition, the process described by Biswas *et al.* [5] does not necessitate the use of costly equipment or a hoover cleaner. However, the deposition of silver (Ag) on the cathode limits the surface area available for particle formation in the electrochemical production of silver NPs. This constraint is one of the factors that need to be considered. Additionally, the presence of silver electrodeposits throughout the region may hinder the formation of NPs [8].

The synthesis of NPs in solution necessitates the use of metal precursors, reducing agents, and stabilizing agents. Common reducing agents used

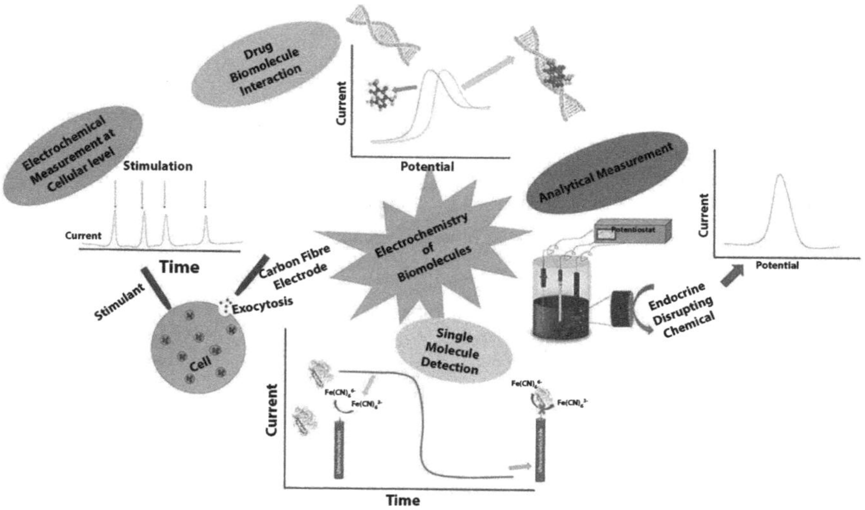

Figure 1.5 Detailed analysis of electrochemical method of nanoparticle synthesis. Source: Singh *et al.* [2].

in aqueous or non-aqueous solutions include sodium citrate, ascorbate, sodium borohydride ($NaBH_4$), elemental hydrogen, polyol process, Tollen's reagent, N,N-dimethylformamide (DMF), and poly (ethylene glycol)–block copolymers. Stabilizing agents such as poly vinyl alcohol (PVA), poly vinylpyrrolidone (PVP), poly ethylene glycol (PEG), poly methacrylic acid, and poly methyl methacrylate have been identified in previous studies [40, 44, 46, 51].

The size and shape of synthesized NPs are significantly affected by the metal precursor, reducing agents, and stabilizing agents [44]. Consequently, it is probable that all nuclei will exhibit similar or comparable sizes, leading to additional expansion. Modulating reaction parameters, including temperature, pH, precursors, reduction agents (such as $NaBH_4$, ethylene glycol, and glucose), and stabilizing agents (such as PVA, PVP, and sodium oleate), enables the manipulation of initial nucleation and subsequent nuclei growth [40, 45].

1.3.1.2.2 Physical Method of Nanoparticle Synthesis

In addition to chemical methods, alternative procedures have been employed for the synthesis of NPs. Physical synthesis methods offer several advantages over chemical approaches, including the absence of solvent contamination in thin films, the homogeneity of NP distribution, and the narrow size distribution of NPs [52]. A notable drawback of these approaches is their increased energy consumption and longer completion times. Physical methods commonly employed for the synthesis of materials include evaporation-condensation, laser ablation, arc-discharge, physical vapor condensation, energy ball milling, electrical irradiation, gamma irradiation, lithography, and direct current magnetron sputtering [3, 48, 52].

1.3.1.2.2.1 Pulse Laser Ablation Method

In addition to conventional chemical processes, new methodologies have been utilized for the creation of NPs. Physical synthesis methods possess various advantages in comparison to chemical alternatives. These advantages encompass the elimination of solvent contamination in thin films, the uniform dispersion of NPs, and the limited range of NP sizes [52]. One significant limitation of these methodologies is their heightened energy consumption and extended durations for completion. Various physical methods are frequently utilized in the synthesis of materials, such as evaporation-condensation, laser ablation (Figure 1.6), arc-discharge, physical vapor condensation, energy ball milling (Figure 1.7), electrical irradiation, gamma irradiation, lithography, and direct current magnetron sputtering [3, 48, 52].

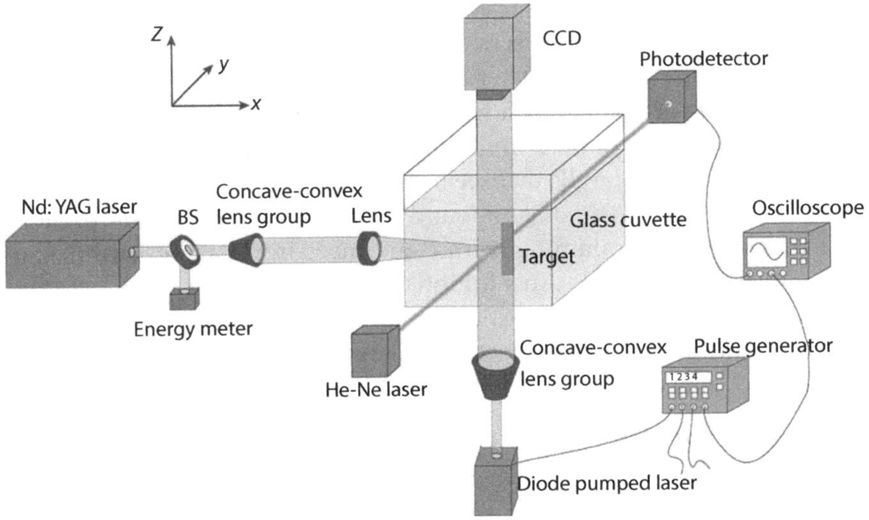

Figure 1.6 Institutionalized procedures for experimentation of pulse laser ablation method. Source: Singh *et al.* [2].

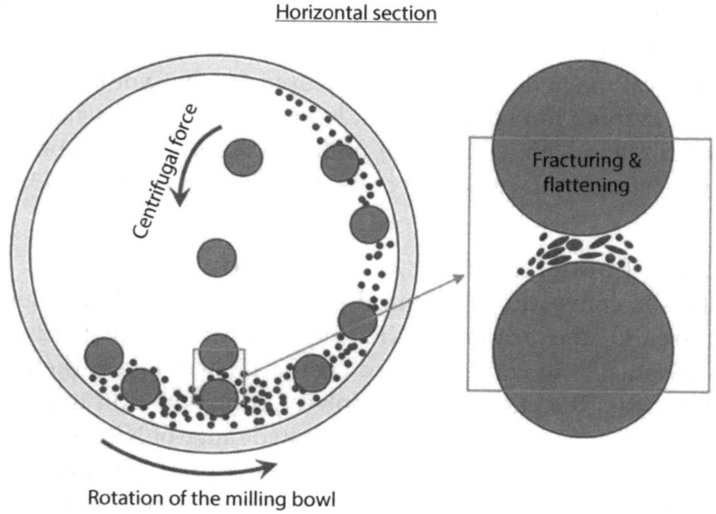

Figure 1.7 Schematic view of high ball-milling approach of nanoparticle synthesis. Source: Singh *et al.* [2].

1.3.1.2.2.2 High Ball-Milling Approach

The technology allows for the fabrication of NPs via a solid-state processing method. Raw materials, such as flakes or powder (size in microns), are placed in containers with tungsten carbide or hardened steel balls, and the containers are spun at high speeds (a few hundreds of revolutions per minute) around their own axis or some central axis [2, 52]. By repeatedly hurling and slamming raw materials against a wall, they are reduced to a powder with a particle size of a few nanometers to a few tens of nanometers. Planetary mills, tumbler mills, vibratory mills, and rod mills are just few of the mechanical mills that see heavy use. By grinding the material in a ball mill, a wide variety of compounds, such as cobalt, chromium, tungsten, nickel-titanium, aluminum-iron, and silver-iron, can be transformed into nanocrystalline forms. In addition to their potential use in active food packaging, some of the nanoparticles mentioned above also have antiviral, antiyeast, antifungal, and antibacterial characteristics [48].

1.3.1.2.2.3 Mechanochemical Approach

Chemical reactions can be initiated through the use of mechanical energy, such as potential or kinetic energy. Typical precursors encompass metallic compounds, oxides, and chlorides. The aforementioned substances exhibit a reaction upon grinding or subjecting them to additional heat, resulting in

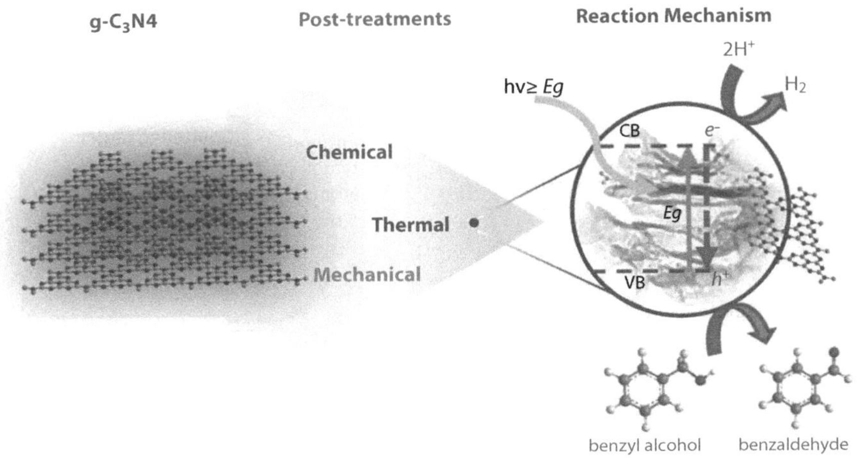

Figure 1.8 Mechanochemical approach of nanoparticle synthesis. Source: Singh *et al.* [2].

the formation of a composite powder. The powder consists of stable, submicron particles that are evenly dispersed inside a salt matrix. The removal of the matrix from the ultrafine particles can be achieved using a selective process subsequent to their washing in a solvent, as stated by Samal [53]. The mechanochemical approach for nanoparticle synthesis is demonstrated in Figure 1.8.

1.3.1.2.2.4 Physical Vapor Deposition (PVD) with Consolidation

Before being blown into a cloud of reactive or inert gas, the source material is vaporized. The scraper is used to clean the cold finger of any accumulated NPs. The powdered NPs are subsequently compressed by the piston-anvil. The entire process takes place in a vacuum chamber to ensure the highest level of purity in the final product (Figure 1.9). Food-safe aluminized film was developed via pulsed vapor deposition for future food packaging [3, 4].

Nanoparticles with uniform size and shape are typically prepared using thermal, alternating current, or arc discharge. The aforementioned methods allow for the simultaneous synthesis of a large number of NPs, reducing production time while preserving NP purity. However, the high cost of the necessary equipment is seen as the primary obstacle to the widespread implementation of such a technique [51, 54].

1.3.1.2.3 Biological (Green) Method of Nanoparticle Synthesis

As an alternative to traditional chemical and physical synthesis methods, the biological synthesis of NPs using plant extracts and/or microbes has

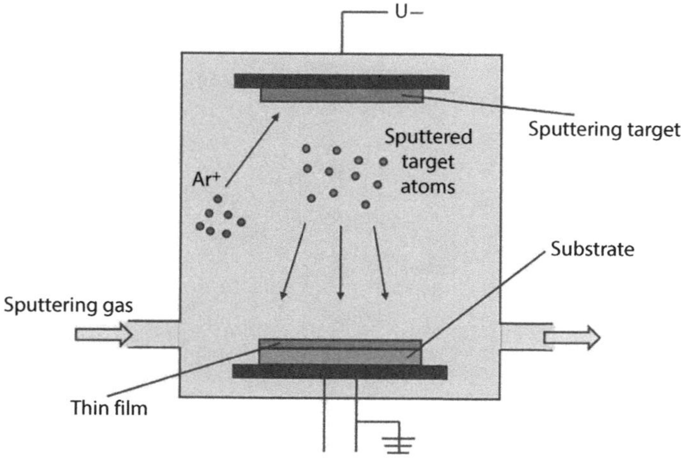

Figure 1.9 Schematic illustration of the physical vapor deposition process. Source: Singh *et al.* [2].

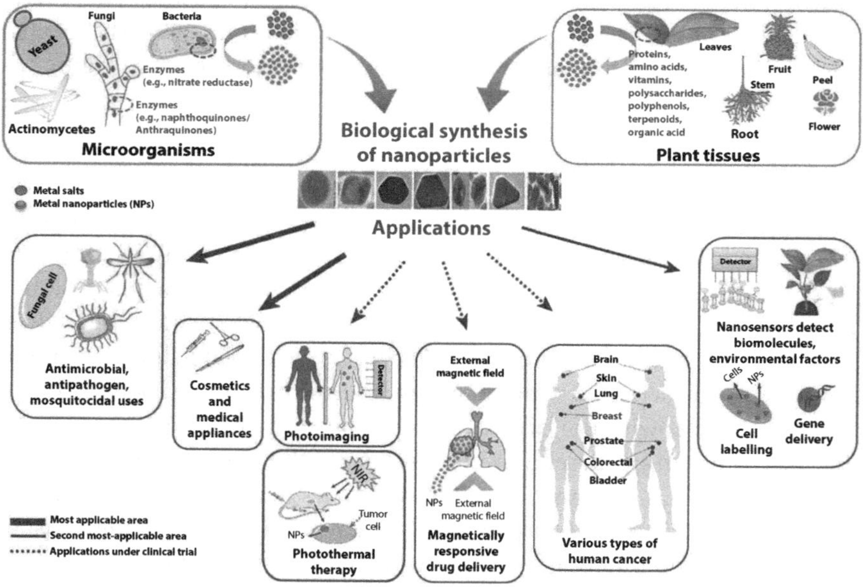

Figure 1.10 Biological synthesis and applications of metal nanoparticles in biomedical and environmental fields. Source: Singh *et al.* [2].

arisen. These methods are also well-known for their simplicity, low cost, environmental friendliness, and scalability [55]. There has been a rise in the utilization of biological agents in the biosynthesis of metal and metal oxide NPs in the field of nanotechnology [40, 44, 46, 55].

In the creation of NPs from metal salts, plants and their parts contain reducing agents in the form of carbohydrates, lipids, proteins, nucleic acids, pigments, and a wide variety of secondary metabolites (Figure 1.10). Enzymes, proteins, and biosurfactants are all examples of microbial macromolecules that perform similar reducing agent roles. Several bacterial species utilize bio-surfactants as capping and/or stabilizing agents [40, 55]. There have also been reports of mixtures of plant extracts. The stem-derived callus of green apples, red apples, egg whites, lemon grass, coffee, and black tea are all used as reductants for $AgNO_3$ [55].

Iravani *et al.* [51], Güzel and Erdal [52], and Ahmad *et al.* [55] utilized various bacterial strains, including *Klebsiella pneumoniae, Bacillus licheniformis, Escherichia coli, Pseudomonas stutzeri, Enterobacter cloacae, Aeromonas* sp., *Lactobacillus* sp., and *Corynebacterium* sp., for NP synthesis. The study also investigated the utilization of various fungal species, such as *Penicillium fellutanum, Aspergillus flavus, Coriolus versicolor, Cladosporium cladosporioides*, and *Penicillium* sp., in the production of NPs. Algae species, including *Tetraselmis gracilis, Chlorella salina, Oscillatoria willei, Spirulina platensis, Isochrysis galbana*, and *Chaetoceros calcitrans*, were employed in the production of NPs. Ahmad *et al.* [55] successfully utilized plant extracts from *Capsicum annuum, Eucalyptus citriodora, Ginko biloba, Diospyros kaki, Platanus orientalis, Euphorbia hirta, Acalypha indica, Moringa oleifera, Pinus desiflora*, and *Magnolia Kobus* for the synthesis of AgNPs.

1.3.2 Application of Nanotechnology in Agriculture

Sustainability, crop enhancement, and greater production are just a few of the agriculture and environmental concerns that nanotechnology in agriculture is being developed to tackle. The potential for agricultural nanotechnology to alleviate world hunger, malnutrition, and infant mortality makes it an appealing investment for developing countries [13]. Changing traditional methods of studying environmental challenges and applying the results to production enhancements, nanotechnology is a promising tool that could revolutionize agricultural sectors [56]. Nanoscale agrochemicals, such as nanofertilizers, nanopesticides, nanosensors, and nanoformulations, have the potential to transform conventional farming methods into ones that are more environmentally friendly and productive [25]. Nanotechnology has many uses in agriculture, including purifying wastewater, restoring degraded soil, protecting crops from pests and diseases, and increasing yields.

1.3.2.1 Application of Nanotechnology in Precision Farming

The long-sought-after goal of precision farming is to tailor input to individual crops' needs, increasing yields while decreasing costs. Computers, Global Positioning System (GPS), GIS, and remote sensing devices are used in precision farming to measure extremely localized environmental conditions, such as whether or not crops are growing at optimal efficiency or precisely identifying the nature and location of crop and environmental problems [12]. Recycling agricultural waste is another way in which precision farming can aid in cleaning up the environment [9]. However, microsensors and monitoring systems provided by nanotechnology will have a major impact on future precision farming methods. One of the key functions of nanotechnology-enabled devices will be the increased use of autonomous sensors connected to a GPS for real-time analysis. Soil conditions and crop development could be tracked using a grid of these nanosensors [13].

1.3.2.2 Nanosensors

The use of nanotechnology to improve soil quality and increase crop yields is another area of active research and development. The use of nanosensors to track crop and animal health and magnetic nanoparticles to remove soil toxins are both possible applications [57]. Developing nations may likewise benefit greatly from the "lab-on-a-chip" concept. Nanoparticles for soil conservation or remediation and nanosensors for plant disease and pesticide detection are two further areas of agriculture that can benefit from nanotechnology. Using enzymes in high-value and low-volume applications is a key component of enzyme immobilization for nanobiosensors utilizing nanomaterials [9].

1.3.2.3 Nanotechnology in Water Management

Surface water, groundwater, and wastewater can all be treated with novel nanomaterials that can remove harmful metal ions, organic and inorganic solutes, and microorganisms. Because NPs have a unique effect against refractory contaminants, they are the subject of extensive study and development for use in water filtration [11]. The public's health requires rapid and precise detection of waterborne pathogens. However, conventional laboratory tests take a long time. Enzymatic, immunological, and genetic diagnostics are being developed as ways to speed up the diagnostic process. Nanofiber membranes and nanobiocides show potential as effective water filtering tools [24]. Biofilms are difficult to cure because the bacteria within them form mats that are covered in natural polymers. They demand a lot of downtime and manual labour for cleaning because they can only be cleaned mechanically. To combat these biofilms, scientists are working on enzyme-based treatments [58].

1.3.2.4 Biosensors to Detect Nutrients and Contaminants

Rapid, sensitive detection of pollutants and pathogens with molecular precision is necessary to safeguard soil health and the environment. The same set of protocols has been used to assess soil fertility for the past 60 years; these may no longer be relevant to modern production systems and precision farming technologies [12]. In-field detection, miniaturized portable devices, and remote sensors for real-time monitoring of large regions all require accurate sensors. Microbiological and immunoassay tests typically take a long time; however, they may be shortened with the help of these equipment. These sensors are utilized to identify contaminants in water, raw ingredients, and finished foods [56].

Enzymes can serve as sensors because to their ability to bind to certain macromolecules with great specificity. An electronic nose (E-nose) employing a pattern of reaction across a set of gas sensors is used to identify various scents. It works similarly to the human nose in identifying odorants, measuring their concentration, and identifying distinctive aspects of the scent [25]. Nanoscale gas sensors, such as ZnO nanowires, are its primary component. When a certain gas passes through them, their resistance shifts, causing an electrical signal shift that serves as a "fingerprint" [25]. Biosensors are high-powered instruments with the potential to identify poisons in food or the environment. They are inexpensive, quick to respond, simple to use, and compact [25], in addition to being very specific and sensitive. Direct enzyme inhibition sensors may be useful as a screening tool to determine when a sample contains one or more contaminants, but they currently lack the analytical capability to distinguish between multiple toxic substances in a sample (such as the simultaneous presence of a heavy metal and a pesticide). Field workers can benefit from these methods because they can be implemented in single-use test strips [25].

1.3.2.5 Nanotechnology to Improve Quality of Soil and Fertilizer Distribution

Farm administration: When it comes to raising crop yields, nanotechnology is indispensable. Examples of nanoscale materials being employed as biosensors in agricultural research include carbon nanotubes, nanofibers, and QDs. The use of NPs has been shown to increase crop quality and production when applied to fertilization practices [9]. For ecological agriculture, the development and use of vermiculite, nanoclay, and zeolite may increase the efficiency of fertilizers and the yield of crops grown on soil with a coarse texture. Soil amendment using inorganic materials decreases NH_4-N passage and increases N fertilizer synthesis in ecological agriculture systems [47].

The usage of fertilizers is crucial to agricultural output. Research shows that increasing soil fertility through the use of fertilizers has a linear effect on harvest yields [16]. Manjunatha *et al.* [9] and Usman *et al.* [16] describe nanofertilizer as a chemical with dimensions on the order of a few nanometers that improves plant nutrition by increasing nutrient delivery and regulating the slow release of nutrients into the soil over time. Increasing crop productivity through the use of nanofertilizers is essential. In order to modify nutrient consumption efficiency and reduce costs for environmental safety, nanotechnology enables the use of nanostructured or NPs for fertilizer transport or limited release routes to generate smooth fertilizer. Nutrient encapsulation within nanoporous structures, coating of thin polymeric films, and delivery as nanoscale particles or suspensions are all methods by which nanofertilizers might increase nutrient efficiency [2, 16, 25].

Nanoscale fertilizers have the potential to improve nutrient delivery because they can access plant surfaces and transport channels. Tomatoes, peppers, and flowers were grown using banana peel nanofertilizer [25]. Many crops have benefited from the use of nano fertilizers, including chickpeas for ZnO NPs, maize for silicon dioxide and iron slag powder, tomatoes for colloidal silica and Nitrogen, Phosphorous and Potassium (NPK), spinach for titanium dioxide, and grapes for gold and sulfur fertilizers [47].

Treatment of fertilizers with nanoscale transporters has the potential to reduce chemical loss and environmental problems by securing the roots of the plant to the surrounding soil contents and organic material. Soil toxicity can be mitigated with nanoscale fertilizers, mitigating the risks of over dosing [48]. These nanofertilizers have a longer fertilizer impact and lessen the rate at which nutrients are released. SiO_2 with TiO_2 NPs increased nitrate action and plant absorption capabilities through controlled use of water and fertilizer, leading to a successful outcome in maize crop development [48].

1.3.2.6 Nanotechnology to Control Plant Diseases

Twenty percent to 40% of annual crop losses are attributed to plant pests and diseases [10]. Insecticides, fungicides, and herbicides are commonly used in modern farming as a means of pest control. It is crucial to create effective, low-cost insecticides that are safer for the environment. By boosting the solubility of weakly water-soluble pesticides, decreasing their toxicity, and extending their shelf life, nanotechnology has the potential to have a positive effect on the environment [4]. Figure 1.11 show the field impact of conventional and nano formulations.

Disease management and crop safety are only two examples of why agricultural nanotechnology is so crucial [10]. Delivery of nutrients and agricultural chemicals to plants can be regulated at a slow and consistent rate with the use of nano-based traditional herbicides and pesticides [15]. Moreover, NPs may play a significant role in the prevention and treatment of human illnesses and the management of insect pests [15, 26]. Polysaccharides such chitosan, alginates, starch, and polyesters have been studied for their potential use in the production of nano-insecticides [14].

There are two main ways in which NPs might be employed to safeguard plants: either directly as crop protectors or as sprayable carriers for

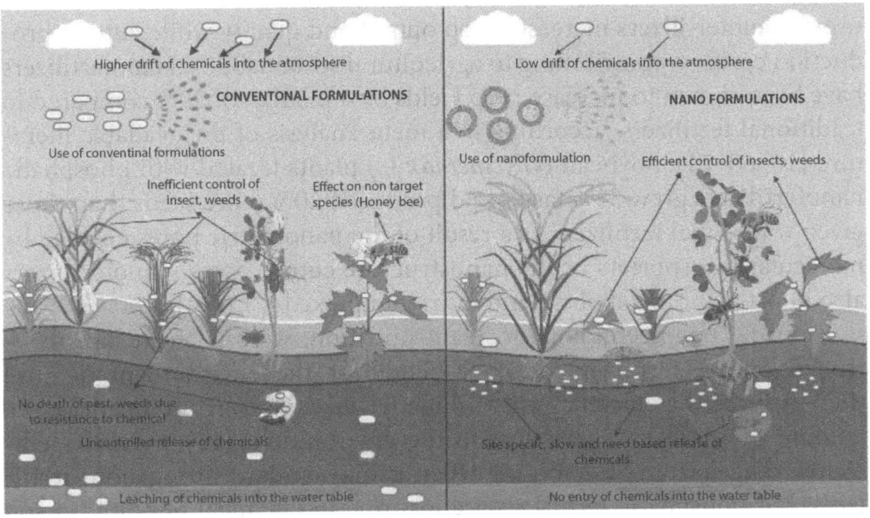

Figure 1.11 Impact of conventional and nanoformulations. Source: Reddy and Chhabra [26].

conventional pesticides. Although NPs have many potential uses, researchers have yet to investigate their potential role in plant protection and food production [26].

1.3.2.7 Nanofertilizers

Over the past 50 years, the world's nutritional needs have been mostly met thanks to the dramatic increase in crop output, especially grain yields. One of the main drivers of improved agricultural productivity is the rising use of chemical fertilizers [56]. Use of artificial fertilizers has increased as farmers have turned to more fertile crop kinds. When compared to organic fertilizers, chemical fertilizers cause more pollution and cost more to produce because so much of the fertilizer is wasted. Fifty percent to 70% of the nitrogen used in conventional fertilizers is lost in the process. Therefore, researchers are focusing on finding new approaches to nutrient management that would not compromise sustainability [59].

Here, nanotechnology is used to create slow-release fertilizers, cut down on ephemeral nutrient losses, and make scarce nutrients more accessible (Figure 1.12). Micronutrients and macronutrients, as well as nanomaterials that serve as carriers/additives for nutrients (by mixing with minerals), are all included in the category of nanofertilizers [12]. Making nanofertilizers by enclosing nutrients in nanomaterials has also been demonstrated [60–62]. Qureshi *et al.* [63], Manjunatha *et al.* [9], Kumar *et al.* [64], Kumar *et al.* [65], Mejias *et al.* [66], and Jakhar *et al.* [67] have all found that the use of nanofertilizers increases crop output and quality while cutting production costs, so contributing to agricultural sustainability. Nanofertilizers have been shown to increase crop yields by a median of 29% compared to traditional fertilizers, according to a meta-analysis of the available literature [16, 68–71]. Soybean (*Glycinemax* L.) plants treated with phosphatic nanofertilizers grew 32% faster and produced 20% more seeds than those given traditional fertilizer. As a result of the nanometric pores enabled by molecular transporters or the nanostructure cuticle pores, nanofertilizers also boost plant metabolism and nutrient uptake [4, 12, 14].

By using nanotechnology to plant nutrition, we can create slow/controlled release fertilizers that are better for the environment because they increase the efficiency with which fertilizer is used and decrease the amount of fertilizer that is lost to the environment [15]. Fertilizer efficiency ranges from 30% to 60% when using standard nitrogenous fertilizers, but phosphatic fertilizers are lost due to chemical bonding in soil and are, therefore, unavailable to plants [13, 15]. Nanocomposites of urea and hydroxyapatite have been shown to give controlled nitrogen release,

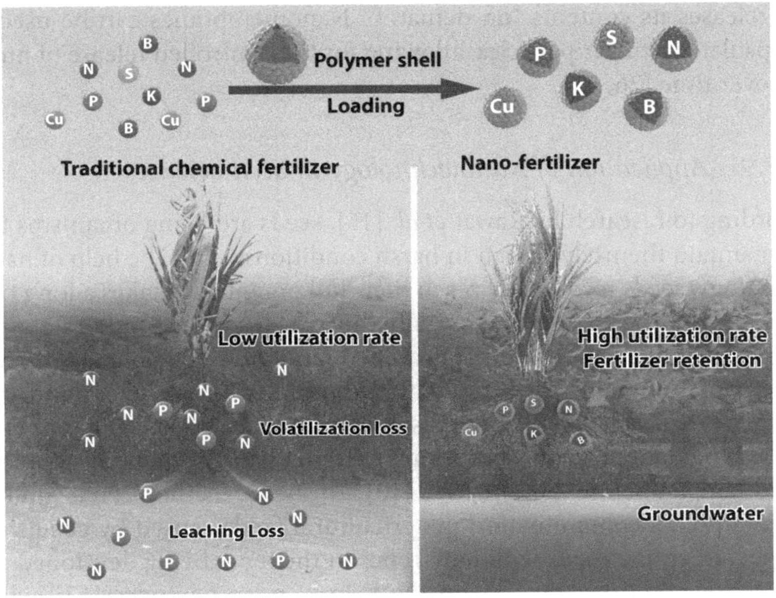

Figure 1.12 Effect of nanomaterials loaded with fertilizers on plant. Source: An *et al.* [56].

decreased NH_3 volatilization, and longer phosphorus availability after only 4 weeks of incubation [15]. Fertilizer consumption can be cut down by using slow-release chemicals. The desired function of nanofertilizers is to release nutrients only when and where plants need them, preventing the wasteful volatilization or leaching away of excess fertilizers. If plant signals control nutrient release, then understanding the signals sent between plant roots and soil microorganisms could lead to the development of smart fertilizers [50].

1.3.2.8 Nanostructured Formulation Reduce Nutrients Loss Into Soil by Leaching

Phosphate fertilizers and other insoluble mineral micronutrients have had their solubility and dispersion vastly improved through nanosizing or nanostructuring processing. Slow nutrient release may be possible with the use of zeolites, a class of naturally occurring minerals having a layered crystal structure resembling a honeycomb. Its complex system of interconnected tunnels and cages may store not just nitrogen and potassium, but also phosphorus, calcium, and a wide variety of other minor and trace nutrients, all of which dissolve slowly. Zeolite is a nutrient storage medium that releases its contents "on demand." Nanomembranes can be used to encapsulate fertilizer particles, allowing for the controlled release of nutrients over time [26, 56].

1.3.2.9 Application of Nanotechnology in Seed Science

According to research by Rawat *et al.* [15], seeds are living organisms that can maintain themselves even in harsh conditions. With the help of nanotechnology, seeds may finally reach their full potential. It takes a long time for plants to produce seeds, especially those that rely on wind pollination. It is possible to ensure genetic purity by identifying pollen loads that will cause contamination. Air temperature, humidity, wind speed, and crop pollen yield all play a role in the dispersal of pollen. Pollen contamination can be avoided if bio-nanosensors are used to identify potentially contaminated samples before they are used [15]. This same method can be utilized to avoid pollen contamination in agricultural fields caused by genetically engineered crops. Seeds with new genes in them are being developed and sold. Seeds could be traced with the help of nano barcodes [13], which are encodable, machine-readable, durable, and a sub-micron-size taggants. Infected seeds can spread disease and often fail to germinate after being kept. Seeds can be preserved and used in considerably lower volumes if they are nano-coated with elemental forms of Zn, Mn, Pt, Au, and Ag. To help differentiate between healthy seeds and those infected with E. coli 0157:H7, QDs have been developed to act as a fluorescent marker in conjunction with immuno-magnetic separation [26].

1.4 Nanofertilizers

Desized nutritional elements can be found in nanoscale fertilizers, which come in powder or liquid form [72]. Although 1–100 nm is considered nanoscale, certain studies have shown an increase in fertilizer action for formulations up to 500 nm in diameter [73]. Improved nutritional quality and other quality aspects, like longer shelf life and dual or multiple roles as pesticides or heavy metal scavenging agents, are among the potential benefits of nanoscale fertilizers [74], which are predicted to reduce the quantity of fertilizer needed and increase use efficiency compared to conventional fertilizers.

One possible conclusion that could significantly benefit agriculture is the use of nanofertilizers. Thanks to their small size and large surface area, nanomaterials can efficiently absorb nutrients and promote growth in crops [75]. There could be significant economic and environmental benefits from incorporating nanotechnology into fertilizer products to enhance release profiles and absorption efficiency [75]. Figure 1.13 shows an overview of nanofertilizer applications in agriculture.

Inefficient formulations of nanofertilizers may be made worse if they dissolve and leach into the environment at a faster rate due to the NPs' increased surface area [76]. The overuse of nanomaterials in any area, meanwhile, would be a purposeful input of nanomaterials into the environment and might have major effects on human and environmental exposure [77].

As conventional fertilizers experience reduced and very low nutrient utilization efficiency, the nanoscale of nutrient components will tend to alter their properties [64, 65]. Improved bioavailability of nanoscale fertilizers leads to their uptake by plant cells immediately following application, resulting in a more favorable growth and nutritional profile with less fertilizer applied [64, 65, 72, 78]. Nanofertilizers will be more effective than traditional fertilizers due to their greater absorption and availability. Nutrient losses can be greatly diminished by switching from conventional to nano-sized N and P manures. Nanoscale micronutrient fertilizers are also anticipated to have significant agronomic benefits [63].

Nanofertilizers also have physicochemical qualities like structure, electrical charge, aggregation, and surface chemistry [79] and can be created

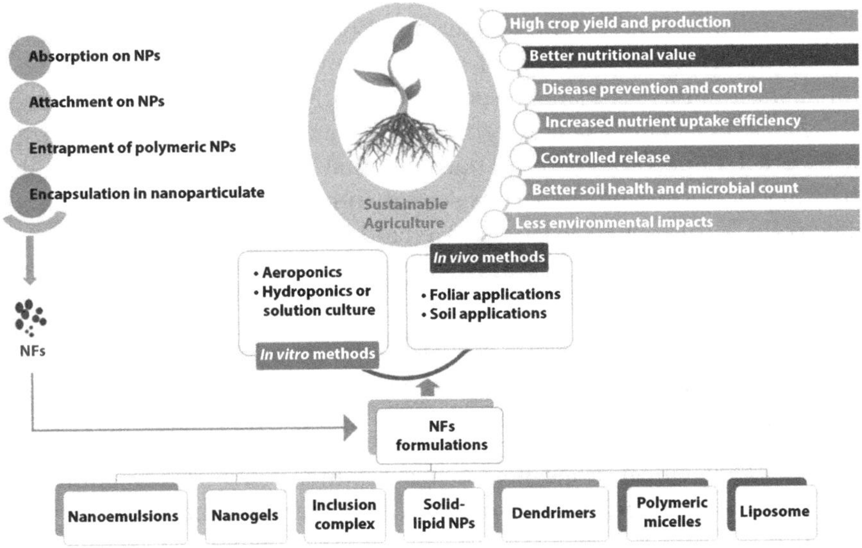

Figure 1.13 Overview of nanofertilizer application in agriculture. Source: Verma *et al.* [81].

from a wide variety of materials such as TiO2, Cu, Si, Al, C, Zn, and N. Current research programmes in nanotechnology-enabled fertilization should be evaluated for future exploration due to the many benefits of nanofertilizers over conventional fertilizers and their potential to help achieve the goals of sustainable agriculture [75]. Protective measures should be employed sparingly until more is known about the toxicity of nanomaterials and their effects on the natural environment, which requires more research and continuous scientific investigations to fill up knowledge gaps. The sustainability of the nanotechnology industry depends on these procedures, which should also improve environmental and human health [80].

1.4.1 Types of Nanofertilizers

Increased crop yields have been linked to the use of nanofertilizers, which include particles so small they may be absorbed by plants [63, 82, 83]. They are the product of cutting-edge technology that may find use in agriculture, but they defy easy categorization. Nanoscale delivery systems and nanobiosensors are two examples of additional products that fall under the purview of some definitions. The scientific community is perplexed by the varying definitions of nanofertilizers. A good example of a hybrid between nanotechnology and fertilizer is nanofertilizers. Because of this ambiguity, it can be difficult to convey the uses and benefits of nanofertilizers without introducing some degree of confusion [84].

Nanofertilizers' active ingredient can also be classified. Nanofertilizers can be created from a variety of materials, including carbon nanotubes, polymers, and metals. Plants respond differently to various nanofertilizers because of their unique characteristics [85]. Nanofertilizers are categorized in Figure 1.14 according to the nutrients that they deliver, the functions that they perform, and their consistency. In order to apply nanofertilizer effectively, it is essential to first understand its nature. Nanofertilizers can be applied to plants in three different ways: foliarly, hydroponically, and topically [83].

1.4.2 Uptake and Accumulation Mechanisms of Nanofertilizers from Soil to Plants

Soil physicochemical properties such as texture, structure, clay minerals, pH, cation exchange capacity, soil organic matter, and microbial population all play a role in how NPs behave in the environment and how they are transported and utilized by living organisms [81]. Because of the surface

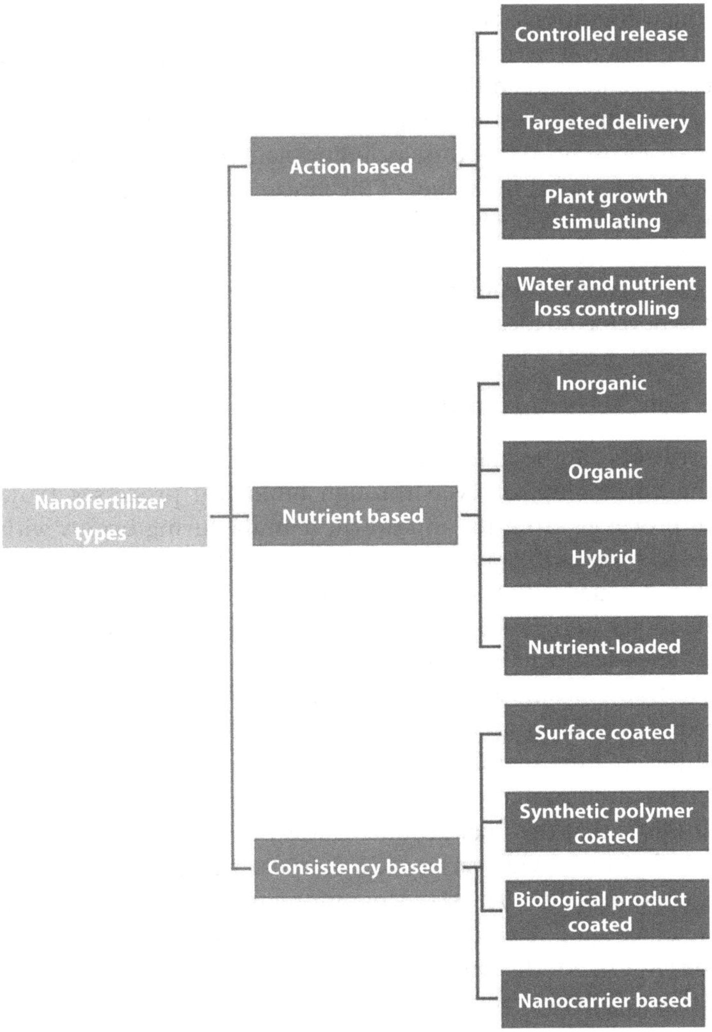

Figure 1.14 Classification and types of nanofertilizers. Source: Yadav *et al.* [83].

charge effect, dissolved organic matter affects the aggregation, mobility, stability, and binding behavior of NPs [86].

Direct application of nanofertilizers to soil or leaves is possible [87]. The foliar entrance comes first, then the root one. Root application gains access through root tips, lateral roots, root hairs, and the rhizodermis, whereas foliar treatment uptake occurs predominantly through the cuticle, stomata, and hydathodes [63, 79]. NF foliar spray is favored in adverse soil and climate conditions. Several empirical models based on the Michaelis–Menten equation describe the movement of water and solutes in soil using the Richards equation and the convection-dispersion equation [81].

Absorption maximizes as the nutrient content rises, although it does so in a nonlinear form. Several factors, including plant species, growing season, and soil temperature, affect the values of the kinetic parameters in the Michaelis–Menten equation. Although the first nutrient transport model for plant tissues was the steady-state source-sink model with flow driven by an osmotically created pressure gradient influenced by diffusive transport factors, more work is needed to develop an ideal uptake model/mechanism for NPs in plants [84].

1.4.3 Synthesis of Nanofertilizers

Nanofertilizers can be made in a variety of ways, from the ground up or from the bottom up, or even through biological processes [85]. Using physical processes from the top down, manufacturing begins with larger particles and works its way down to the nanometric scale.

There are issues with the methods used to implement this idea, such as a lack of ability to regulate factors like particle size and homogeneity [88]. Another chemical reaction–based technique is the bottom-up generation of nanofertilizers. Nanostructure size and impurity reduction can be better managed with bottom-up approaches [89]. Biological synthesis is another method for creating nanofertilizers, and it makes use of microorganisms like bacteria and fungi. An advantage of obtaining nanofertilizers through biosynthesis is that the final product is less cytotoxic [90]. Nanofertilizers can be made from a wide variety of materials, including organic and inorganic compounds [85, 91, 92]. Most inorganic nanostructures are made from metal oxides like zinc oxide (ZnO), magnesium oxide (MgO), or silver oxide (AgO). For this purpose, we use nanomaterials like polymers and carbon that originate from organic sources [91].

Consequently, there are a number of potential benefits and drawbacks associated with the numerous routes available for manufacturing nanofertilizers, including reduced energy expenditure, increased crop yields, and more effective material synthesis. Yadav *et al.* [83] found that, by combining these properties, effective and long-lasting agrochemicals might be developed.

Therefore, there are so many different ways to create nanofertilizers, it is important to pick the method that will work best for a certain application. Manufacturing, use, preparation, and ultimate application costs must all be considered before a conclusion can be reached [83]. Understanding the methods of nutrient absorption in the culture under research is also important for ensuring the success of the end products [85].

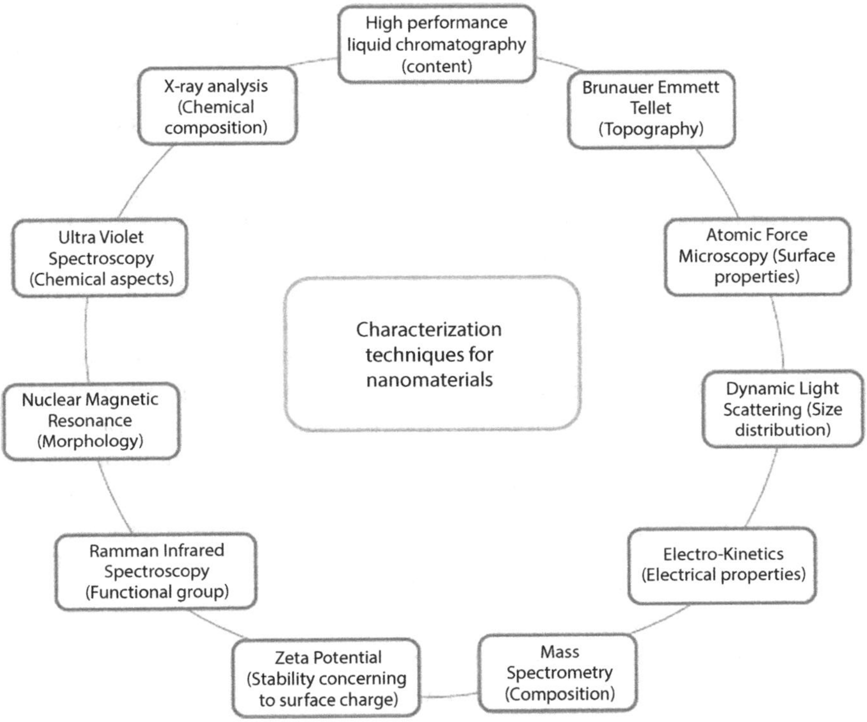

Figure 1.15 Characterization techniques for nanomaterials. Source: Nilmini [82].

1.4.4 Characterization of Nanofertilizers

The synthesized nanofertilizers are characterized with instruments like particle size analyzers, X-ray diffraction, Fourier transform infrared spectroscopy, Raman spectroscopy, Zeta analyzers, scanning electron microscopes, transmission electron microscopes, energy-dispersive x-ray spectroscopy, and AFMs. Several methods of categorization are shown in Figure 1.15. Nanofertilizers have been the subject of several characterization studies [82]. These studies have focused on those containing nitrogen, phosphate, potassium, sulfate, and zinc.

1.4.5 Advantages of Nanofertilizers

Sustainable food production on a worldwide scale may be possible with the use of nanofertilizers in agriculture. Less nutritious food consumption and a lack of fruit and vegetable consumption are major contributors to nutritional inadequacies in human populations [83, 93]. Nanofertilizers' nutrient distribution method is vastly superior to that of traditional chemical fertilizers. To manage crop yields, farmers implement slow/control release strategies for nitrogen. This delayed nutrition delivery has been linked to the coating or cementing of nutrients with NPs. Because nutrients are constantly provided to plants over time, growers can take use of this method to speed up crop development. Instead of waiting 4–10 days for nutrients to be released from traditional fertilizers, slow-release fertilizers can do so in as few as 40–50 days [77].

Half of the fertilizer given in conventional systems either leaches away or is inaccessible to the plant because of its high availability, which can hinder root absorption and even have negative effects on the plant [80, 94]. Nanofertilizers cut down on costs associated with delivery and use. The soil is also spared the negative effects of salts, which are sometimes over-applied when conventional fertilizers are used, either temporarily or permanently [80]. Nanofertilizers can be synthesized to meet the nutrient needs of future crops [94]. Soil nutrient status, crop growth season, and environmental conditions can all be taken into account when determining how much fertilizer to apply [80, 85]. Biosensors can be combined to this unique form of fertilizer to achieve this goal.

The availability of micronutrients to plants is critical for crop growth, and a lack of these nutrients can lead to nutritionally inadequate fruits and vegetables [80, 94]. Nanofertilizers allow growers to supply optimal levels of nutrients, whereas, in a conventional nutrient management system, it is difficult to control the distribution of micronutrients to a given crop [64, 65]. For instance, nanofertilizers can be used as efficient and effective fortification solutions for agricultural and fresh food items because most horticultural growing areas throughout the world are deficient in certain micronutrients (Zn and Fe) [82, 83]. Nanofertilizers boost nutrient absorption because of their large specific surface area, small size, and strong reactivity. In contrast, the balanced nutrition provided by nanofertilizers helps the plant endure a wide range of biotic and abiotic stresses, with clear benefits for the plant as a whole [83].

1.4.6 Limitations of Nanofertilizers

Increased crop yields thanks to the use of various nanofertilizers are a solid sign that sustainable agriculture is progressing. However, there may be unintended, long-lasting effects of implementing this technology on farms on purpose. Henrique da Silva Júnior *et al.* [85] noted that this scenario raises new environmental and unexpected health safety concerns that may slow the widespread use of this technology to increase horticultural crop yields. Another issue in this field is the potential phytotoxicity of nanomaterials, which might vary from plant to plant and from dose to dose. The advantages and disadvantages of nanofertilizers must be weighed carefully before they are introduced to the market [77]. Nanomaterials have a large surface area in relation to their size, making them very reactive. The reactivity and unpredictability of such materials also raises concerns. Concerns have been raised about the well-being of farm workers who may be exposed to xenobiotics from the manufacturing to the application process [74, 77, 94].

Given the advantages, research into the usefulness and viability of these innovative smart fertilizers is essential. Unfortunately, their adoption and application in environmentally responsible farming and gardening is hampered by worries about their transport, toxicity, bioavailability, and the potential for unintended environmental impacts when they come into contact with biological systems [85].

Toxicological research goals must be established in addition to risk assessment and hazard identification for nanomaterials, which includes evaluating the life cycle of nanomaterials or fertilizers. The accumulation of NPs in plants poses serious health hazards, making this point all the more salient. Indeed, there are serious worries about food safety, human and food security, due to the use of nanomaterial-derived nanofertilizers [94].

Several studies have shown that NPs can be hazardous to plants. The uptake, translocation, transformation, and accumulation (phytotoxicity) of NPs in plants varies with species, dose, application method, and NP type (composition, size, shape, and surface characteristics). Understanding the uptake and translocation of nanofertilizers, the transformation of NPs through interactions with soil and plant compounds, and NP accumulation in various plant tissues all require knowledge of the toxicity of individual NPs to the crop in question [85, 92].

1.5 Conclusion

The potential for increased agricultural productivity and sustainability due to the use of nanofertilizers is substantial. The controlled release mechanisms and increased nutrient bioavailability of nanofertilizers are huge advancements over traditional chemical fertilizers. However, it is necessary to be cautious of the potential risks associated with nanomaterials, including phytotoxicity and environmental impact. Recognizing and tackling these difficulties through the use of nanofertilizers could significantly impact global food security and sustainable agricultural practices.

References

1. Fadhil, Y.S., Applications of Nanotechnology in Food Processing and Packaging. *Mesopotamia J. Agric.*, 50, 3, 27–36, 2022.
2. Singh, T., Shukla, S., Kumar, P., Wahla, V., Bajpai, V.K., Rather, I.A., Application of Nanotechnology in Food Science: Perception and Overview. *Front. Microbiol.*, 8, 1501, 2017, Doi: 10.3389/fmicb.2017.01501.
3. Sahoo, M., Vishwakarma, S., Panigrahi, C., Kumar, J., Nanotechnology: Current applications and future scope in food. *Food Front.*, 2, 1, 3–22, 2020, Doi: 10.1002/fft2.58.
4. Sahoo, M., Panigrahi, C., Vishwakarma, S., Kumar, J., A Review on Nanotechnology: Applications in Food Industry, Future Opportunities, Challenges and Potential Risks. *J. Nanotechnol. Nanomater.*, 3, 1, 28–33, 2022.
5. Biswas, R., Alam, M., Sarkar, A., Haque, I., Hasan, M., Hoque, M., Application of nanotechnology in food: processing, preservation, packaging and safety assessment. *Heliyon*, 8, 11, e11795, 2022, Doi.org/10.1016/j.heliyon. 2022. e11795.
6. Sriram, P. and Suttee, A., Nanotechnology advances, benefits, and applications in daily life, in: *Nanotechnology*, pp. 23–44, CRC Press, Boca Raton, Florida, 2020.
7. Kumar, P., Mahajan, P., Kaur, R., Gautam, S., Nanotechnology and its challenges in the food sector: a review. *Mater. Today Chem.*, 17, 100332, 2020, Doi.org/10.1016/j.mtchem.2020.100332.
8. Ansari, M.Z., Nanotechnology in Food and Plant Science: Challenges and Future Prospects. *Plants*, 12, 13, 2565, 2023, Doi.org/10.3390/plants12132565.
9. Manjunatha, R.L., Naik, D., Usharani, K.V., Nanotechnology application in agriculture: A review. *J. Pharmacogn. Phytochem.*, 8, 3, 1073–1083, 2019.
10. Singh, S., Kumar, A., Arora, P., Sahu, S., Nanotechnology and its applications in Crop Improvement. *Bull. Environ. Pharmacol. Life Sci.*, 8, 10, 1–6, 2019.
11. Abd-Elrahman, S.H. and Mostafa, M.A.M., Applications of Nanotechnology in Agriculture: An Overview. *Egypt. J. Soil Sci.*, 55, 2, 197–214, 2015.
12. Joshi, H., Choudhary, P.S., Mundra, S.L., Future Prospects of Nanotechnology in Agriculture. *Int. J. Chem. Stud.*, 7, 2, 957–963, 2019.
13. Neme, K., Nafady, A., Uddin, C., Tola, Y.B., Application of nanotechnology in agriculture, postharvest loss reduction and food processing: food security implication and challenges. *Heliyon*, 7, 12, e08539, 2021, Doi.org/10.1016/j.heliyon.2021.e08539.
14. Pathak, S.O., Pandey, B.K., Singh, S., Hasanain, M., Shukla, G., Maurya, D.K., Nanotechnology in Agriculture: benefits, risks and challenges. *Food Sci. Rep.*, 1, 4, 89–92, 2020.
15. Rawat, A., Kumar, R., Bhatt, B., Ram, P., Nanotechnology in Agriculture- A Review. *Int. J. Curr. Microbiol. Appl. Sci.*, 7, 8, 969–978, 2018.

16. Usman, M., Farooq, M., Wakeel, A., Nawaz, A., Cheema, S.A., Rehman, H., Sanaullah, M., Nanotechnology in agriculture: Current status, challenges and future opportunities. *Sci. Total Environ.*, *721*, 137778, 2020, Doi.org/10.1016/j.scitotenv.2020.137778.
17. Prasad, R., Bhattacharyya, N., D. Nguyen, Q.D., Nanotechnology in Sustainable Agriculture: Recent Developments, Challenges, and Perspectives. *Front. Microbiol.*, 8, 1014, 2017, Doi: 10.3389/fmicb.2017.01014.
18. Chellaram, C., Murugaboopathi, G., John, A.A., Sivakumar, R., Ganesan, S., Krithika, S., Priya, G., Significance of nanotechnology in food industry. *APCBEE Procedia*, *8*, 109–113, 2014.
19. Prasad, R., Kumar, V., Prasad, K.S., Nanotechnology in sustainable agriculture: present concerns and future aspects. *Afr. J. Biotechnol.*, *13*, 6, 705–713, 2014.
20. Kim, D.Y., Kadam, A., Shinde, S., Saratale, R.G., Patra, J., Ghodake, G., Recent developments in nanotechnology transforming the agricultural sector: a transition replete with opportunities. *J. Sci. Food Agric.*, *98*, 3, 849–864, 2018.
21. Ashraf, S.A., Siddiqui, A.J., Abd Elmoneim, O.E., Khan, M., II, Patel, M., Alreshidi, M., Adnan, M., Innovations in nanoscience for the sustainable development of food and agriculture with implications on health and environment. *Sci. Total Environ.*, *768*, 144990, 2021.
22. Manjunatha, S.B., Biradar, D.P., Aladakatti, Y.R., Nanotechnology and its applications in agriculture: A review. *J. Farm Sci.*, *29*, 1, 1–13, 2016.
23. Lowry, G.V., Avellan, A., Gilbertson, L.M., Opportunities and challenges for nanotechnology in the agri-tech revolution. *Nat. Nanotechnol.*, *14*, 6, 517–522, 2019.
24. Abobatta, W.F., Nanotechnology Application in Agriculture. *Acta Sci. Agric.*, 2, 6, 99–102, 2018.
25. Hob-Allah, E., Saber, M., Zaghloul, A., Nanotechnology Applications in Agriculture. *Int. J. Environ. Pollut. Environ. Modell.*, 2, 4, 196–211, 2019.
26. Reddy, S.S. and Chhabra, V., Nanotechnology: its scope in agriculture. *J. Phys. Conf. Ser.*, 2267, 1, 012112, 2022, Doi:10.1088/1742-6596/2267/1/012112.
27. Nouailhat, A. (Ed.), *An Introduction to Nanoscience and Nanotechnology*, ISTE Ltd and John Wiley & Sons, Inc, France, 2020.
28. Jeevanandam, J., Barhoum, A., Chan, Y.S., Dufresne, A., Danquah, M.K., Review on nanoparticles and nanostructured materials: history, sources, toxicity and regulations. *Beilstein J. Nanotechnol.*, *9*, 1, 1050–1074, 2018.
29. Chaturvedi, S. and Dave, P.N., Nanomaterials: Environmental, human health risk, in: *Handbook of nanomaterials for industrial applications*, pp. 1055–1062, Elsevier, Amsterdam, 2018.
30. Darweesh, H.H.M., Nanomaterials: classification and properties-Part I. *Nanoscience*, *1*, 1, 1–11, 2018.

31. Harish, K.K., Venkatesh, N., Bhowmik, H., Anuttam, K.,.A., Metallic Nanoparticle: A Review. *Biomed. J. Sci. Tech. Res.*, 4, 2, 3765–3775, 2018, Doi: DOI: 10.26717/BJSTR.2018.04.001011.
32. Gilbertson, L.M., Zimmerman, J.B., Plata, D.L., Hutchison, J.E., Anastas, P.T., Designing nanomaterials to maximize performance and minimize undesirable implications guided by the Principles of Green Chemistry. *Chem. Soc. Rev.*, *44*, 16, 5758–5777, 2015.
33. Kumar Teli, M., Mutalik, S., Rajanikant, G.K., Nanotechnology and nanomedicine: going small means aiming big. *Curr. Pharm. Des.*, *16*, 16, 1882–1892, 2010.
34. Guerra, F.D., Attia, M.F., Whitehead, D.C., Alexis, F., Nanotechnology for environmental remediation: materials and applications. *Molecules*, *23*, 7, 1760, 2018.
35. Gatto, F. and Bardi, G., Metallic Nanoparticles: General Research Approaches to Immunological Characterization. *Nanomaterials*, 8, 10, 753, 2018, Doi:10.3390/nano8100753.
36. Salman, H.D., Evaluation and Comparison the Antibacterial Activity of Silver Nan Particles (AgNPs) and Silver Nitrate ($AgNO_3$) on Some Pathogenic Bacteria. *J. Global Pharma Technol.*, 10, 9, 238–248, 2017.
37. Huynh, K., Pham, X., Kim, J., Lee, S.H., Chang, H., Rho, W., Jun, B., Synthesis, Properties, and Biological Applications of Metallic Alloy Nanoparticles. *Int. J. Mol. Sci.*, 21, 14, 5174, 2020, doi: 10.3390/ijms21145174.
38. Zhang, D., Ma, X., Gu, Y., Huang, H., Zhang, G., Green Synthesis of Metallic Nanoparticles and Their Potential Applications to Treat Cancer. *Front. Chem.*, 8, 799, 2020, Doi: 10.3389/fchem.2020.00799.
39. Gavanji, S., The effects of silver Nano particles on microorganisms: A review. *Appl. Sci. Rep.*, 1, 2, 50–56, 2019.
40. Siddiqi, K.S., Husen, A., Rao, R.A.K., A review on biosynthesis of silver nanoparticles and their biocidal properties. *J. Nanobiotechnol.*, 16, 14, 1–28, 2018.
41. Chouhan, N., Silver Nanoparticles: Synthesis, Characterization and Applications. *Intech Open*, 1, 1–39, 2018. dx.doi.org/10.5772/intechopen.75611.
42. Gharibshahi, L., Saion, E., Gharibshahi, E., Shaari, A., Matori, K.A., Structural and Optical Properties of Ag Nanoparticles Synthesized by Thermal Treatment Method. *Materials*, 10, 5, 1–13, 2017.
43. Kolhatkar, A.G., Jamison, A.C., Litvinov, D., Willson, R.C., Lee, T.R., Tuning the magnetic properties of nanoparticles. *Int. J. Mol. Sci.*, *14*, 8, 15977–16009, 2013.
44. Gamboa, S.M., Rojas, E.R., Martínez, V.V., Vega-Baudrit, J., Synthesis and characterization of silver nanoparticles and their application as an antibacterial agent. *Int. J. Biosens. Bioelectron.*, 5, 5, 166–173, 2019.
45. Lee, S. and Jun, B., Silver Nanoparticles: Synthesis and Application for Nanomedicine. *Int. J. Mol. Sci.*, 20, 4, 865, 2019, Doi: 10.3390/ijms20040865.

46. Firdhouse, M.J. and Lalitha, P., Biosynthesis of Silver Nanoparticles and Its Applications. *J. Nanotechnol.*, 2015, 1, 829526, 2020. http://dx.doi.org/10.1155/2015/829526.
47. Mohammad, Z.H., Ahmad, F., Ibrahim, S.A., Zaidi, S., Application of nanotechnology in different aspects of the food industry. *Discover Food*, 2, 1, 12, 2022, Doi.org/10.1007/s44187-022-00013-9.
48. Singh, R., Dutt, S., Sharma, P., Sundramoorthy, A.K., Dubey, A., Singh, A., Arya, S., Future of Nanotechnology in Food Industry: Challenges in Processing, Packaging, and Food Safety. *Global Challenges*, 7, 4, 2200209, 2023, Doi: 10.1002/gch2.202200209.
49. Inobeme, A. and Adetunji, C.O., Advances, Prospects and Concerns in Nanomaterials for Food Processing. *Acta Sci. Nutr. Health*, 7, 3, 96–97, 2023.
50. Abbas, K.A., Saleh, A.M., Mohamed, A., MohdAzhan, N., The recent advances in the nanotechnology and its applications in food processing: A review. *J. Food Agric. Environ.*, 7, 3&4, 14–17, 2019.
51. Iravani, S., Korbekandi, H., Mirmohammadi, S.V., Zolfaghari, B., Synthesis of silver nanoparticles: chemical, physical and biological methods. *Res. Pharm. Sci.*, 9, 6, 385–406, 2014.
52. Güzel, R. and Erdal, G., *Synthesis of Silver Nanoparticles*, 20182020, Retrieved from https://www.intechopen.com/books/silver-nanoparticles-fabrication-characterization-and-applications/synthesis-of-silver-nanoparticles on 8th July.
53. Samal, D., Use of Nanotechnology in Food Industry: A review. *Int. J. Environ. Agric. Biotechnol.*, 2, 4, 2270–2280, 2017.
54. Keat, C.L., Aziz, A., Eid, A.M., Elmarzugi, N.A., Biosynthesis of nanoparticles and silver Nanoparticles. *Bioresour. Bioprocess.*, 2, 1–11, 2015, Doi 10.1186/s40643-015-0076-2.
55. Ahmad, S., Munir, S., Zeb, N., Ali, S., Green nanotechnology: a review on green synthesis of silver nanoparticles-an eco-friendly approach. *Int. J. Nanomed.*, 1, 4, 5087–5107, 2019.
56. An, C., Sun, C., Li, N., Wang, Y., Nanomaterials and nanotechnology for the delivery of agrochemicals: strategies towards sustainable agriculture. *J. Nanobiotechnol.*, 20, 1, 11, 2022, Doi.org/10.1186/s12951-021-01214-7.
57. Mousavi, S.R. and Rezaei, M., Nanotechnology in Agriculture and Food Production. *J. Appl. Environ. Biol. Sci.*, 1, 10, 414–419, 2018.
58. Fraceto, L.F., Grillo, R., Medeiros, G.A., Scognamiglio, V., Rea, G., Bartolucci, C., Nanotechnology in Agriculture: Which Innovation Potential Does It Have? *Front. Environ. Sci.*, 4, 186737, 2016, Doi: 10.3389/fenvs.2016.00020.
59. Joseph, T. and Morrison, M., Nanotechnology in Agriculture and Food. *Eur. Nanotechnol. Gateway*, 5, 7, 1–15, 2020.
60. Panpatte, D.G., Jhala, Y.K., Shelat, H.N., Vyas, R.V., Nanoparticles: the next generation technology for sustainable agriculture. *Microb. Inoculants Sustainable Agric. Productivity Vol. 2 Funct. Appl.*, 289–300, 2016. https://doi.org/10.1007/978-81-322-2644-4_18

61. Mahanta, N., Dambale, A., Rajkhowa, M., Mahanta, C., Mahanta, N., Nutrient use efficiency through nano fertilizers. *Int. J. Chem. Stud.*, *7*, 3, 2839–2842, 2019.
62. Okey-Onyesolu, C.F., Hassanisaadi, M., Bilal, M., Barani, M., Rahdar, A., Iqbal, J., Kyzas, G.Z., Nanomaterials as nanofertilizers and nanopesticides: An overview. *Chem. Select*, *6*, 33, 8645–8663, 2021.
63. Qureshi, A., Singh, D.K., Dwivedi, S., Nano-fertilizers: A Novel Way for Enhancing Nutrient Use Efficiency and Crop Productivity. *Int. J. Curr. Microbiol. Appl. Sci.*, *7*, 2, 3325–3335, 2018.
64. Kumar, Y., Tiwari, K.N., Nayak, R.K., Rai, A., Singh, S.P., Singh, A.N., Kumar, Y., Tomar, H., Singh, T., Raliya, R., Nanofertilizers for Increasing Nutrient Use Efficiency, Yield and Economic Returns in important Winter Season Crops of Uttar Pradesh. *Indian J. Fertilizers*, 16, 8, 772–786, 2020.
65. Kumar, Y., Tiwari, K.N., Singh, T., Sain, N.K., Laxmi, S., Verma, R., Raliya, R., Nanofertilizers for enhancing nutrient use efficiency, crop productivity and economic returns in winter season crops of Rajasthan. *Ann. Plant Soil Res.*, *22*, 4, 324–335, 2020.
66. Mejias, J.H., Salazar, F., Pérez Amaro, L., Hube, S., Rodriguez, M., Alfaro, M., Nanofertilizers: A cutting-edge approach to increase nitrogen use efficiency in grasslands. *Front. Environ. Sci.*, *9*, 52, 2021.
67. Jakhar, A.M., Aziz, I., Kaleri, A.R., Hasnain, M., Haider, G., Ma, J., Abideen, Z., Nano-fertilizers: A sustainable technology for improving crop nutrition and food security. *NanoImpact*, 27, 100411, 2022.
68. Badawy, M.T., Mostafa, M., Khalil, M.S., Abd-Elsalam, K.A., Agri-food and environmental applications of bionanomaterials produced from agri-waste and microbes, in: *Agri-waste and microbes for production of sustainable nanomaterials*, pp. 441–463, Elsevier, Amsterdam, 2022.
69. Zhang, Y. and Goss, G.G., Nanotechnology in agriculture: Comparison of the toxicity between conventional and nano-based agrochemicals on non-target aquatic species. *J. Hazard. Mater.*, *439*, 129559, 2022.
70. Jha, A., Pathania, D., Damathia, B., Raizada, P., Rustagi, S., Singh, P., Chaudhary, V., Panorama of biogenic nano-fertilizers: A road to sustainable agriculture. *Environ. Res.*, 235, 116456, 2023.
71. Vithanage, M., Zhang, X., Gunarathne, V., Zhu, Y., Herath, L., Peiris, K., Siddique, K.H., Plant nanobionics: Fortifying food security via engineered plant productivity. *Environ. Res.*, 229, 115934, 2023.
72. Kalia, A., Sharma, S.P., Kaur, H., Nanoscale fertilizers: harnessing boons for enhanced nutrient use efficiency and crop productivity. *Nanobiotechnol. Appl. Plant Prot.*, *2*, 191–208, 2019.
73. Kanjana, D., Evaluation of Foliar Application of Different Types of Nanofertilizers on Growth, Yield and Quality Parameters and Nutrient Concentration of Cotton under Irrigated Condition. *Int. J. Curr. Microbiol. Appl. Sci.*, 9, 7, 429–441, 2020.

74. El-Ghamry, A.E., Mosa, A.A., Alshaal, T.A., El-Ramady, H.R., Nanofertilizers vs. Biofertilizers: New Insights. *Environ. Biodivers. Soil Secur.*, 2, 1, 51–72, 2018.
75. Alrbaihat, M.R., Agricultural Nano Fertilizers: Macronutrient Types and Applications Review. *Curr. Trends Geotech. Eng. Constr.*, 306–316, 2023, Doi.org/10.1007/978-981-19-7358-1_26.
76. Mastronardi, E., Tsae, P., Zhang, X., Monreal, C., DeRosa, M.C., Strategic role of nanotechnology in fertilizers: potential and limitations. *Nanotechnol. Food Agric.*, 25–67, 2015. https://doi.org/10.1007/978-3-319-14024-7_2
77. Kale, S.K., Parishwad, G.V., Husainy, A.D.N., Patil, A.S., Emerging Agriculture Applications of Silver Nanoparticles. *ES Food Agrofor.*, 3, 17-22, 2021, Doi: DOI: doi.org/10.30919/esfaf438.
78. Kalia, A., Sharma, S.P., Kaur, H., Kaur, H., Novel nanocomposite-based controlled-release fertilizer and pesticide formulations: Prospects and challenges. *Multifunct. Hybrid Nanomater. Sustain. Agri-food Ecosyst.*, 99–134, 2020. https://doi.org/10.1016/B978-0-12-821354-4.00005-4
79. Ilmudeen, S., Farween, F., Mahaletchumi, S., Kanchana, D.K.D., II, Advances of Nanofertilizers in Modern Agriculture; A review. *J. Res. Technol. Eng.*, 3, 2, 19–29, 2022.
80. Meghana, K.T., Wahiduzzaman, M., Vamsi, G., Nanofertilizers in Agriculture. *Acta Sci. Agric.*, 5, 3, 35–46, 2021.
81. Verma, K.K., Song, X., Joshi, A., Tian, D., Rajput, V.D., Singh, M., Arora, J., Minkina, T., Li, Y., Recent Trends in Nano-Fertilizers for Sustainable Agriculture under Climate Change for Global Food Security. *Nanomaterials*, 12, 1, 173, 2022, Doi.org/10.3390/ nano12010173.
82. Nilmini, W.K.A., Nanofertilizers: A Novel Approach Towards a Sustainable Agriculture. *J. Res. Technol. Eng.*, 2, 1, 92–98, 2021.
83. Yadav, A., Yadav, K., Abd-Elsalam, K.A., Nanofertilizers: Types, Delivery and Advantages in Agricultural Sustainability. *Agrochemicals*, 2, 2, 296–336, 2023, Doi.org/10.3390/ agrochemicals2020019.
84. Nongbet, A., Mishra, A.K., Mohanta, Y.K., Mahanta, S., Ray, M.K., Khan, M., Baek, K.H., Chakrabartty, I., Nanofertilizers: A Smart and Sustainable Attribute to Modern Agriculture. *Plants*, 11, 19, 2587, 2022, Doi.org/10.3390/plants11192587.
85. Henrique da Silva Júnior, A., Mulinari, J., Júnior, F.W.R., Silva de Oliveira, C.R., Nanofertilizers: An Overview. *Int. Agrobusiness Congr.*, 14, 2020, Doi.org/10.31692/ICIAGRO.2020.0041.
86. Raliya, R., Saharan, V., Dimkpa, C., Biswas, P., Nanofertilizer for Precision and Sustainable Agriculture: Current State and Future Perspectives. *J. Agric. Food. Chem.*, 66, 26, 6487–6503, 2017, doi: DOI: 10.1021/acs.jafc.7b02178.
87. Yomso, J. and Menon, S., Impact of nanofertilizers on growth and yield parameters of rice crop; A Review. *Pharm. Innov. J.*, 10, 6, 249–253, 2021.
88. Rahman, H., Hasan, N., Nigar, S., Ma, F., Aly Saad Aly, M., Khan, Z.H., Synthesis and Characterization of a Mixed Nanofertilizer Influencing the

Nutrient Use Efficiency, Productivity, and Nutritive Value of Tomato Fruits. *ACS Omega*, 6, 41, 27112–27120, 2021, Doi.org/10.1021/acsomega.1c03727.

89. Shebl, A., Hassan, A.A., Salama, D.M., Abd El-Aziz, M.E., Abd Elwahed, M.S.A., Green Synthesis of Nanofertilizers and Their Application as a Foliar for *Cucurbita pepo L. J. Nanomater.*, 2019, 1, 3476347, 2019, Doi.org/10.1155/2019/3476347.
90. Patil, S.S., Balpande, S.S., Sajid, M., Pandao, M.R., Green Synthesis and Characterization of Nano Phosphorus Fertilizer for Wheat. *Int. J. Curr. Microbiol. Appl. Sci.*, 9, 12, 19–23, 2020.
91. Preetha, P.S. and Balakrishnan, N., A Review of Nano Fertilizers and Their Use and Functions in Soil. *Int. J. Curr. Microbiol. Appl. Sci.*, 6, 12, 3117–3133, 2017.
92. Al-Juthery, H.W.A., Lahmod, N.R., Al-Taee, A.H.G., Intelligent, Nano-fertilizers: A New Technology for Improvement Nutrient Use Efficiency (Article Review). *Earth Environ. Sci.*, 735, 1, 012086, 2021, Doi:10.1088/1755-1315/735/1/012086.
93. Bairwa, P., Kumar, N., Devra, V., Abd-Elsalam, K.A., Nano-Biofertilizers Synthesis and Applications in Agroecosystems. *Agrochemicals*, 2, 1, 118-134, 2023, Doi.org/10.3390/ agrochemicals2010009.
94. Zulfiqar, F., Navarro, M., Ashraf, M., Akram, N.A., Munné-Bosch, S., Nanofertilizer use for sustainable agriculture: Advantages and limitations. *Plant Sci.*, 289, 110270, 2019, Doi.org/10.1016/j.plantsci.2019.110270.

2

Synthesis of Nano-Chitosan Using Agricultural Waste

Konjerimam Ishaku Chimbekujwo[1*], Japhet Gaius Yakubu[2], Isibor Patrick Omoregie[3], Oluwafemi Adebayo Oyewole[2], Wuna Muhammad Muhammad[2], Charles Oluwaseun Adetunji[4], Eniola K. I. T.[4], Mohammed Bello Yerima[5] and John Tsado Mathew[6]

[1]*Department of Microbiology, Modibbo Adama University, Yola, Nigeria*
[2]*Department of Microbiology, Federal University of Technology, Minna, Nigeria*
[3]*Department of Biological Sciences, Covenant University, Ota, Ogun State, Nigeria*
[4]*Department of Microbiology, Edo State University, Uzairue, Nigeria*
[4]*Department of Biological Sciences, Joseph Ayo Babalola University, Ikeji Arakeji, Osun State, Nigeria*
[5]*Department of Microbiology, Sokoto State University, Sokoto, Sokoto State, Nigeria*
[6]*Department of Chemistry, Ibrahim BadamasiBabangida University, Lapai, Niger State, Nigeria*

Abstract

Industrial agro-waste has the potential to harm soil and groundwater, which is causing significant environmental pollution. The slower degradation is a result of the increased waste volume over time. Interestingly, these wastes proved to have the potential to be useful in various applications such as composts, adsorbents for pollutant removal, and organic fertilizers (chitosan), which will gradually reduce negative environmental impacts. Waste products, including shrimp shells, squid, crab, and lobster, are used to synthesize nano-chitosan (N-acetyl derivative) of chitin. The superior polymeric cationic nature, non-toxicity, and biodegradability of nano-chitosan make them a suitable choice for use. They are commonly used as promoters of growth in agriculture, immobilizing enzymes and providing bio-medical controlled drug delivery through encapsulation or carrier. This review outlines chitosan traits like the structural, physical, and chemical. The synthesis of chitosan nanoparticles is also described in detail. Chitosan nanoparticles are

**Corresponding author*: chimbekonje@gmail.com

Charles Oluwaseun Adetunji, Maulin P. Shah, Yerima Mohammed Bello, Daniel Hefft, Jay Singh, Shyam S. Pandey and Ravindra Pratap Singh (eds.) Nanochitosan Applications for Enhanced Crop Production and Food Security, (39–68) © 2025 Scrivener Publishing LLC

also employed in various contexts. Nano-chitosan applications comprise plant growth-promoting agents, drug transporter, encapsulation, antimicrobial agent, and plant preserver. In addition, its distinct traits also make it useful in wastewater treatment, serving as a fungicide, wound healer, and framework for engineering tissue. Nano-chitosan is an efficient solution for toxic chemical usage in various industries. Thus, production of nano-chitosan nanoparticles is advantageous in most industries.

Keywords: Nano-chitosan, nanotechnology, nanoparticles synthesis, biopolymer, waste management, agro-waste

2.1 Introduction

Agricultural wastes are remains obtained from the growth and clarification of raw agricultural produce. They are the non-useful yield obtained from the manufacturing and clarification of produce from agriculture [1]. Humans rely on edible parts for energy, while agro-waste is made up of non-consumable plant portions [2]. Increased production in agriculture has experience rapid growth of domestic animals, industrial product, and crop residues. Increasing farming systems in developing countries may lead to accumulation of agricultural waste greatly. Agricultural waste generated annually is estimated to be 998 million tons [3]. Obi *et al.* [1] suggest that organic wastes make up as much as 80% of the trash generated in a farm. Inadequate dumping of these waste products can negatively impact the lithosphere and water from the ground, leading to all types of pollution. Many trashes are left unused and discarded improperly, which only serves to increase the problem. Beneficial products formed from leftover materials are aided by this waste [2]. The bio-chemicals, such as lignin and cellulose, and the polyphenolic compounds found in agricultural wastes (such as horticultural, aquatic, or culinary waste) are essential [4]. Elemike *et al.* [5] reported that certain compounds can generate carbon dots through pyrolysis, whereas some balances, caps, and serves as a reductant when producing nanoparticle (NP) [5]. NPs are rapidly becoming more popular as agro-waste is used as substrate for the production of nanomaterials because of advantages like innovations and profit making [5]. The production of NPs is done via crushing, grinding, alkali and acid treatment, and bleaching, which are blend of mechanical-chemical processes. The physical appearance of NPs was ascertain using techniques like atomic force microscopy, transmission electron microscopy (TEM), and scanning electron microscopy (SEM) [6], Crystallographic shape and modification of the functional group is looked into using diffraction of X-ray and Fourier-transform infrared spectroscopy

(FTIR) [7]. Nanostructures have been produced from orange, banana, and pomegranate peel, which serve as agricultural residues, furthermore, the excrement or waste of sea creatures such as shrimp, lobster, crab, oysters (yam), krill, and squid. Chitin, proteins, and calcium carbonate are found in most marine crustacean shells, with a dry weight of approximately 15%–20%. Guleria *et al.* [8] stated many potential uses of NPs, such as the production of molecular biology products, synthesis of NPs, and manure formation and biofuels, among others. The establishment of nano-innovations and NPs to increases the productivity of various sectors in recent years. Nano-innovations focus on particle size between 1 nm and 100 nm, as identified by Afolalu *et al.* [9]. There are several potential applications of NPs, one of the main goals is to manage particle dimension, form, and even crystallization characteristics of synthesized NPs [10]. Large-scale materials are linked to atomic or molecular structures through the use of NPs. This has led to its wide usage in biological innovation, medicals, and electrochemistry and in detection of trace element [10].

2.2 Different Sources of Agricultural Waste

The process of processing and manufacturing agricultural products usually results to a residual output known as agro-waste [11]. Saini *et al.* [12] identified four primary categories of waste from agriculture: manure from livestock and agricultural sector, crustacean shells, vegetables and waste from crop, and other by-products.

2.2.1 Shell Wastes

Shellfish is a significant contributor to fisheries, which have incredibly high economic value. During shrimp processing, the shell and head portions are produced as waste and are consumed in large quantities. A substantial amount of biowaste causes environmental issues, resulting in an unpleasant air and unpleasant smell. Protein, minerals, and chitin are the mainstay of these shrimp shells. Shell waste deproteinization and demineralization can yield chitin [13]. The removal of acetyl group from chitin can result in polysaccharide chitosan, with numerous uses. For example, plant protection can be enhanced by chitosan. In agriculture, chitosan as an anti-transpirant may be used to conserve water, as suggested by Biltelli *et al.* [14]. Biodegradable, biocompatible, and non-toxic polymer chitosan has diverse uses in biomedical applications. Biomedical use is facilitated by the low-immunogenicity, microbial growth-inhibiting ability of chitosan.

It has wide application in engineer of tissue, gene delivery, and lesion healing, as reported by Jayakumar *et al.* [15], and Jayakumar *et al.* [16] reported on the delivery of drugs.

2.2.2 Livestock Wastes

Rural households worldwide generate income from livestock and crops. In certain societies, animal dung is a viable option for producing biogas or creating manure cakes to burn. Examples commonly used are chicken litter burned as fuel, animal slurry integrated for energy and power, and animal residue breakdown [17].

2.2.3 Crop Residues

Bulk agro-debris is developed in field, which includes stovers, straws and leaves, and seed pods. Organic wastes converted to variety valuable goods are often affordable [18]. Bioethanol is produced using rice and wheat straw and corn stovers as by-products that are available year-round. The majority of it is burned down, with the remaining portion being burned solely for biofuels or animal feed. Residues from wheat are derived from straws, with 355.35 million metric tons being produced worldwide each year [19]. Agricultural waste from crop wastes like oatmeals, sisorghum, and grains have also been employed as feedstock for biofuels [20].

2.2.4 Agricultural Industry Wastes

Waste resulting from vegetable and fruit peels and sugarcane bagasse, are extracts of oil or juice. The list also comprises of fat and meat of animals, eggs, and skin of chicken [21]. Sugarcane bagasse is the most common waste from the agro-industry. After extracting juice, enterprises produce dry pulpy fibrous residue known as bagasse [22]. Approximately 180 million tons of bagasse of sugarcane is estimated to be available worldwide. Other examples of agro-industry wastes (fruit, orange peel, and apple pomace) are produced in foods and other processing units. The non-food agro-industries produce accumulated waste, like seeds of de-oiled cakes from oil plants that are non–edible (*Pongamia pinnata* and *Jatropha curcas*) [23].

2.2.5 Nanomaterial Synthesis Using Agro-Waste

Nanoparticles have been triumphantly produced by various agricultural wastes, including wheat bran, corn cob, peels, rice bran, and effluent from palm oil mill [24]. NPs can be generated within the cell or outside of it. Dimension and the kind of NPs heavily are dependent upon time of synthesis, temperature, and pH of growth [25]. Phenolics, proteins, and flavonoids are biological molecules formed from these agricultural wastes that are environmentally friendly as well as metalllic NP production. Production of nanomaterials employs the use of top-down technique. Larger materials are broken down with mechanical, chemical, or other energy to create the desired structure. The bottom-up approach acknowledges that the essential components of life, such as enzymes and other elements in each living cell, act as machines at the nanoscale. Scientists aimed at discovering affordable alternative NPs synthesizes, which resulted in the establishment of a relatively new nanomaterials innovation [26].

The primary mechanism of biosynthesis of NPs is reduction/oxidation, which takes a step up from the top. Various techniques such as deposition of chemical vapor, pyrolysis, sol-gel process, plasma, and spinning are employed. Take a step from the bottom up. Nanofactories that can synthesize metals intra- and extra-cellularly have been proposed already as microorganisms. The process of generating NPs using plants is a novel and under-researched approach. The use of plant extracts for metallic NPs production is economical and reliable choice, as demonstrated [27]. The well-organized bio-synthesis of silver, copper, and gold NPs is produced from broth extracts of several plants [27].

2.2.6 Chitin

The primary polysaccharide chitin is enough to hold up commercially chitin or chitosan industry and is found in crustaceans, like crabs, shrimp, crawfish, and lobster (dry basis, 20%–30%) [27]. Glycos-aminoglycans are joined with negatively charged polysaccharides with poly-β-(1,4)-N-acetyl-D glucose-2-amine linear [28]. It is found in three polymorphic forms alpha, beta and gamma chitin. The polymorphic, alpha, beta, and gamma chitin are the of chitosan. The powerful interactions between the amide and carbonyl groups in chitins leads to the formation of hydrogen bonds and crystalline fibrils. In deacetylated chitosan, chitin property has shown to change by similar and dis-similar N-deacetylation, alkali concentration, polymer chemical form, single or multiple methods, type and variety of chitin, particle dimension, and incubation time. All the above characteristics pave way for greater use in addition to their biological breakdown and compatibility and mechanical traits [29]. Chitin is gotten from white shrimp shell waste when demineralized, deproteinized, and decolorized [30]. Nevertheless, insoluble nature of organic solvents and the hydrogen bond present has reduced its field of application [31]. Crustacean shell in which chitin is obtained from is coated with calcium carbonate for the formation of hard shell. The detachment of chitin from shells entails the removal inorganic calcium carbonate and proteins [32]. Different kinds of chemicals [NaOH, KOH, Na_2CO_3, K_2CO_3, $NaHCO_3$, Ca $(OH)_2$, and Na_3PO_4] have been tired as chemical agent for the removal of proteins, although NaOH is often used at different temperature and varied time. Also, molecular weight of the biopolymer is reduced when NaOH hydrolyzes and leads to chitin partial deacetylation [30]. Chitin contains $CaCO_3$, which is removed alongside minerals a chemically requires usage with hydrogen chloride, HNO_3, H_2SO_4, and CH_3COOH. Calcium carbonate is broken down into dissolvable calcium salt and CO_2 gas. Some factors such as temperature, rate of mineralization extraction time, particle dimension, and contents of acid solute to solvent ratio affects treatment rate [33]. Moreso, enzymes (proteases) for chitin extraction are very necessary as proteases help to get rid of proteins from shells and also lessen the separation processes of chitin during depolymerization and deacetylation. The removal of protein is done by the use of unmixed and crudely protease and treatment use prior or after the removal of minerals from the materials and helps change the availability of reactant [34].

The protein removal process by enzymes has been ascertained. For example, *Bacillus licheniformis* that produced alcalase, a serine endopeptidase, is used to remove proteins because it effectively selects terminal hydrophobic amino acids and permits the rate of control of hydrolysis [32]. Enzyme alcalase was used for the recovery of chitin, protein hydrolysate, and astaxanthin [35]. Chitin separated by enzymatic protein removal had a significantly higher residual protein concentration than chitin isolated from alkali solutions [36]. Removal of protein shells by untreated alkaline enzyme (proteases) was said to occur in marine invertebrates.

2.2.7 Chitosan Nanoparticles

Chitosan, a polyelectrolyte polysaccharide, is produced by partially deacetylating chitin. Chitosan is a heteropolymer consisting of linear β-(1-4) glycosidic linkages of N-acetyl glucosamine and D-glucosamine unit. Chitosan has a primary amine group and two free hydroxyl groups for each monomer with a formula of C6H11O4N with molar mass of 3.8 [37]. Crustaceans, including shrimps and crabs, frequently gather their exoskeletons along with yeast cell walls, fungi. Due to its lack of affinity for water, it is unsolvable in other carbon-based liquid [38]. Dry wastes from lobster and shrimps' shells are estimated to make up 1.44 million of total annual production. The level of acetyl group removal is crucial biotic and chemical properties of substances [39]. The higher chelating amino group count of chitin, due to its atomic weight and deacetylated level, makes chitosan superior to chitin [40]. Different metallic ions can be chelated by negatively or neutral D-glucosamine charged amino groups [41]. Nanoparticles range between 1 nm and 1,000 nm, as suggested by Zhao *et al.* [42]. Chitin prepared into NPs is simply called "chitosans." The transformation of chitosan into nano-chitosans is tiredlessly researched for various purposes [43]. The polymeric cationic character of chitin, along with its non-poisonous properties and unique eco-friendly ability, muco-adhesiveness, and sorption-increasing effects, makes it an intriguing biopolymer for NPs preparation [44]. According to Pereira *et al.* [40], small-scale chitosan particles have more functional groups that enhance their sorption procedure. As a linear polyamine that has numerous free amine groups and can be easily linked [45], ionic interplay with the anionic sodium tripolyphosphate (STPP) is eco-friendlier than glutaraldehyde and is an important element that makes chitosan a significant contributor to the production of NP materials [45]. The nano-chitosan biopolymer, as described by Pereira *et al.* [40] is considered to be fit for the removal of organic contaminate like phenol [40, 43]. Nano-chitosan possesses unique characteristics that enable them to target biologically anionic membranes with greater specificity in living organism [46]. Therefore, effectively utilization to encapsulate catalyst, DNA, and drugs can be achieved [47, 48]. Release control is effective done using well-suited NPs [49]. As conveyor frame of engineered tissue, chitosan NPs' biological activity depends on preparation process and system conditions.

Transformation of chitin to chitosan is through catalytic blend/chemical process [50]. Chitosan economical and large production scale is key to general usage for commercialized production through chemical methods [51]. Chitosan's cation makes it insolvable in acidic or neutral surroundings, which is one of its advantages. The responsiveness of glycosidic link to acids makes alkali deacetylation more prevalent [51]. Chitin can undergo N-deacetylation either diversely or similarly [52]. Hot concentration of sodium hydroxide has several been used to treat chitin, leading to the formation of an insolvable residue deacetylated up to 85% to 99%. In addition, dissolubility of chitosan can also be weighed by the dissemination of its N-acetic acid groups [53]. In heterogeneous situations, N-acetate D-glucosamine and D glucosamine remnant deacetylation reactions yield an unequal dispersal noted by Aiba [54] who also observed some block-wise dispersal along polymeric chains. The average attribute of chitosans may vary due to differences in their dissolvable and level of disintegration in liquid solutions [55]. Chitin deacetylases are involved in enzyme reactions that change chitin into a specific, explicit, and controlled form of chitosan [56]. By hydrolyzing the N-acetamido connections in chitin, deacetylase produces chitosan. Chitin deacetylases that are glycoproteins are set free and then enter periplasmic space or medium. Additionally, heat stable N-acetyl–D-glucosamine (1,4) polymer exhibits high precision for every catalyst. Acetate, a deacetylation product, does not affect deacetylases proteins of chitin by *A. nidulans* and *C. Lindemuthianum* that can be utilized in molecular biology processes [57, 58]. Fungus *M. rouxii* uses chitin as nutrient to biologically synthesize chitosan [59]. Insoluble chitins do not support the effective activity of enzymes because acetyl group removal is low.

2.2.8 Properties of Nano-Chitosan

The prevalent of chitin as natural polysaccharides succeeds behind cellulose in global scale [60]. It was reported that chitosan standard is influenced by the origin of chitin, its separation, and level of removal of acetyl group [61]. Chitin and chitosan initial sources are numerous in nature, including both water and earthly organism. Marine life's processing (e.g., biowaste, which is generated from seafood such as shrimp, lobster, crab, and squid) is converted into chitosan and chitin [62]. Displacing raw materials, unpredictability, and processing situations have hindered the manufacture of chitosan from water organisms in industrial settings. This led to the production of chitosan, a partial deacetylated polymer from land crustaceans [63]. Microbial cells like yeasts molds, bacteria, and some algae contain chitin [64]. It appears to be a viable option to produce chitosan from microorganisms, as the innovation can be modified to create standardized and pure products [65]. Despite the ability of micro- and macro-organisms to produce chitosan and chin from land and sea, their economic application is restricted to select few species [66]. Chitosan is known for its ability to not trigger immune stimulatory responses or induce severe inflammation. Various deacetylation rates and chitosan molecular weights demonstrate little toxic rate [67].

Chitosan's cationic nature is responsible for its bactericidal properties, as it can prevent growth of microorganism by adhering to bacterial surfaces and altering membrane wall permeability [68]. Chitosan bacteria inhibiting ability enhanced lower pH and reduced deacetylation. Decreasing molar weight of Gram-negative bacteria, the antibacterial activity can be increased and reduced [69]. Chitosan's hydrophilic nature requires the extracellular wall to be highly charged, which is why it appears to have minimal impact on human cells [70]. The adsorption of compounds lacking an admixture in mucus can be facilitated by the mucus adherent composition of chitosan, as demonstrated by Salamat-Miller *et al.* [71]. Chitosan enhances drug penetration through epidermis by facilitating the clearance of tighter epithelial intersections [68]. Okamoto *et al.* [72] acknowledge that chitosan's surface amino group and platelet and amino group interaction has made it a popular study for wound healing through coagulation studies [72]. Healing of injury facilitates hemostatic characteristics of chitosan for a long time. Vaccines can activate macrophage and neutrophils, quicken granulation tissue repair, reduce scar formation and contracture, and induce hemostasis [73].

2.3 Synthesis of Nano-Chitosan

In 1994, the homogenizing and cross-linking process was employed to produce nano-chitosan as drug conveyor [74]. Subsequently, ionic gelation, reversed micellar technique, precipitation, and straining were developed, along with emulsion droplet coalescence and spray drying. Nano-chitosan are produced using "bottom-up" methods like ultra-sonication, microemulsion, polymerization, grinding, and high-pressure homogenization [75].

2.3.1 Nano-Precipitation

This method, also known as solvent displacement, is more favored than others. Precipitation of NPs involves stirring with mix of water and oil phase, which forms particles suspension of colloid. Process of forming NPs is quick and easy, as it requires a single step. The rate of injection of the water phase stirring rate, organic phase, and oil phase/water phase proportion are important parameters that determine the nano-precipitation method when producing [76]. The absence of shearing stress enables the creation of particles with narrow sizes. Water-soluble solvent dissolve polymers and drugs. A drip-by-drip surfactant is added to the solution in a drop of an absorbing water. The NPs are rapidly generated through the diffusion of solvent. The solvents are subsequently eliminated with decreased pressure. A phase term diffusion is formed by solubilizing and introducing chitosan dispersal phase, i.e., a constant flux of 0. 8 mL/min was used to pump the methanol through the membrane. The diffusion phase is where Tween 80 mixes to create NPs [77]. Nano-precipitation can formed particles of 50 nm and 300 nm, these small sizes form more surface contact. Characteristic indispensable application occurs in desorption and adsorption systems [78, 79]. By producing particles as tiny as 170 nm, this technique is not only efficient but also expands the range of application.

2.3.2 Drying Through Spraying

Chitosan NP is produced through dry spraying. This method makes use of a nano spray dryer. Chitosan is dissolved by acetic acid when stored through the night. By creating droplets in an atomizer, the solution is oxidized. Chitosan mixed with a gas dryer is evaporated to form NP. The standard size for a spray-drying valve is 4.0–5 5 μm or 7 μm. A drying gas flows at a rate of 1.3 m^3 per minute when 2 mL per minute is used. A temperature range present at the inlet is 120°C and at the outlet is 80°C. The characteristics and production yield of nano-chitosan in the spray-drying process are dependent on the initial feed and operational variables, including flow rate [80]. The pharmaceutical industry often employs spray-drying as a method for micro-encapsulating antibiotics like amoxicillin, vancomycin, and ampicillin [80]. Spray-drying is an easy process that occurs continuously, but its solubility is determined by the polymer and drug. The drug can be used with hydrophilic or hydrophobic polymers and other heat-resistant, water-dissolvable, or water pharmaceutical dissolvable [81]. Protein-packed nano-chitosan can be quickly and easily treated with spray drying, which is an easy technique [82]. Using spray-drying, dexketoprofen was added to nano-chitosan and tested their anti-inflammatory agility [83]. A possible method for delivering low dosages and better efficacy of dexketoprofen-packed nano-chitosan coupled with trometamol may be explored.

2.3.3 The Gelation Ionotropic Technique

Chitosan's backbone contains several amine groups, which can be protonized to generate NH^{3+} in acid solutions. Nanoparticles can be prepared by physically cross-linking both [84]. Physical cross-linking is more prepared than chemical to avoid possible toxicity of reagents, unpleasant effects and enhance biocompatible. Linking of chitosan cation with anions ions by Na tripolyphosphate is established physically [84]. Gelation ionotropic process was confirmed first by Calvo *et al.* [86], and it has since been extensively studied and clarified. Small ionic molecules, such as sulfate, phosphate, or citrate, are utilized in the gelation of chitosan through the process of labeled anionic gelation. Instead of using small molecules, ionic macromolecules are thought to induce a polyelectrolyte appearance [87]. This technique employs the electrostatic interplay between the anion polyanion tripolyphosphate and the chitosan amine group [88]. To create an acidic chitosan and tripolyphosphate solution, the method introduced in 1989 was used [84]. Chitosan becomes solution in an acid that can be either malic, acetic, or citric. It may also be combined with a stabilizer such as poloxamer [84, 89]. Positively charged of chitosan was formed by adding tripolyphosphate solution and continuous agitation. As a result of the fusing between opposingly charged species, chitosan undertakes anionic gelation and becomes particle in spherical shape [90]. When the negatively charges particles are dispersed into the chitosan molecules, they link to form NPs with sizes between 200 nm and 1000 nm. Nano-chitosan is received by freeze-drying or by oven after centrifugation and washing with water. The chitosan-stabilizer proportion in this technique can employ the modification of NPs' exterior charge and dimension [91]. The size of particles increased with increase proportion chitosan back to polyanion [92]. The small-scale size of particles in sodium chloride resulted in an increased stability of NPs dispersed in a normal saline solution. By adding sodium chloride to the solvent, electrostatically resistance of chitosan backbone and amine groups decreases. Polymer stability increases as well [93]. The combination of radical polymerization and ionotropic gelation can result in the formation of chitosan, whereas acrylic or methacrylic acid also undergoes polymerization [80]. In this method, particles are removed by means of dialysis or washing. Application of gelatin ionotropic in serum albumin, insulin packing has been employed. From table-to-commercial scale-up, this approach is both the easiest and most economical.

2.3.4 Droplet Emulsion Coalescence and Solvent Emulsion Diffusion

To produce a stable aqueous-in-oil emulsion during etching, the drug is homogenized with chitosan in liquor paraffin and aided by SpanTM 83, an additive that can be applied at high speeds. The combination of emulsions will result in the formation of NPs with chitosan drops, which will create at a distance of approximately 452 nm when they collide during mixing [94]. El-Shabouri [95] adapted an adjusted approach of solvent dispersion in the publication of Niwa *et al.* [96]. The process of forming an oil-in-water emulsion begins with introducing drug in a water affinity solution with the addition of methylene chloride/acetone as an organic phase and then blends with excessive pressure to volatilize methylene chloride. An organic solvent causes precipitation of NPs when acetone is dispersed in water phase, resulting to 100-nm to 500-nm polymer that initially solubilize chitosan [97]. To achieve complete acetone dispersion, surplus water is added, and the NPs are sort out by centrifugation. This new method decreases both the dimension of the synthesized nano-chitosan particles and its dissemination. The use of emulsification results in preferred control particle dimension, but it often involves the use and removal of remaining agents of cross-linking [98]. Both water insoluble and solubility drugs are prepared using the solvent oil blend dispersal technique. In the instant of hydrophilic drugs, a mixture of water, oil, and water–oil blend with drug's dissolution in the beginning of the liquid phase must be determined through double emulsion. This technique has two drawbacks: extensive shear forces and the solvents needed to form NPs.

2.3.5 Reverse Micelles

Reverse micelles are a combination of water, oil, and surfactant that combine to create reversible liquid mixtures. Highly fine small-dimension particles are the reason for its dynamic behavior, unlike other traditional methods of oil blend polymerization [90]. To make reverse micelles, dissolution of the surfactant in an organic solvent is done using this method. They move the micellar drops at random and perform a random movement [99]. The chitosan and drugs are introduced to the emulsion through a liquid phase with constant agitation, wherein NPs form at the center. Liquid controlled phase maintains a virtually clear micro-emulsion phase throughout the mixture [90]. To generate NPs of greater size, Shi *et al.* [100] suggest using water content rise. The dimension of oil/water drops increased in series. Micellar concentration reduces as the water content rises. The solution that is clear is made by adding a cross-linking agent and agitating continuously through the night. Drug quantity in solution differs from one another in reverse micelles; it must be gently added until solution becomes clear. After vaporization, NPs are obtained while purifying the solvent [90].

2.3.6 Polyelectrolyte Complex (PEC)

A PEC is formed when the positively charged polymer and plasmid DNA self-assemble due to lack of affinity for water. It is believed that the use of negatively charged macromolecules results in a polymer electrolyte appearance [86]. Chitosan is manufactured automatically by adding acetic acid solution in combination with DNA, agitating it. Different sizes of chitosan are achieved through appearance of cationic/anionic species at room temperature [101]. Fluorine exterior change of carboxymethyl nano-chitosan was used to produce them for effective drug conveyance, and they are sensitive to pH. To produce the sensitive-pH NPs, N-(3-aminopropyl)-imidazole was affixed onto carboxymethyl nano-chitosan and then combined with perfluorobutyric anhydride to form NPs that are fluorinated. The study indicated that carboxymethyl nano-chitosan could be an effective means of transporting drugs for cancer chemotherapy [102].

2.3.7 Biological Synthesis

Microorganisms, including bacteria, fungi actinomycete, and plants, are utilized in biological synthesis [103, 104]. A method of synthesizing NPs that links nano-innovation and bio-innovations takes a technique called bottom-up. Biological NP synthesis is most preferred because it is safe and economical and because of its energy consumption and converts large molecules into nanoscale by harnessing the maximum oxidation-reduction reactions of living organisms than physical and chemical processes [105, 106].

NP synthesis done biologically via catalytic action in cell involves taking metallic ions and detoxifying them through oxidation-reduction reactions. This occurs both inside and outside the cell. It intracellularly involves transportation of metallic ions into the cell and it is decreased within the cytoplasm, periplasmatic space, or cell wall to produce NPs that develop inside the cells. Conversely, NP created from the latter is form on cell exterior through reductase activity [107]. The process of obtaining NPs from the cell network is time-consuming, making external production of nanomaterials preferred. Microorganisms are commonly regarded as biological nano-factories that can produce NPs on a wide scale with conditional restraint over their dimension and structure, to easier production process [108].

Plant extracts are used in phytofabrication, with the addition of metal ions to form NPs. Typically, the process of phytoproduction of NPs involves adding metallic ions to extract from plant and agitating it in a magnetic turner. Color change results in NPs formation in solution [105]. Polyphenol, flavonoids, and phytoalexins, which are antioxidants that convert metallic ions into non-poisonous metallic ion in the make of nano, play a crucial role in this. A plant with a greater proportion of natural decreasing elements is preferred for phyto-production of NPs. A plant known as rose geranium is rich in natural antioxidizing agents. Its leaves and essential oil have various therapeutic uses. A mixture of chitosan in solution and extract of *P. graveolens* was mixed in equal ratio and was nurtured at 50°C in a shaker [106]. In a turbid solution that represents oxidation-reduction reaction in progress, centrifugation is carried out at 10,000 x g for 10 min. The chitosan was left inactive and then subjected to freeze drying to remove nano-chitosan. Biologically, created nano-chitosan from fungal cell contains chitosan. Chitosan treatment with alkali and acid was done to eliminate fungal biomass using reserved conditions. NaOH is used to treat biomass of fungal and then applied as an alkali-insolvable material (AIM) that was subsequently extracted by acid treat. A 1% TPP solution was agitated in a magnetic stirrer to produce nano-chitosan [109]. Chitosan solutions are made by adding negatively charged proteins from *Penicillium oxalicum* culture to the solution. The biogenic procedure yields nano-chitosan with powerful antifungal properties [110]. Moreover, the use of curcumin-packed NPs with dextran sulfate and nano-chitosan preparations led to the selective destroying cells of cancer. According to Anitha *et al.* [111], curcumin-packed nano-preparations have the potential to be a good candidate for cancer treatment.

2.3.8 Biogenic Synthesis of Nano-Chitosan Over Other Nanoparticles

Chitosan is a more adaptable and biocompatible NP than most other NPs, with less toxic and can be broken down in the same time span [112]. Microbial inhibition by nano-chitosan is as a result of antioxidation ability, antifungal, and other antiviral properties, response to immune system, and ability to manage plant stress. Chitosan's less toxic makeup makes it an ideal choice of restricted pesticide release, fertilization, and promoters of plant growth. Its utilization in wastewater treatment, cosmetics, adhesive, drug delivery, and bioimaging applications is also possible [113]. The nano-chitosan and nonchemical chitosan employ various techniques, such as spray drying, cross-linking in the emulsion, ionic gelation, and precipitation [114]. The disadvantage of chemical/physical techniques like the sieving technique can produce uneven structure and dimension of NPs. Although, cross-linking emulsion procedure is stressful and needs toxic chemicals formaldehyde, which has a tendency of causing obstacles due to its incompatibility with the ingredients active in agrochemicals. The dimension of the formed particles is mainly determined by the dimension [113], agitation speed of an emulsion, cross-linking, surface active agent type, and molecular mass of chitosan. The thermodynamically stable reversal micellar technique produces the tiny, evenly distributed nanoparticulate chitosan technique. The process requires a poisonous, exorbitant, and tedious surfactant (acetyl trimethyl ammonium bromide). Chitosan NPs (ChNPs) were created through precipitation processes without stability, uneven structure, and less strength. Chitosan produced using spray drying techniques, including dry granules, pellets, and powders. The methods involve the inclusion of acetic acid to chitosan dispersed in solution and cross-receptor agents. Nanoparticles are produced by the vaporization of the precursor solution using hot air steam. Several disadvantages of the chemical and physical methods discussed above justify biogenic processes. Advantages of biogenic synthesis of nano-chitosan include their firmness, uniformity, structure and dimension, bio-accessibility as biogenic compounds involved in coating, decrease procedure, non-toxic, no extra medium or hard work needed, and more. Thus, in recent times, the scientific inquiry has shifted its focus to biological synthesis of NPs by biogenic means. The primary aim of nano-preparation is to increase the gain of agricultural chemicals and chitosan while simultaneously minimizing unwanted effects [115].

2.3.9 Chitosan Nanoparticle Characterization

Particle size of nano-chitosan is a vital physical attribute for measuring, analyzing, and characterizing it with devices that fall within the sub-nanometer-to-millimeter range. SEM is commonly analyzed morphologically; NP properties are focused on the sample. The sensitive detector picks up on various signals and presents an image of the sample with no diffraction limit. Samples are covered with conductive matter (gold) and placed on metal butt below drying vaccum before being processed for SEM. SEM can show exterior coarseness/regular shape or permeable shape, which is useful in future analysis like insolubility behavior or intracellular reactions [116]. SEM used to study chitosan NPs alginate has been observed [117]. Chitosan NPs are analyzed using the transmission electron microscope system. Here, nanomaterial sample allows electrons from an energy beam to pass through [118]. The first step in sample preparation using TEM is to dry and observe NPs over a layer of carbon. The smallest TEM value of 0.2 nm is resolved under absolute conditions [119]. Nanomaterials have been extensively characterized using cryogenic-SEM/TEM. This technique totally moistened chemically altered nanomaterials at very low temperatures. It is employed to analyze the exterior appearance, structure, dimension, and inside structure of nano-chitosan, just like SEM and TEM. Due to the speed of sample preparation, ease, and efficiency, it is now popular [120]. By using the critical point drying method, there are no prerequisites for preparing a standard sample. Cryo-SEM/TEM is a more effective method than SEM and TEM due to its resolution power, speed, fully hydrated state, and ability to relocate dispersed materials. Zeta potential's strength fluctuates with the amount of static electric attraction between particles that are charged. Particle density and stability in the colloidal system are indicated by high electro-kinetic potential, not minding its electrical [121].

X-ray photoelectron spectroscopy (XPS) is used to identify chemical states and patterns that change with particle size. The exterior particle properties are analyzed in time-of-flight secondary ion mass spectrometry (ToF-SIMS), which also presents the chemical compositions. The surface chitosan was studied by using ToF-SIMS and electron microscopy through XPS [122]. Morphology and chemical constituent of chitosan was examined using ToF-SIMS/XPS in other studies bearing in mind stabilizers of NPs [123]. Electro-kinetic potential and emulsifier were measured by laser Doppler anemometry for nano-chitosan. Packing and enveloped efficacy of NP earnings and the contents of drug are all important parameters that should be roughly calculated before use in nanotechnology [123].

2.3.10 Nano-Chitosan Applications

2.3.10.1 In Agriculture

Nano-chitosan is microbial inhibitory agent in agriculture. The use of ChNP preparations is said to strengthen the plant mechanisms for protection, as such it recognizes the defense catalytic operations. Polyacrylic acid–based ChNP preparations offer significant control over certain common pests, used in soybean farming [124]. Several studies established nano-chitosan preparation improves protection system in plants by obtaining the protection catalytic functions.

Chitosan failed to inhibit mycelia as effectively as the nano-chitosan. ChNPs' decreases dimension, permeability, and electro-kinetic potential, which make them very firm, which has been shown to have an impact on fungal pathogens that have been tested. Oleoyl-chitosan nano-preparation was found to have a significant antifungal consequence against various pathogens [125]. Like tomato plants, they are able to resist *Phytophtora infestans* and *A. solani* by being treated with chitosan and then applying it from the foliar [126]. Anthracnose disease in cucumbers is brought about by *Colletotrichum* spp. The application chitosan through foliar was efficient in the control [127]. Maize seeds were subjected to copper nano-chitosan for 4 h (chitosan, 50–190 kDa; deacetylation degree, 80%); then, the sprouting plants were sprayed daily for 35 days; the results indicate more plant height, root extent and amount, content of chlorophyll, plot/extent of ear, and high yield of grain [128]. Waste of silica with chitosan by Indonesian farmers could potentially decrease the need for NKP fertilizers, which would have environmental and economic benefits [129].

The packing, enveloping capability, discharge profile, particle dimension, and structure are crucial factors in agrochemicals development of nano-chitosan. Multiple studies examined and reported all the parameters mentioned while creating and developing nano-preparation. An herbicide (diuron) nano-conveyor system was created to control weed growth by linking chitosan carboxymethyl/2-nitro benzyl as a photosynthetic inhibitor. These nano-preparations were able to be released through a photo-controlled release mechanism [130].

In the review of Maluin and Hussein [113], the protection of crops using nano-chemicals was a sustainable approach, and they did identify variable forms of these chemicals in various applications such as gradual release, boost for plant growth, and biocidal. The use of ChNPs and their nano-preparations as booster of plant was studied. Protonated chitosan with a high charge shows increased affinity with regard to cell membrane, leading to increased plant activity. Chitosan's content of nitrogen is roughly 9%–10%, which makes it a crucial supplement for the plant [114]. The use of HRTEM reveals nano-chitosan of different dimensions, including an increased molecular weight chitosan oligomer. The nutrient absorption of Mg, Ca, N, K, and P was boost when seedling of Robusta coffee was dispersed on leaves, different dimension, such as 420 nm for small, 50 nm for medium, or 970 ppm for large. Nano-chitosans in all dimensions are superior to chitosan oligomers. Absorption of nutrients was not notably impacted by the size [131]. The treatment of ChNPs has resulted in an increase of 61% and 29% in small-sized and 82% and 55% in medium-sized NPs. Bio-accessibility hormones could be improved by using chitosan nano-conveyors to efficiently release regulator for plant growth-containing hormone ingredients through nano-effective delivery system. Hormones of plants, including their class II and class III classes I, are the only known plant growth regulators. French beans showed a 37% increase development in root [132].

2.3.10.2 *Biomedicals*

Nano-chitosan observed certain areas to regulate rate of drug release; boost therapeutic efficacy; and convey drugs, DNA, proteins, or antigens to a target site [133, 134]. Nano-chitosan has been employed in pulmonary inhalations, mucosal adhesion systems, oral conveyance systems for peptic ulcer treatment, and drug targeting intratumoral. Physical trapping of layer polymeric binds to spine could be employed as the mechanism for loading. Physical trapping is an effective method in the delivery of drug. Two distinct methods exist for physically loading of drugs, when formulating and incubating phases of nano-particles. The drug is either fixed into the network or adsorbed onto the exterior in these systems. Stacking relays on formulation process for drug compositions and various techniques have been formed to achieve this aim. Both soluble and insoluble drugs can be stacked using these processes. Its primary focus is on introducing biomolecules without decreasing its biogenic activity [90, 135]. Drugs can be delivered through NPs either actively or passively. Leaks in capillary aperture are responsible for the transportation of nutrients in and through passive delivery. Through passive diffusion, the cells and tumor interstitium become exposed. The use of active moiety target allows for the increase delivery to a specific site, which is determined by molecular recognition site. The NP is coated with antibody externally, and interaction between them and target antigenic site occurs [135, 136]. Release drug generally involves a few mechanisms: exterior surfaces release, dispersal via network, and by erosion of films of polymer. Majority of times, the release of drug occurs through multiple pathways [90]. Chitosan-based particulate systems release drugs outside living cells differently based on cross-linking, physical appearance, and adjuvant presence. Factors like insolubility in medium pH, polarity, and catalytic presence affect the release done inside the cell. A sonication bath used to mix 5 mg/100 μL quantity of drug soluble in chemical (chloroform) with corresponding amounts of NPs in varying solutions until the opaque solution became clear. The water affinity core of NPs was used to ascertain the free drug's absorbance and calculate its entrapment efficacy using a ultraviolet ray at 249 nm [137].

2.3.10.3 Industry

Chitosan was developed as a packaging material that could be used to combat bacteria and fungi, including microorganisms, and increase shelf life. A chitosan NP edible film made from hydroxypropyl methylcellulose was developed [138]. They used of hydroxypropyl methylcellulose in chitosan/tripolyphosphate NPs. The film's mechanical compositions, like permeable to water vapor, and heat stability, were examined. Nano-chitosans have been found to increase the film barrier characteristics, as well as heat stability, in this film. Moreover, the empty spaces in the matrix are filled with nano-chitosan, which boost the tensile strength of the film and also permeability of water vapor. For food and beverage packaging, carboxymethylcellulose film with nano-chitosan has been explored as a good material, given their stability/properties [139]. Additionally, nano-chitosans have created opening mechanisms for reducing components of food lost during processing. L-ascorbic acid's stability is modified through chitosan NPs, as suggested by Jang and Lee [140]. Chitosan preparation, which was packed with ascorbic acid, involved the ionic gelation with tripolyphosphate that is negatively charged. The release profile was assessed by testing different temperatures and concentrations of NPs stacked with ascorbic acid. Increasing the temperature at release profiles was found to improve the stability of burst release on the exterior. As the release time increased, so did the internal ascorbic acid. The internal stability of chitosan NPs stacked with ascorbic acid was found to be higher than their exterior stability, which was dependent on temperature. The antioxidant effects were noted to be increased by the continual nano-chitosan release of ascorbic acid that contains ascorbic acids. Nanoparticles for food industry applications as intelligent packaging were explored [139]. Quaternization reaction yielded chitosan N,N,N-trimethyl trimethylated chitosan chloride (TMC) and then produced (tripolyphosphate) TPP NPs by combining chitosan with TMP through ionic gelation. The effect of TMC NPs on their formulation conditions and characteristics was later examined. It is argued that the optimization of TMC-TPP NP production for food industry is crucial due to the ability of nano-innovation to enhance the performance of ready to eat films used in food applications.

2.4 Conclusion

In conclusion, agricultural waste disposal has led to the accumulation of waste, resulting to environmental problem. This waste can be managed effectively as resources for the production of nanoscale materials. Shell waste deproteinization and demineralization yield chitin, and the removal of acetyl group from chitin results in polysaccharide chitosan. Chitosan is an organic wonder drug with fascinating antimicrobial and stimulating properties. The recent focus on nano-chitosan is due to its increased effectiveness, biodegradability, biocompatibility, and non-toxicity. Nano-chitosans are synthesized using both physical and biological techniques. Biogenic synthesis and phytofabrication of nano-chitosan have more advantages due to its bio-accessibility, economical, and eco-friendly features. Chitosan NPs in combination with the other metallic and metal oxide NPs show a considerable range of activities. Nano-chitosan is effectively employed in biomedical, agriculture, and waste water treatment; can increase drug and gene delivery; remove pollutants from the environment; and also maintain quality and yield of crops. Despite the potential advantages of nano-chitosan, more studies are required on the developments of products and optimization procedure required for large production and *in vivo* use. Further investigation on new products based on nano-chitosan and long-term dependability is a significant barrier to scaling-up chitosan formulations in pharmaceuticals. Moreso, studies need to be channelized to reveal the exact mode of action of nano-chitosan specific to plants and pet control.

References

1. Obi, F.O., Ugwuishiwu, B.O., Nwakaire, J.N., Agricultural waste concept, generation, utilization and management. *Niger. J. Technol.*, 35, 4, 957–64, 2016 Oct 11.
2. Yuan, N., Tan, K., Zhang, X., Zhao, A., Guo, R., Synthesis and adsorption performance of ultra-low silica-to-alumina ratio and hierarchical porous ZSM-5 zeolites prepared from coal gasification fine slag. *Chemosphere*, 303, 134839, 2022 Sep 1.
3. Agamuthu, P., Challenges and opportunities in agro-waste management: An Asian perspective, in: *Inaugural meeting of first regional 3R forum in Asia*, vol. 11, University of Malasiya, p. 12, 2009 Nov 11.
4. Lu, H., Yadav, V., Bilal, M., Iqbal, H.M., Bioprospecting microbial hosts to valorize lignocellulose biomass–Environmental perspectives and value-added bioproducts. *Chemosphere*, 288, 132574, 2022 Feb 1.
5. Elemike, E.E., Ekennia, A.C., Onwudiwe, D.C., Ezeani, R.O., Agro-waste materials: Sustainable substrates in nanotechnology, in: *Agri-Waste and Microbes for Production of Sustainable Nanomaterials*, pp. 187–214, Elsevier, Amsterdam, the Netherlands, 2022 Jan 1.
6. Venkateshaiah, A., Padil, V.V., Nagalakshmaiah, M., Waclawek, S., Černík, M., Varma, R.S., Microscopic techniques for the analysis of micro and nano-structures of biopolymers and their derivatives. *Polymers*, 12, 3, 512, 2020 Feb 27.
7. Khan, S.A., Khan, S.B., Khan, L.U., Farooq, A., Akhtar, K., Asiri, A.M., Fourier transform infrared spectroscopy: fundamentals and application in functional groups and nanomaterials characterization, in: *Handbook of materials characterization*, pp. 317–44, 2018.
8. Guleria, G., Thakur, S., Shandilya, M., Sharma, S., Thakur, S., Kalia, S., Nanotechnology for sustainable agro-food systems: The need and role of nanoparticles in protecting plants and improving crop productivity. *Plant Physiol. Biochem.*, 194, 533–549, 2022 Dec 9.
9. Afolalu, S.A., Ikumapayi, O.M., Oloyede, O.R., Ogedengbe, T.S., Ogundipe, A.T., Advances in Nanotechnology and Nanoparticles in the 21st Century–An Overview, in: *Proceedings of the 3rd African International Conference on Industrial Engineering and Operations Management*, pp. 5–7, 2022 Apr.
10. Sharma, D., Kanchi, S., Bisetty, K., Biogenic synthesis of nanoparticles: a review. *Arabian J. Chem.*, 12, 8, 3576–600, 2019 Dec 1.

11. Pattanaik, L., Pattnaik, F., Saxena, D.K., Naik, S.N., Biofuels from agricultural wastes, in: *Second and third generation of feedstocks*, pp. 103–142, Elsevier, New Delhi, India, 2019 Jan 1.
12. Saini, J.K., Saini, R., Tewari, L., Lignocellulosic agriculture wastes as biomass feedstocks for second-generation bioethanol production: concepts and recent developments. *3 Biotech*, 5, 337–53, 2015 Aug.
13. Alshehri, M.A., Aziz, A.T., Trivedi, S., Panneerselvam, C., Efficacy of chitosan silver nanoparticles from shrimp-shell wastes against major mosquito vectors of public health importance. *Green Process. Synth.*, 9, 1, 675–84, 2020 Nov 19.
14. Bittelli, M., Flury, M., Campbell, G.S., Nichols, E.J., Reduction of transpiration through foliar application of chitosan. *Agric. For. Meteorol.*, 107, 3, 167–75, 2001 Apr 2.
15. Jayakumar, R., Chennazhi, K.P., Srinivasan, S., Nair, S.V., Furuike, T., Tamura, H., Chitin scaffolds in tissue engineering. *Int. J. Mol. Sci.*, 12, 3, 1876–87, 2011 Mar 15.
16. Jayakumar, R., Prabaharan, M., Reis, R.L., Mano, J., Graft copolymerized chitosan—present status and applications. *Carbohydr. Polym.*, 62, 2, 142–58, 2005 Nov 10.
17. Awogbemi, O. and Von Kallon, D.V., Valorization of agricultural wastes for biofuel applications. *Heliyon*, 2022 Oct 18.
18. Beuchat, L.R., Oncom (fermented peanut press cake), in: *Legume-based fermented foods*, N.R. Reddy, M.D. Pierson, D.K. Salunkhe (Eds.), pp. 135–44, CRC Press, Boca Raton, 1986.
19. Sarkar, N., Ghosh, S.K., Bannerjee, S., Aikat, K., Bioethanol production from agricultural wastes: an overview. *Renew. Energy*, 37, 1, 19–27, 2012 Jan 1.
20. Jahirul, M.I., Rasul, M.G., Chowdhury, A.A., Ashwath, N., Biofuels production through biomass pyrolysis—a technological review. *Energies*, 5, 12, 4952–5001, 2012 Nov 23.
21. Jogi, K. and Bhat, R., Valorization of food processing wastes and by-products for bioplastic production. *Sustain. Chem. Pharm.*, 18, 100326, 2020 Dec 1.
22. Putri, D.N., Sahlan, M., Montastruc, L., Meyer, M., Negny, S., Hermansyah, H., Progress of fermentation methods for bio-succinic acid production using agro-industrial waste by Actinobacillus succinogenes. *Energy Rep.*, 6, 234–9, 2020 Feb 1.
23. Duran, M., Aday, M.S., Zorba, N.N., Temizkan, R., Büyükcan, M.B., Caner, C., Potential of antimicrobial active packaging 'containing natamycin, nisin, pomegranate and grape seed extract in chitosan coating'to extend shelf life of fresh strawberry. *Food Bioprod. Process.*, 98, 354–63, 2016 Apr 1.
24. Meral, R., Kose, Y.E., Ceylan, Z., Cavidoglu, I., The potential use of agro-industrial by-products as sources of bioactive compounds: A nanotechnological approach. *Stud. Nat. Prod. Chem.*, 73, 435–66, 2022 Jan 1.
25. Zamare, D., Vutukuru, S.S., Babu, R., Biosynthesis of nanoparticles from agro-waste: a sustainable approach. *Int. J. Eng. Appl. Sci. Technol.*, 1, 12, 85–92, 2016.

26. Yadav, M. and Kaur, P., A review on exploring phytosynthesis of silver and gold nanoparticles using genus Brassica. *Int. J. Nanopart.*, 10, 3, 165–77, 2018.
27. Faqir, Y., Ma, J., Chai, Y., Chitosan in modern agriculture production. *Plant Soil Environ.*, 67, 12, 679–99, 2021 Dec 10.
28. Vani, R. and Stanley, S.A., Studies on the extraction of chitin and chitosan from different aquatic organisms. *Adv. BioTech.*, 12, 12, 12–5, 2013.
29. Kardas, I., Struszczyk, M.H., Kucharska, M., van den Broek, L.A., van Dam, J.E., Ciechańska, D., Chitin and chitosan as functional biopolymers for industrial applications, in: *The European Polysaccharide Network of Excellence (EPNOE) Research Initiatives and Results*, pp. 329–73, 2012.
30. Younes, I. and Rinaudo, M., Chitin and chitosan preparation from marine sources. Structure, properties and applications. *Mar. Drugs*, 13, 3, 1133–74, 2015 Mar 2.
31. Patria, A., Production and characterization of Chitosan from shrimp shells waste. *Aquacult. Aquarium Conserv. Legislation*, 6, 4, 339–44, 2013 Jul 1.
32. Joseph, J., Manigundan, K., Shamya Arokia Rajan, M., Radhakrishnan, M., Gopikrishnan, V., Kumaran, S., Thirunavukkarasu, R., Aruni, W., Shanmugam, V., Conversion of aquaculture waste into biomedical wealth: Chitin and chitosan journey, in: *Advances in Materials Science and Engineering*, 2022 Apr 22, 2022.
33. Shahidi, F. and Synowiecki, J., Isolation and characterization of nutrients and value-added products from snow crab (Chionoecetes opilio) and shrimp (Pandalus borealis) processing discards. *J. Agric. Food. Chem.*, 39, 8, 1527–32, 1991 Aug.
34. Pal, J., Verma, H.O., Munka, V.K., Maurya, S.K., Roy, D., Kumar, J., Biological method of chitin extraction from shrimp waste an eco-friendly low cost technology and its advanced application. *Int. J. Fish. Aquat. Stud.*, 1, 6, 104–7, 2014.
35. Gildberg, A. and Stenberg, E., A new process for advanced utilisation of shrimp waste. *Process Biochem.*, 36, 8-9, 809–12, 2001 Mar 1.
36. Manni, L., Ghorbel-Bellaaj, O., Jellouli, K., Younes, I., Nasri, M., Extraction and characterization of chitin, chitosan, and protein hydrolysates prepared from shrimp waste by treatment with crude protease from Bacillus cereus SV1. *Appl. Biochem. Biotechnol.*, 162, 345–57, 2010 Sep.
37. Mati-Baouche, N., Elchinger, P.H., de Baynast, H., Pierre, G., Delattre, C., Michaud, P., Chitosan as an adhesive. *Eur. Polym. J.*, 60, 198–212, 2014 Nov 1.
38. Nguyen, V.B. and Wang, S.L., Reclamation of marine chitinous materials for the production of α-glucosidase inhibitors via microbial conversion. *Mar. Drugs*, 15, 11, 350, 2017 Nov 7.
39. Wassila, A., Adour, L., Amrane, A., Chitin extraction from crustacean shells using biological methods, review. *Food Technol. Biotechnol.*, 51, 1, 12–25, 2013.

40. Pereira, A.G., Muniz, E.C., Hsieh, Y.L., Chitosan-sheath and chitin-core nanowhiskers. *Carbohydr. Polym.*, 107, 158–66, 2014 Jul 17.
41. Dash, M., Chiellini, F., Ottenbrite, R.M., Chiellini, E., Chitosan—A versatile semi-synthetic polymer in biomedical applications. *Prog. Polym. Sci.*, 36, 8, 981–1014, 2011 Aug 1.
42. Zhao, L.M., Shi, L.E., Zhang, Z.L., Chen, J.M., Shi, D.D., Yang, J., Tang, Z.X., Preparation and application of chitosan nanoparticles and nanofibers. *Braz. J. Chem. Eng.*, 28, 353–62, 2011.
43. Hajji, S., Younes, I., Ghorbel-Bellaaj, O., Hajji, R., Rinaudo, M., Nasri, M., Jellouli, K., Structural differences between chitin and chitosan extracted from three different marine sources. *Int. J. Biol. Macromol.*, 65, 298–306, 2014 Apr 1.
44. Kunjachan, S., Jose, S., Lammers, T., Understanding the mechanism of ionic gelation for synthesis of chitosan nanoparticles using qualitative techniques. *Asian J. Pharm. (AJP)*, 4, 2, 148–153, 2010.
45. Wijesena, R.N., Tissera, N., Kannangara, Y.Y., Lin, Y., Amaratunga, G.A., de Silva, K.N., A method for top down preparation of chitosan nanoparticles and nanofibers. *Carbohydr. Polym.*, 117, 731–8, 2015 Mar 6.
46. Qi, L., Xu, Z., Jiang, X., Hu, C., Zou, X., Preparation and antibacterial activity of chitosan nanoparticles. *Carbohydr. Res.*, 339, 16, 2693–700, 2004 Nov 15.
47. Bowman, K. and Leong, K.W., Chitosan nanoparticles for oral drug and gene delivery. *Int. J. Nanomed.*, 1, 2, 117–28, 2006 Dec 1.
48. Colonna, C., Conti, B., Perugini, P., Pavanetto, F., Modena, T., Dorati, R., Genta, I., Chitosan glutamate nanoparticles for protein delivery: development and effect on prolidase stability. *J. Microencapsulation*, 24, 6, 553–64, 2007 Jan 1.
49. Corradini, E., De Moura, M.R., Mattoso, L.H., A preliminary study of the incorparation of NPK fertilizer into chitosan nanoparticles. *eXPRESS Polym. Lett.*, 4, 8, 509–515, 2010 Aug 1.
50. Kafetzopoulos, D., Martinou, A., Bouriotis, V., Bioconversion of chitin to chitosan: purification and characterization of chitin deacetylase from Mucor rouxii. *Proc. Natl. Acad. Sci.*, 90, 7, 2564–8, 1993 Apr 1.
51. No, H.K. and Meyers, S.P., Preparation and characterization of chitin and chitosan-A review. *J. Aquat. Food Prod. Technol.*, 4, 2, 27–52, 1995.
52. Chang, K.L., Tsai, G., Lee, J., Fu, W.R., Heterogeneous N-deacetylation of chitin in alkaline solution. *Carbohydr. Res.*, 303, 3, 327–332, 1997.
53. Rinaudo, M. and Domard, A., A solution property of chitosan, in: *editorsChitin and Chitosan. Sources, Chemistry, Biochemistry, Physical properties and applications*, G. Skajak-Braek, T. Anthonsen, P. Sanford (Eds.), pp. 71–86, Elsevier, London and New York, 1989.
54. Aiba, S.I., Studies on chitosan: 4. Lysozymic hydrolysis of partially N-acetylated chitosans. *Int. J. Biol. Macromol.*, 14, 225–228, 1992.

55. Berger, J., Reist, M., Chenite, A., Felt-Baeyens, O., Mayer, J.M., Gurny, R., Pseudo-thermosetting chitosan hydrogels for biomedical application. *Int. J. Pharm.*, 288, 2, 197–206, 2005.
56. Tsigos, I., Martinou, A., Kafetzopoulos, D., Bouriotis, V., Chitin deacetylases: new, versatile tools in biotechnology. *Trends Biotechnol.*, 18, 7, 305–312, 2000.
57. Alfonso, C., Nuero, O., Santamaria, F., Reyes, F., Purification of a heat-stable chitin deacetylase from Aspergillus nidulans and its role in cell wall degradation. *Curr. Microbiol.*, 30, 1, 49–54, 1995.
58. Tsigos, I. and Bouriotis, V., Purification and characterization of chitin deacetylase from Colletotrichum lindemuthianum. *J. Biol. Chem.*, 270, 44, 26286–26291, 1995.
59. Gao, X.D., Katsumoto, T., Onodera, K., Purification and characterization of chitin deacetylase from Absidia coerulea. *J. Biochem.*, 117, 2, 257–263, 1995.
60. Dutta, P.K., Dutta, J., Tripathi, V., Chitin and chitosan: Chemistry, properties and applications. *J. Sci. Ind. Res.*, 63, 20–31, 2004.
61. Younes, I., Sellimi, S., Rinaudo, M., Jellouli, K., Nasri, M., Influence of acetylation degree and molecular weight of homogeneous chitosans on antibacterial and antifungal activities. *Int. J. Food Microbiol.*, 185, 57–63, 2014.
62. Nwe, N. and Stevens, W., Production of chitin and chitosan and their applications in the medical and biological sector. *Recent Res. Biomed. Aspect Chitin Chitosan*, 978, 161–176, 2008.
63. Shante, P., Yadav, P.N., Adhikari, R., Application of chitin and chitosan in industry and medical sciences: A review. *Nepal J. Sci. Technol.*, 16, 1, 99, 2016, DOI: 10.3126/njst.v16i1.14363.
64. Gooday, G.W., The ecology of chitin degradation, in: *Advances Microb Ecol*, pp. 387–430, Springer, Berlin/Heidelberg, Germany, 1990.
65. Santos, V.P., Marques, N.S., Maia, P.C.B., Lima, M.A.B.D., Franco, L.D.O., Seafood waste as attractive source of chitin and chitosan production and their applications. *Int. J. Mol. Sci.*, 21, 4290, 2020.
66. Hayes, M., Carney, B., Slater, J., Brück, W., Mining marine shellfish wastes for bioactive molecules: Chitin and chitosan - Part A: Extraction methods. *Biotechnol. J. Healthcare Nutr. Technol.*, 3, 871–877, 2008.
67. Bugnicourt, L. and Ladavière, C., Interests of chitosan nanoparticles ionically cross-linked with tripolyphosphate for biomedical applications. *Prog. Polym. Sci.*, 60, 1–17, 2016.
68. Yamamoto, H., Kuno, Y., Sugimoto, S., Takeuchi, H., Kawashima, Y., Surface-modified PLGA nanosphere with chitosan improved pulmonary delivery of calcitonin by mucoadhesion and opening of the intercellular tight junctions. *J. Control. Release*, 102, 373–381, 2005.
69. Fernández-Pan, I., Maté, J.I., Gardrat, C., Coma, V., Effect of chitosan molecular weight on the antimicrobial activity and release rate of carvacrol-enriched films. *Food Hydrocolloids*, 51, 60–68, 2015.

70. Kong, M., Chen, X.G., Xing, K., Park, H.J., Antimicrobial properties of chitosan and mode of action: A state of the art review. *Int. J. Food Microbiol.*, 144, 51–63, 2010.
71. Salamat-Miller, N., Chittchang, M., Johnston, T.P., The use of mucoadhesive polymers in buccal drug delivery. *Adv. Drug Deliv. Rev.*, 57, 1666–1691, 2005.
72. Okamoto, Y., Yano, R., Miyatake, K., Tomohiro, I., Shigemasa, Y., Minami, S., Effects of chitin and chitosan on blood coagulation. *Carbohydr. Polym.*, 53, 337–342, 2003.
73. Busilacchi, A., Gigante, A., Mattioli-Belmonte, M., Manzotti, S., Muzzarelli, R.A., Chitosan stabilizes platelet growth factors and modulates stem cell differentiation toward tissue regeneration. *Carbohydr. Polym.*, 98, 665–676, 2013.
74. Ohya, Y., Shiratani, M., Kobayashi, H., Ouchi, T., Release behavior of 5-fluorouracil from chitosan-gel nanospheres immobilizing 5-fluorouracil coated with polysaccharides and their cell specific cytotoxicity. *J. Macromol. Sci., Pure Appl. Chem.*, 31, 629–642, 1995.
75. Rizeq, B.R., Younes, N.N., Rasool, K., Nasrallah, G.K., Synthesis, bio applications, and toxicity evaluation of chitosan-based nanoparticles. *Int. J. Mol. Sci.*, 16, 101–112, 2015.
76. Luque-Alcaraz, A.G., Lizardi-Mendoza, J., Goycoolea, F., Higuera-Ciapara, I., Argüelles-Monal W. Preparation of chitosan nanoparticles by nanoprecipitation and their ability as a drug nanocarrier. *RSC Adv.*, 6, 59250–59256, 2016.
77. Grenha, A., Chitosan nanoparticles: A survey of preparation methods. *J. Drug Target*, 20, 291–300, 2012.
78. Barreras-Urbina, C.G., Ramírez-Wong, B., López-Ahumada, G.A., Burruel-Ibarra, S.E., Martínez-Cruz, O., Tapia-Hernández, J.A., Rodriguez, F., Nano- and micro-particles by nanoprecipitation: Possible application in the food and agricultural industries. *Int. J.*, 19, 1912–1923, 2019.
79. Khan, I.U., Serra, C.A., Anton, N., Vandamme, T.F., Production of nanoparticle drug delivery systems with microfluidics tools. *Expert Opin. Drug Deliv.*, 12, 547–562, 2015.
80. Ilium, L., Chitosan and its use as a pharmaceutical excipient. *Pharm. Res*, 15, 1326–1331, 1998.
81. Ngan, L.T., Wang, S.L., Hiep, ETHM., Luong, P.M., Vui, N.T., Đinh, T.M., Dzung, N.A., Preparation of chitosan nanoparticles by spray drying, and their antibacterial activity. *Res. Chem. Intermed.*, 40, 2165–2175, 2014.
82. Mikušová, V. and Mikuš, P., Advances in chitosan-based nanoparticles for drug delivery. *Int. J. Mol. Sci.*, 22, 9652, 2021.
83. Öztürk, A.A. and Kıyan, H.T., Treatment of oxidative stress-induced pain and inflammation with dexketoprofen trometamol loaded different molecular weight chitosan nanoparticles: Formulation, characterization and

anti-inflammatory activity by using *in vivo* HET-CAM assay. *Microvasc. Res.*, 128, 103961, 2020.
84. Kafshgari, M.H., Khorram, M., Khodadoost, M., Khavari, S., Reinforcement of chitosan nanoparticles obtained by an ionic cross-linking process. *Iran. Polym. J.*, 20, 445–456, 2011.
85. Bodmeier, R., Oh, K.H., Pramar, Y., Preparation and evaluation of drug-containing chitosan beads. *Drug Dev. Ind. Pharm.*, 15, 1475–1494, 1989.
86. Calvo, P., Remuñan-López, C., Vila-Jato, J.L., Alonso, M.J., Chitosan and chitosan ethyleneoxide-propyleneoxide block co-polymer nanoparticles as novel carriers for proteins and vaccines. *Pharm. Res.*, 14, 1431–1436, 1997a.
87. Calvo, P., Remunan-Lopez, C., Vila-Jato, J.L., Alonso, M., Novel hydrophilic chitosan-polyethylene oxide nanoparticles as protein carriers. *J. Appl. Polym. Sci.*, 63, 125–132, 1997b.
88. Patil, J., Kamalapur, M., Marapur, S., Kadam, D., Ionotropic gelation and polyelectrolyte complexation: The novel techniques to design hydrogel particulate sustained, modulated drug delivery system: A review. *Dig. J. Nanomater. Biostruct.*, 5, 241–248, 2010.
89. Bodnar, M., Hartmann, J.F., Borbely, J., Preparation and characterization of chitosan-based nanoparticles. *Biomacromolecules*, 6, 2521–2527, 2005.
90. Agnihotri, S.A., Mallikarjuna, N.N., Aminabhavi, T.M., Recent advances on chitosan-based micro- and nanoparticles in drug delivery. *J. Control. Release*, 100, 5–28, 2004.
91. Idrees, H., Zaidi, S.Z.J., Sabir, A., Khan, R.U., Zhang, X., Hassan, S.U., A review of biodegradable natural polymer-based nanoparticles for drug delivery applications. *Nanomaterials*, 10, 1970, 2020.
92. Gan, Q., Wang, T., Cochrane, C., McCarron, P., Modulation of surface charge, particle size and morphological properties of chitosan–TPP nanoparticles intended for gene delivery. *Colloids Surf. B Biointerfaces*, 44, 65–73, 2005.
93. Jonassen, H., Kjøniksen, A.L., Hiorth, M., Stability of chitosan nanoparticles cross-linked with tripolyphosphate. *Biomacromolecules*, 13, 3747–3756, 2012.
94. Tokumitsu, H., Ichikawa, H., Fukumori, Y., Chitosan-gadopentetic acid complex nanoparticles for gadolinium neutron-capture therapy of cancer: Preparation by novel emulsion-droplet coalescence technique and characterization. *Pharm. Res.*, 16, 1830–1835, 1999.
95. El-Shabouri, M.H., Positively charged nanoparticles for improving the oral bioavailability of cyclosporin-A. *Int. J. Pharm.*, 249, 101–108, 2002.
96. Niwa, T., Takeuchi, H., Hino, T., Kunou, N., Kawashima, Y., Preparations of biodegradable nanospheres of water-soluble and insoluble drugs with D, L-lactide/glycolide copolymer by a novel spontaneous emulsification solvent diffusion method, and the drug release behavior. *J. Control. Release*, 25, 89–98, 1993.

97. Karnchanajindanun, J., Srisa-ard, M., Baimark, Y., Genipin-cross-linked chitosan microspheres prepared by a water-in-oil emulsion solvent diffusion method for protein delivery. *Carbohydr. Polym.*, 85, 674–680, 2011.
98. Perera, U. and Rajapakse, N., Chitosan nanoparticles: Preparation, characterization, and applications, in: *Seafood Processing By-Products*, pp. 371–387, Springer, Berlin/Heidelberg, Germany, 2014.
99. Pileni, M.P., Reverse micelles used as templates: a new understanding in nanocrystal growth. *J. Exp. Nanosci.*, 1, 13–27, 2006.
100. Shi, L.E., Fang, X.J., Xing, L.Y., Chen, M., Zhu, D.S., Guo, X.F., Zhao, L.M., Tang, Z.X., Chitosan nanoparticles as drug delivery carriers for biomedical engineering. *J. Chem. Soc. Pak.*, 33, 929–934, 2011.
101. Alishahi, A., Mirvaghefi, M.R., Tehrani, H., Farahmand, S.A., Shojaosadati, F.A., Dorkoosh, B., Maher, Z., Elsabee, D., Chitosan nanoparticle to carry vitamin C through the gastrointestinal tract and induce the non-specific immunity system of rainbow trout (Oncorhynchus mykiss). *Carbohydr. Polym.*, 86, 1, 142–146, 2011.
102. Xu, C., Zeng, Y., Zheng, X., Wang, R.T., Surface-fluorinated and pH-sensitive carboxymethyl chitosan nanoparticles to overcome biological barriers for improved drug delivery *in vivo*. *Carbohydr. Polym.*, 208, 59–69, 2018.
103. Das, A. and Dutta, P., Antifungal activity of biogenically synthesized silver and gold nanoparticles against sheath blight of rice. *J. Nanosci. Nanotechnol.*, 21, 6, 3547–3555, 2021.
104. Ahmed, A.A. and Dutta, P., Effect of green synthesized silver nanoparticles on soil properties. *J. Med. Plant Stud.*, 8, 1, 2320–3862, 2020.
105. Sharma, A., Dutta, P., Mahanta, M., Kumari, A., Yasin, A., Botanicals as a source of nanomaterial for pest and disease management. *Plant Health Arch.*, 1, 5–9, 2022.
106. El-Naggar, N.E.A., Saber, W.I., Zweil, A.M., Bashir, S.I., An innovative green synthesis approach of chitosan nanoparticles and their inhibitory activity against phytopathogenic Botrytis cinerea on strawberry leaves. *Sci. Rep.*, 12, 1, 1–20, 2022.
107. Mohanpuria, P., Rana, N.K., Yadav, S.K., Biosynthesis of nanoparticles: Technological concepts and future applications. *J. Nanopart. Res.*, 10, 3, 507–517, 2008.
108. Li, X., Xu, H., Chen, Z.S., Chen, G., Biosynthesis of nanoparticles by microorganisms and their applications. *J. Nanomater.*, 2011, 1–16, 2011, doi: 10.1155/2011/270974.
109. Boruah, S. and Dutta, P., Fungus-mediated biogenic synthesis and characterization of chitosan nanoparticles and its combined effect with Trichoderma asperellum against Fusarium oxysporum, Sclerotium rolfsii, and Rhizoctonia solani. *Indian Phytopathol.*, 74, 1, 81–93, 2021.

110. Sathiyabama, M. and Parthasarathy, R., Biological preparation of chitosan nanoparticles and its *in vitro* antifungal efficacy against some phytopathogenic fungi. *Carbohydr. Polym.*, 151, 321–325, 2016.
111. Anitha, A., Deepagan, V.G., Divya, V.V., Menon, R.D., Nair, S.V., Jayakumar, R., Preparation, characterization, *in vitro* drug release, and biological studies of curcumin-loaded dextran sulphate–chitosan nanoparticles. *Carbohydr. Polym.*, 84, 3, 1158–1164, 2011.
112. Karlsson, H.L., Toprak, S.M., Fadeel, B., Toxicity of metal and metal oxide nanoparticles, in: *Handbook of Toxicology*, pp. 75–112, 2015.
113. Maluin, F.N. and Hussein, M.Z., Chitosan-based agronanochemicals as a sustainable alternative in crop protection. *Molecules*, 25, 7, 1611, 2020.
114. Agarwal, M., Nagar, D., Srivastava, N., Agarwal, M., Chitosan nanoparticles based drug delivery: an update. *Int. J. Adv. Multidiscip.*, 2, 1–13, 2015.
115. Campos, E.V.R., Oliveira, J.L., Fraceto, L.F., Singh, B., Polysaccharides as safer release systems for agrochemicals. *Agron. Sustain. Dev.*, 35, 47–66, 2015.
116. Khalid, M., El-Helw, A.R.M., Ahmed, O.A., Hosny, K.M., Ahmed, T.A., Novel controlled release of antihypertensive drug: Preparation, *in-vitro* and *in-vivo* evaluation. *Afr. J. Pharm. Pharmacol.*, 7, 1744–1756, 2013.
117. Ma, Z., Yeoh, H.H., Lim, L.Y., Formulation pH modulates the interaction of insulin with chitosan nanoparticles. *J. Pharm. Sci.*, 91, 1396–1404, 2002.
118. Choudhary, R.C., Kumari, S., Kumaraswamy, R., Pal, A., Raliya, R., Biswas, P., Saharan, V., Characterization methods for chitosan-based nanomaterials. *Plant Nanobionics*, pp. 103–116, Springer, Nature, Switzerland, 2019.
119. Kumar, N.V., Basavegowda, V.R., Murthy, A.N., Synthesis and characterization of copper-chitosan based nanofungicide and its induced defense responses in Fusarium wilt of banana. *Inorg. Nano-Metal Chem.*, 1–9, 2022.
120. Kuntsche, J., Horst, J.C., Bunjes, H., Cryogenic transmission electron microscopy (cryo-TEM) for studying the morphology of colloidal drug delivery systems. *Int. J. Pharm.*, 417, 120–137, 2011.
121. Cover, N.F., Lai-Yuen, S., Parsons, A.K., Kumar, A., Synergetic effects of doxycycline-loaded chitosan nanoparticles for improving drug delivery and efficacy. *Int. J. Nanomed.*, 7, 2411, 2012.
122. Casettari, L., Castagnino, E., Stolnik, S., Lewis, A., Howdle, S.M., Illum, L., Surface characterization of bioadhesive PLGA/chitosan microparticles produced by supercritical fluid technology. *Pharm. Res.*, 28, 1668–1682, 2011.
123. Rodrigues, S., da Costa, A.M.R., Grenha, A., Chitosan/carrageenan nanoparticles: Effect of cross-linking with tripolyphosphate and charge ratios. *Carbohydr. Polymer.*, 89, 282–289, 2012.
124. Sahab, A.F., Waly, A.I., Sabbour, M.M., Nawar, L.S., Synthesis, antifungal and insecticidal potential of Chitosan (CS)-g-poly (acrylic acid) (PAA) nanoparticles against some seed-borne fungi and insects of soybean. *Int. J. Chem. Technol. Res.*, 8, 2, 589–598, 2015.
125. Xing, K., Shen, X., Zhu, X., Ju, X., Miao, X., Tian, J., Feng, Z., Peng, X., Jiang, J., Qin, S., Synthesis and *in vitro* antifungal efficacy of oleoyl-chitosan

nanoparticles against plant pathogenic fungi. *Int. J. Biol. Macromol.*, 82, 830–836, 2016.
126. Kiprushkina, E.I., Shestopalova, I.A., Pekhotina, A.M., Kuprina, E.E., Nikitina, O.V., Protective-stimulating properties of chitosan in the vegetation and storing tomatoes. *Prog. Chem. Appl. Chitin Derivative*, 22, 77–81, 2017.
127. Dodgson, J.L. and Dodgson, W., Comparison of effects of chitin and chitosan for control of Colletotrichum sp. on cucumbers. *J. Pure Appl. Microbiol.*, 11, 1, 87–94, 2017.
128. Siddaiah, C.N., Prasanth, K.V.H., Satyanarayana, N.R., Mudili, V., Gupta, V.K., Kalagatur, N.K., Satyavati, T., Dai, X., Chen, J., Mocan, A., Singh, B.P., Srivastav, R.K., Chitosan nanoparticles having higher degree of acetylation induce resistance against pearl millet downy mildew through nitric oxide generation. *Sci. Rep.*, 8, 2485, 2018, doi: 10.1038/s41598-017-19016-z.
129. Gumilar, J.A., Prihastanti, E.H., Haryanti, S., Subagio, A., Ngadiwiyana, A., Utilization of waste silica and chitosan as fertilizer nanochisil to improve corn production in Indonesia. *Adv. Sci. Letter*, 23, 2447–2449, 2017.
130. Ye, Z., Guo, J., Wu, D., Tan, M., Xiong, X., Yin, Y.G., He. Photo-responsive shell cross-linked micelles based on carboxymethyl chitosan and their application in controlled release of pesticide. *Carbohydr. Polym.*, 132, 520–528, 2015.
131. Minh, H.D. and Anh, N.D., Study on chitosan nanoparticles on biophysical characteristics and growth of Robusta coffee in the greenhouse. *Biocatalyst Agric. Biotechnol.*, 2, 289–294, 2013.
132. Pereira, A.E.S.I., Sandoval-Herrera, S., Zavala-Betancourt, H., Oliveira, A., Ledezma-Perez, J., Romero, L., Fraceto, L.F., γ-Polyglutamic acid/chitosan nanoparticles for the plant growth regulator gibberellic acid: characterization and evaluation of biological activity. *Carbohydr. Polym.*, 157, 1862–1873, 2017.
133. Ritthidej, G.C., Pichayakorn, W., Kusonwiriyawong, N., Lipipun, V., Preparation of chitosan nanoparticles for protein delivery by w/o/w emulsion solvent evaporation and simple ionotropic gelation techniques. *Solid State Phenom.*, 121, 751–754, 2007.
134. Papadimitriou, S., Bikiaris, D., Avgoustakis, K., Karavas, E., Georgarakis, M., Chitosan nanoparticles loaded with dorzolamide and pramipexole. *Carbohydr. Polym.*, 73, 44–54, 2008.
135. Gonçalves, C., Pereira, P., Gama, M., Self-assembled hydrogel nanoparticles for drug delivery applications. *Materials*, 3, 1420–1460, 2010.
136. Zambaux, M.F., Bonneaux, F., Gref, R., Dellacherie, E., Vigneron, C., Preparation and characterization of protein C-loaded PLA nanoparticles. *J. Control. Release*, 60, 177–189, 1999.
137. Kaloti, M. and Bohidar, H.B., Kinetics of coacervation transition versus nanoparticle formation in chitosan–sodium tripolyphosphate solutions. *Colloids Surf. Biointerfaces*, 81, 165–173, 2010.

138. Campos, C.A., Gerschenson, L.N., Flores, S.K., Development of edible films and coatings with antimicrobial activity. *Food Bioprocess Technol.*, 4, 6, 849–875, 2010, DOI: 10.1007/s11947-010-0434-1.
139. De Moura, M.R., Aouada, F.A., Avena-Bustillos, R.J., McHugh, T.H., Krochta, J.M., Mattoso, L.H.C., Improved barrier and mechanical properties of novel hydroxypropyl methylcellulose edible films with chitosan/tripolyphosphate nanoparticles. *J. Food Eng.*, 92, 448–453, 2009.
140. Jang, K.I. and Lee, H.G., Stability of chitosan nanoparticles for L-ascorbic acid during heat treatment in aqueous solution. *J. Agric. Food Chem.*, 56, 1936–1941, 2008.

3

Reduction of Agricultural Greenhouse Gas Emissions by Nanochitosan

Isibor Patrick Omoregie[1*], Samuel Adeniyi Oyegbade[1], Oluwafemi Adebayo Oyewole[2] and Charles Oluwaseun Adetunji[3]

[1]*Department of Biological Sciences Covenant University, Ota, Nigeria*
[2]*Department of Microbiology, Federal University of Technology, Minna, Nigeria*
[3]*Department of Microbiology, Edo State University, Uzairue, Nigeria*

Abstract

Agriculture is a significant contributor to greenhouse gas emissions, including methane (CH_4), nitrous oxide (N_2O), and carbon dioxide (CO_2). These emissions result from practices such as fertilizer use, pesticide application, and soil management. Key greenhouse gases emitted in agriculture include N_2O from nitrogen-based fertilizers and CH_4 from enteric fermentation in livestock. Greenhouse gas emissions from agriculture contribute to climate change and environmental degradation. They can also impact crop productivity and food security. Nanochitosan, derived from chitosan, has gained attention for its potential to reduce emissions and improve sustainability in agriculture. Nanochitosan can reduce emissions by improving nutrient management, enhancing pesticide efficacy, and promoting soil health. Emission reductions after nanochitosan application can vary widely, but studies have reported reductions in nitrous oxide (N_2O) emissions, among others. Nanochitosan can enhance crop yield and quality by improving nutrient uptake, stress tolerance, and disease resistance. Nanochitosan improves nutrient availability, water use efficiency, and soil health, resulting in higher crop yields and improved crop quality. Research is ongoing to address the long-term environmental impact, safety, and regulation of nanochitosan. Further study is thus needed to optimize application methods and dosages and address region-specific challenges. Stakeholders are encouraged to support ongoing research, promote education and outreach, and collaborate to facilitate the responsible adoption of nanochitosan in agriculture. Nanochitosan is crucial

**Corresponding author*: patrick.isibor@covenantuniversity.edu.ng

Charles Oluwaseun Adetunji, Maulin P. Shah, Yerima Mohammed Bello, Daniel Hefft, Jay Singh, Shyam S. Pandey and Ravindra Pratap Singh (eds.) Nanochitosan Applications for Enhanced Crop Production and Food Security, (69–88) © 2025 Scrivener Publishing LLC

in reducing emissions by improving resource use efficiency, reducing synthetic chemical use, and enhancing soil health. Nanochitosan offers significant potential to mitigate greenhouse gas emissions in agriculture by improving nutrient management, enhancing crop productivity, and promoting sustainable practices. There are challenges and research gaps that can be filled by collaborative efforts among stakeholders. It is thus essential to harness the full benefits of nanochitosan in reducing emissions and advancing sustainable agriculture.

Keywords: Nitrogen-based fertilizers, environmental impact, leachate, sustainable agriculture, soil health, crop productivity

3.1 Introduction

Agricultural greenhouse gas (GHG) emissions refer to the release of GHGs into the atmosphere as a result of various agricultural practices. These emissions are significant due to their contribution to global climate change and the potential for adverse impacts on food security, the environment, and human well-being [1].

Agriculture is a major contributor to global GHG emissions, accounting for a substantial portion of the total emissions. The primary GHGs associated with agriculture include carbon dioxide (CO_2) from land-use changes such as deforestation, methane (CH_4) from enteric fermentation in livestock and rice paddies, and nitrous oxide (N_2O) from fertilizer application and soil management practices [2]. GHGs like methane and nitrous oxide are particularly potent in terms of their warming potential. Methane, for instance, has a much higher heat-trapping capacity than CO_2, although it has a shorter atmospheric retention. These gases contribute to the greenhouse effect, leading to global warming and associated climate change impacts, such as extreme weather events, sea-level rise, and altered precipitation patterns. Agriculture is essential for global food production [3]. Climate change caused by agricultural emissions can have significant implications for food security. Changes in temperature and precipitation patterns can disrupt crop yields, reduce livestock productivity, and increase the risk of pests and diseases, affecting food availability and affordability [4].

Agricultural emissions can have adverse effects on ecosystems and biodiversity. Excessive application of nitrogen fertilizer may result in nitrate pollution of water bodies, causing eutrophication and reduced biodiversity. Such polluted aquatic habitat can expose humans who rely on the water for domestic uses or consume the native fish. Livestock wastes contribute to the nitrogenous load in soil and, hence, impact land arability [5]. Recognizing the significance of agricultural emissions, various international agreements and national policies have aimed to address this issue. The Paris Agreement, for instance, encourages countries to reduce emissions from agriculture while increasing food production sustainably. Mitigation strategies may include adopting climate-smart agricultural practices, improving livestock management, reducing food waste, and enhancing soil carbon sequestration.

In recent times, research efforts have been focused on developing innovative technologies and practices to reduce agricultural emissions [6]. These efforts involve developing low-emission crop varieties, precision agriculture techniques, and sustainable land management practices that sequester carbon and reduce the release of methane and nitrous oxide. Addressing agricultural emissions often involves balancing environmental goals with economic interests, as agriculture is a significant industry in many countries. Transitioning to more sustainable practices may require investment and support from governments and organizations to ensure that farmers can adapt and thrive while reducing emissions [7, 8].

Agricultural GHG emissions are a critical aspect of the climate change challenge. Their significance lies in their contribution to global warming, their impact on food security and the environment, and the need for coordinated efforts to mitigate these emissions while ensuring the sustainability of agriculture. Addressing agricultural emissions is a complex task that requires a multidisciplinary approach involving farmers, policymakers, scientists, and the broader community to develop and implement effective solutions. Nanochitosan is a nanoscale derivative of chitosan, a biopolymer derived from chitin, which is found in the shells of crustaceans like shrimp and crabs [9]. It has gained attention for its potential role in various applications, including agriculture, where it may contribute to reducing GHG emissions through several mechanisms.

Nanochitosan can be used as a soil amendment to improve nutrient and water retention in the soil. This can reduce the need for synthetic fertilizers, which are a source of nitrous oxide (N_2O) emissions, a potent GHG. Carbon Sequestration: Chitosan-based nanomaterials can enhance soil organic carbon sequestration. Increased carbon sequestration in soils helps mitigate atmospheric CO_2 levels, a primary GHG. Nanochitosan can be used as a carrier for delivering pesticides and herbicides. This targeted delivery system allows for more efficient use of agrochemicals, reducing their overall application [10–12]. This, in turn, can minimize the production and release of GHGs associated with pesticide production and application. Nanochitosan has been shown to enhance stress tolerance in plants, making them more resilient to adverse environmental conditions. This can result in higher crop yields and reduced pressure on land conversion, helping to mitigate emissions associated with deforestation and land-use changes.

Nanochitosan has been explored as a potential livestock feed additive. It may help reduce methane emissions from enteric fermentation in the digestive systems of ruminant animals. Mitigating methane emissions, nanochitosan can indirectly contribute to reducing agricultural emissions [11]. Chitosan and nanochitosan can be used in water treatment processes to remove pollutants and nutrients from agricultural runoff, preventing water pollution and the release of GHGs associated with water contamination. Plastic Alternatives: Nanochitosan-based materials can be used as biodegradable mulch films in agriculture. These films can replace traditional plastic mulches, reducing plastic pollution and its associated environmental impacts. It is important to note that, while nanochitosan holds promise in reducing emissions in agriculture, more research is needed to fully understand its effectiveness, safety, and long-term impacts on ecosystems [13]. Additionally, the adoption of nanochitosan-based technologies should consider potential socioeconomic and regulatory factors. Nanochitosan represents an innovative avenue for sustainable agriculture practices that can contribute to reducing agricultural emissions while improving soil health, crop productivity, and environmental sustainability. However, its implementation should be part of a broader strategy that includes other mitigation measures and sustainable farming practices.

3.2 Types of Greenhouse Gases Emitted in Agriculture

Agriculture is a significant source of several GHGs, each of which contributes to global warming to varying degrees. The primary GHGs emitted in agriculture include carbon dioxide (CO_2), methane (CH_4), nitrous oxide (N_2O), carbon monoxide (CO), volatile organic compounds (VOCs), sulfur hexafluoride, and hydrofluorocarbons.

Carbon Dioxide (CO_2): Agriculture contributes to CO_2 emissions primarily through land-use changes, such as deforestation and the conversion of forests and grasslands into agricultural land. When forests are cleared or wetlands drained for farming, the carbon stored in vegetation and soil is released into the atmosphere as CO_2. While CO_2 emissions from agriculture are substantial, they are often considered separately from emissions associated with the combustion of fossil fuels, as they result from changes in land use.

Methane (CH_4): Methane emissions in agriculture primarily result from enteric fermentation in the digestive systems of ruminant animals (e.g., cattle, sheep, and goats). Microbes in the stomachs of these animals produce methane during digestion, which is then released when they belch.
Manure Management: Methane is also produced when livestock manure decomposes under anaerobic (low-oxygen) conditions, such as in lagoons and pits.
Nitrous Oxide (N_2O): N_2O emissions in agriculture mainly come from the use of synthetic and organic fertilizers. Nitrogen-based fertilizers, in particular, release nitrogen oxides into the atmosphere, which can lead to the production of N_2O. Soil management practices, such as tillage and irrigation, also affect the nitrogen cycle and contribute to N_2O emissions. N_2O is a potent GHG with a much higher heat-trapping capacity than CO_2.
Carbon Monoxide (CO): CO emissions can occur in agriculture from incomplete combustion of organic matter, such as crop residues, during open-field burning or inefficient use of equipment powered by fossil fuels.
Volatile Organic Compounds (VOCs): Agriculture can emit VOCs from a variety of sources, including plants and soil. VOCs can be released during plant growth, decay, and certain agricultural practices, such as the use of certain pesticides and herbicides.
Sulfur Hexafluoride: Although relatively rare, SF6 can be used in some agricultural applications, such as electrical equipment used in precision agriculture. SF6 is a synthetic GHG with a high global warming potential.
Hydrofluorocarbons (HFCs): HFCs are synthetic compounds used in some agricultural refrigeration and air conditioning systems. They are potent GHGs, but their emissions from agriculture are relatively low compared to other sectors.

It is important to note that the significance of these GHGs varies depending on regional agricultural practices, livestock populations, and the use of agricultural inputs like fertilizers. Agricultural emissions can vary widely between countries and regions, and they are influenced by factors such as crop types, land management practices, and livestock systems. Reducing agricultural GHG emissions often involves adopting more sustainable farming practices, optimizing resource use, and improving the management of livestock and organic matter.

3.3 Environmental and Economic Consequences of Greenhouse Gases

Greenhouse gases (GHGs) have significant environmental and economic consequences. They are a primary driver of global climate change, and their effects are wide-ranging and complex. The primary environmental consequence of GHGs is global warming, leading to higher average temperatures on Earth's surface [14]. This warming is directly linked to the increase in GHG concentrations in the atmosphere, primarily carbon dioxide (CO_2), methane (CH), and nitrous oxide (N_2O).

Global warming alters weather patterns and contributes to climate change. This includes more frequent and severe heatwaves, changes in precipitation patterns, increased frequency of extreme weather events (such as hurricanes and droughts), and rising sea levels. Higher temperatures lead to the melting of polar ice caps, glaciers, and permafrost, contributing to sea-level rise. This can result in coastal erosion, loss of habitat, and increased risk of flooding in low-lying areas [15].

CO_2 is absorbed by the world's oceans, leading to ocean acidification. This can harm marine ecosystems, including coral reefs, shellfish, and other organisms with calcium carbonate shells or skeletons. Changes in temperature and weather patterns can disrupt ecosystems and threaten species with extinction. The altered climate can affect migration patterns, breeding seasons, and habitats, which can result in reduced biodiversity. Increased frequency and intensity of extreme weather events, such as hurricanes, floods, and wildfires, are linked to climate change driven by GHGs. These events can lead to loss of life, property damage, and infrastructure destruction, leading to high economic costs for affected communities, insurance companies, and governments [16, 17].

Furthermore, climate change can affect crop yields, livestock productivity, and food production. Farmers may face challenges due to changing weather patterns, leading to crop failures and reduced agricultural incomes. Heatwaves and the spread of infectious diseases influenced by climate change can increase healthcare costs and place additional burdens on healthcare systems.

Efforts to mitigate and adapt to climate change often require investments in renewable energy sources, energy efficiency measures, and infrastructure upgrades. These not only can result in initial costs but also can lead to long-term energy savings. Climate-related events and regulations can disrupt markets and supply chains. Companies may face increased costs due to climate-related risks, supply chain interruptions, or shifting consumer preferences. Climate change poses financial risks to various sectors, including insurance companies, banks, and investors. These entities may face losses due to climate-related liabilities and stranded assets [16].

On the positive side, efforts to reduce GHG emissions and transition to a low-carbon economy can create job opportunities in renewable energy, clean technology, and sustainable agriculture. Innovation in green industries can drive economic growth. Greenhouse gases have profound environmental consequences, including climate change, ocean acidification, and biodiversity loss. These environmental changes, in turn, have significant economic ramifications, including property damage, healthcare costs, disruptions to markets, and financial risks. Addressing GHG emissions through mitigation and adaptation measures is essential to reduce these negative impacts and promote a more sustainable and resilient future [18].

3.4 Nanochitosan as a Potential Mitigation Strategy

Nanochitosan is a nanoscale derivative of chitosan, which is a natural biopolymer derived from chitin, a substance found in the shells of crustaceans such as shrimp and crabs. Chitosan has a long history of use in various industrial and biomedical applications due to its biocompatibility, biodegradability, and non-toxicity. Nanochitosan is created by breaking down chitosan into nanoparticles, which are extremely small particles with dimensions typically in the nanometer range (1 nanometer = 1 billionth of a meter).

Nanochitosan exhibits unique properties and characteristics that make it useful in a wide range of applications. Nanochitosan nanoparticles have a significantly larger surface area compared to bulk chitosan, making them useful in adsorption and binding applications, such as water purification and drug delivery systems. In pharmaceutical and biomedical applications, nanochitosan can improve the solubility and bioavailability of poorly water-soluble drugs, enhancing their therapeutic effectiveness. Chitosan, and, by extension, nanochitosan, has natural antimicrobial properties. It can be used in food packaging materials to extend the shelf life of products and reduce the growth of spoilage microorganisms. Nanochitosan has been investigated for its potential use in wound dressings and tissue engineering due to its biocompatibility and ability to promote tissue regeneration [19].

In agriculture, nanochitosan-based formulations have been explored as crop protectants, soil conditioners, and carriers for agricultural inputs such as pesticides and fertilizers. They may help improve crop yields and reduce the environmental impact of farming. Nanochitosan can be used for environmental remediation purposes, including the removal of heavy metals and pollutants from water and soil through adsorption. Like chitosan, nanochitosan is biodegradable, making it environmentally friendly and suitable for applications where non-toxic, sustainable materials are desired. It is important to note that nanotechnology, including nanochitosan, is a rapidly evolving field, and research continues to explore new applications and benefits. However, as with any emerging technology, the safety and potential environmental impacts of nanochitosan should be carefully considered and studied to ensure its responsible use in various applications. Regulatory agencies and research institutions are actively monitoring and assessing nanomaterials like nanochitosan to understand their potential risks and benefits.

Nanochitosan has gained attention in agricultural research due to its potential benefits in enhancing crop productivity, improving soil health, and reducing environmental impacts. Previous studies have explored various applications of nanochitosan in agriculture. Nanochitosan has been used as a carrier for pesticides and herbicides, helping to improve their adhesion to plant surfaces and enhance their efficacy. This can result in reduced pesticide use and minimized environmental contamination. Nanochitosan has shown antifungal properties, making it a potential alternative to synthetic fungicides. It has been studied for its ability to control plant diseases caused by fungi. Nanochitosan-based formulations have been explored to improve the efficiency of nutrient delivery to plants [20]. They can help reduce nutrient leaching and enhance the uptake of essential elements by plant roots. Nanochitosan can be incorporated into slow-release fertilizer formulations, allowing for a gradual release of nutrients over time, which improves nutrient use efficiency and minimizes environmental pollution. Nanochitosan has been used as a soil conditioner to enhance soil structure, water retention, and nutrient availability. It promotes beneficial microbial activity in the soil, which can lead to improved crop growth. Furthermore, in contaminated soils, nanochitosan has been studied for its ability to immobilize heavy metals and organic pollutants, reducing their mobility and potential harm to crops. Nanochitosan can be applied as a seed coating to protect seeds from pathogens and environmental stress. This can improve germination rates and early plant growth. Nanochitosan has been investigated for its potential to stimulate plant growth by enhancing root development and nutrient uptake, resulting in improved crop yields. Nanochitosan has shown promise in enhancing plant tolerance to abiotic stresses such as drought and salinity. It can help plants maintain water balance and reduce stress-related damage. Nanochitosan-based materials have been explored as biodegradable mulch films in agriculture, offering an eco-friendly alternative to traditional plastic mulches [21, 22].

Nanochitosan holds promise in various agricultural applications. Further research is, however, needed to optimize formulations, dosage, and application methods. The efficacy, safety, and environmental impact of nanochitosan in agriculture must be carefully evaluated to ensure its responsible use and minimize potential risks. Additionally, regulatory agencies in different countries may have specific requirements and guidelines for the use of nanomaterials in agriculture, and researchers and practitioners should adhere to these regulations [23].

3.5 Mechanisms of Action for Emission Reduction

Nanochitosan can play a role in reducing GHG emissions in agriculture through various mechanisms. While the specific mechanisms can vary depending on the application, nanochitosan's actions generally revolve around improving nutrient use efficiency, reducing the need for synthetic chemicals, and enhancing soil and crop health.

Nanochitosan can be used as a soil conditioner to improve nutrient retention in the soil. It binds with nutrients such as nitrogen, phosphorus, and potassium, reducing nutrient leaching and runoff into water bodies. This, in turn, minimizes the production of nitrous oxide (N_2O), a potent GHG associated with fertilizer use. Nanochitosan can be incorporated into slow-release fertilizer formulations. These formulations release nutrients gradually over time, matching plant uptake needs and reducing the risk of nutrient loss and emissions [24].

Nanochitosan can serve as a carrier for pesticides and herbicides. By enhancing the adhesion of these chemicals to plant surfaces and protecting them from degradation, nanochitosan can reduce the amount of chemicals needed for effective pest and weed control, thereby decreasing emissions associated with pesticide production and application. Nanochitosan can stimulate beneficial microbial activity in the soil. This enhanced microbial activity can improve nutrient cycling and organic matter decomposition, potentially reducing emissions of methane (CH_4) and nitrous oxide (N_2O) from the soil. Nanochitosan's ability to improve soil structure, water retention, and aeration can lead to healthier soil conditions. This can promote root growth, increase plant health, and reduce the need for irrigation, ultimately decreasing energy-intensive water pumping [25]. Nanochitosan has demonstrated antifungal properties and can help plants resist diseases. Healthier plants are less likely to suffer from stress-related issues, reducing the need for remedial treatments and associated emissions. By improving nutrient and pesticide uptake by crops and minimizing runoff, nanochitosan can help protect water bodies from contamination. This mitigates the environmental impact of agricultural runoff, including eutrophication, which can contribute to GHG emissions from water bodies [26].

Nanochitosan-based materials can serve as biodegradable alternatives to traditional plastic mulch films. Reducing the use of plastic in agriculture can decrease the environmental footprint associated with plastic production and disposal. It is essential to note that the effectiveness of nanochitosan in emission reduction depends on factors such as application methods, dosage, crop types, and local agricultural practices. Further research and field trials are necessary to optimize the use of nanochitosan for emission reduction and to ensure that it aligns with sustainable agricultural practices and environmental goals. Additionally, the specific impact on emissions can vary depending on the GHG of interest (e.g., CO_2, CH_4, and N_2O) and the specific agricultural context [27].

Quantitative data on GHG emissions before and after nanochitosan application in agriculture can vary significantly depending on the specific application, crop type, soil conditions, climate, and other factors. Emission reductions are often context-specific and may not follow a uniform pattern. Nevertheless, we can provide some general insights and examples of studies that have assessed emission reductions after nanochitosan application in certain agricultural contexts.

It is important to emphasize that the effectiveness of nanochitosan in reducing emissions can vary widely depending on the specific conditions of each agricultural system. Moreover, quantitative data may not be readily available for all scenarios, and results can be influenced by numerous factors, including the type of nanochitosan formulation used, application rates, and the duration of the study [28]. Assessing the potential for emission reduction through nanochitosan application is best conducted through site-specific studies and considers the overall environmental and economic impacts. Additionally, researchers and practitioners should take into account regional and crop-specific variations to optimize nanochitosan applications for emission reduction while ensuring sustainable agricultural practices [29].

3.6 Crop Yield and Quality

Nanochitosan application has shown promise in enhancing crop yield and quality by improving various aspects of plant growth and health. Nanochitosan can chelate and stabilize essential nutrients in the soil, making them more available to plant roots. This results in improved nutrient uptake, leading to increased plant growth and higher crop yields [30].

Nanochitosan has been reported to enhance drought tolerance in crops. It can help plants maintain water balance, reduce water stress, and continue growing during dry periods, ultimately leading to improved yield stability. Nanochitosan has demonstrated antifungal properties, which can protect crops from various fungal diseases. Reduced disease incidence can result in higher crop quality and yield [31]. Nanochitosan-based formulations can help manage certain pests by acting as a natural pesticide. This can reduce damage to crops and minimize yield losses. Nanochitosan can be used as a seed coating material, enhancing germination rates and early seedling growth. This leads to better crop establishment and higher overall yields. Nanochitosan has been shown to improve salt tolerance in plants. It can help mitigate the negative effects of salinity on crop growth, enabling better yields in saline soils. Some studies suggest that nanochitosan can help crops adapt to temperature stress, including extreme heat or cold. This can contribute to improved yield stability [32].

Nanochitosan improves soil aggregation, leading to better soil structure. This facilitates root growth, nutrient uptake, and water retention, all of which positively impact crop yield and quality. Nanochitosan can stimulate beneficial microbial activity in the soil, enhancing nutrient cycling and organic matter decomposition. This can improve nutrient availability to plants. Nanochitosan-based biodegradable mulch films can suppress weed growth, reducing competition for resources and allowing crops to thrive. Nanochitosan's ability to improve nutrient and pesticide use efficiency and reduce runoff can help minimize environmental pollution and its adverse effects on crop quality. It is important to note that the effectiveness of nanochitosan in enhancing crop yield and quality may vary depending on factors such as crop type, soil conditions, climate, and application methods. Site-specific trials and experiments are essential to determine the optimal use of nanochitosan in specific agricultural contexts. Additionally, while nanochitosan offers potential benefits, its long-term environmental impacts should be carefully considered to ensure sustainable agricultural practices [33].

In the pursuit of sustainable and climate-resilient agricultural practices, the encouragement of ongoing innovation and research in the field of nanochitosan applications is imperative. This report emphasizes the need to expand the scope of nanochitosan applications in agriculture to address region-specific and crop-related challenges. Such endeavors are vital for stakeholders to collaboratively harness the full potential of nanochitosan, with the overarching goals of reducing emissions, enhancing food security, and promoting sustainable farming practices, especially in the context of climate change and broader environmental challenges.

Nanochitosan, derived from chitosan, has demonstrated significant potential in enhancing crop production, soil health, and environmental sustainability. However, to fully realize these benefits, ongoing research and innovation are essential. Researchers must explore and develop novel applications of nanochitosan that cater to the unique challenges faced by different regions and crops. This entails investigating the adaptability of nanochitosan formulations to varying climatic conditions, soil types, and agricultural systems. Agriculture is a diverse and dynamic sector, with distinct challenges across regions and crops. Therefore, the customization of nanochitosan applications to address these specific challenges is paramount. For example, in regions prone to water scarcity, nanochitosan can be tailored to improve drought tolerance in crops. In areas with high pest pressure, nanochitosan-based pest management solutions can be developed. Crop-specific research should guide the formulation and application of nanochitosan to ensure maximum efficacy and minimal environmental impact [34].

3.6.1 Collaborative Stakeholder Engagement

The success of nanochitosan adoption in agriculture hinges on collaborative efforts among stakeholders. These stakeholders include farmers, agricultural researchers, policymakers, industry representatives, and regulatory agencies. Engaging in open dialogues and knowledge-sharing forums is crucial for aligning research and development efforts with the practical needs of farmers and the regulatory requirements of governments. Furthermore, such collaboration fosters a holistic approach to sustainable agriculture, promoting the responsible and safe use of nanochitosan.

Nanochitosan's potential to improve nutrient management, reduce synthetic chemical use, and enhance crop resilience can contribute significantly to emissions mitigation and food security. By optimizing nutrient uptake, mitigating stress factors, and bolstering disease resistance, nanochitosan aids in increasing crop yields. Moreover, its ability to reduce nutrient runoff and leaching minimizes environmental pollution, thereby reducing GHG emissions.

3.6.2 Promoting Sustainable Farming Practices

Sustainable farming practices are essential for ensuring the long-term viability of agriculture in the face of climate change and environmental challenges. Nanochitosan, as a tool for sustainable agriculture, aligns with the broader goal of reducing the carbon footprint of food production. By improving resource use efficiency and soil health, nanochitosan facilitates sustainable farming practices that prioritize environmental stewardship.

The encouragement of ongoing innovation and research in nanochitosan applications is essential for the advancement of sustainable agriculture. Addressing region-specific and crop-related challenges through tailored nanochitosan solutions can lead to reduced emissions, improved food security, and the promotion of environmentally responsible farming practices. The collaborative engagement of stakeholders, including farmers, researchers, and policymakers, is vital to realizing the full potential of nanochitosan in mitigating climate change impacts on agriculture while ensuring food security and sustainability [35].

3.7 Limitations and Future Research Directions

Nanochitosan holds promise in agriculture for its potential to improve crop yield, quality, and sustainability. However, like any emerging technology, it also has limitations and areas where further research is needed. The long-term environmental and ecological impacts of nanochitosan need further investigation. This includes assessing its persistence in the environment and potential effects on non-target organisms. Nanochitosan-based products may require regulatory approval before widespread adoption in agriculture. Research and testing are necessary to meet regulatory standards for safety and efficacy. The most effective application methods and dosages of nanochitosan can vary for different crops and conditions. Research is needed to optimize these parameters.

The cost of producing nanochitosan and nanochitosan-based formulations may be a limitation for some farmers. Research into cost-effective production methods is essential to make nanochitosan more accessible. The interactions between nanochitosan and other agricultural inputs, such as fertilizers and pesticides, need further investigation to maximize their combined benefits and minimize potential conflicts. Comprehensive studies should assess the long-term environmental impact of nanochitosan, including its fate in soil and water, potential accumulation, and effects on ecosystems. Research should focus on crop-specific applications of nanochitosan to determine its effectiveness, optimal application methods, and benefits for various crops; investigate dose–response relationships to identify the ideal nanochitosan dosage for different crops and growth stages; conduct rigorous safety and toxicity studies to ensure that nanochitosan does not pose risks to human health, animals, or beneficial organisms in the environment; and assess the economic viability of nanochitosan applications in agriculture by conducting cost-benefit analyses and evaluating the return on investment for farmers. It is essential for research institutes to work with regulatory agencies to establish guidelines and standards for nanochitosan-based products to ascertain safety and efficacy precision. It is imperative to conduct large-scale field trials on diverse farms to validate the effectiveness of nanochitosan in real-world agricultural settings [36].

Collaboration is required across disciplines, including agronomy, chemistry, biology, and environmental science, to holistically assess nanochitosan's potential and limitations in agriculture. Researchers need to explore how nanochitosan can be integrated into sustainable agricultural practices, including organic and regenerative farming systems. Overall, nanochitosan offers exciting possibilities for sustainable agriculture, but its safe and effective use requires continuous research, regulatory oversight, and a commitment to responsible and sustainable farming practices. Aiming future studies on these recommended considerations may unlock the full potential of nanochitosan in agriculture.

3.8 Conclusion and Recommendations

Nanochitosan can enhance crop yield and quality by improving nutrient uptake, water use efficiency, and stress tolerance. It has demonstrated antifungal properties, reducing the incidence of diseases and pests. Seed coating with nanochitosan improves germination and seedling growth.

Nanochitosan helps reduce environmental impacts by improving nutrient and pesticide use efficiency, reducing runoff, and potentially lowering GHG emissions. It can be used in biodegradable mulch films to suppress weeds, reducing the need for herbicides. Research is ongoing to understand the long-term environmental impact, safety, and toxicity of nanochitosan.

Regulatory approvals and cost-effectiveness are important considerations for widespread adoption. Further research is needed to optimize application methods, dosages, and crop-specific use. Multi-disciplinary collaboration is essential to assess nanochitosan's potential and limitations in diverse agricultural settings. Sustainable integration of nanochitosan into farming practices should be explored. Nanochitosan offers promise in improving crop production, sustainability, and environmental protection in agriculture, but its responsible use and safety require ongoing research and careful consideration. Nanochitosan holds significant importance in mitigating agricultural emissions due to its potential to enhance resource use efficiency, reduce the need for synthetic chemicals, and improve soil health. Nanochitosan can improve nutrient retention in the soil, reducing the leaching of nutrients such as nitrogen into water bodies. This helps prevent nutrient pollution, which is a major contributor to GHG emissions like nitrous oxide (N_2O) and can result in water quality issues [32].

By enhancing nutrient uptake by crops, nanochitosan reduces the need for excessive fertilizer application. Over-fertilization can lead to higher emissions of N_2O, a potent GHG. Nanochitosan can improve the efficacy of pesticides and herbicides, reducing the need for their excessive use. This minimizes emissions associated with the production and application of these chemicals. Nanochitosan enhances soil structure, microbial activity, and organic matter content. Healthy soils with improved structure can sequester more carbon dioxide (CO_2) and reduce the emission of nitrous oxide (N_2O) from soil.

Nanochitosan's ability to reduce the environmental impact of agriculture, including GHG emissions, aligns with global efforts to reduce the carbon footprint of food production. It promotes more sustainable and climate-friendly farming practices. Nanochitosan can help crops become more resilient to climate change-induced stresses, such as drought and temperature fluctuations. This resilience reduces the need for reactive measures that may lead to increased emissions. By improving nutrient uptake, stress tolerance, and disease resistance, nanochitosan can lead to higher crop yields. Producing more food per unit of land can help meet global food demands while minimizing the expansion of agriculture into carbon-rich ecosystems like forests.

Nanochitosan-based biodegradable mulch films can replace traditional plastic mulches, reducing plastic waste and the emissions associated with plastic production and disposal. Nanochitosan's role in mitigating agricultural emissions is crucial for promoting more sustainable and climate-resilient farming practices. By improving resource use efficiency, reducing the environmental impact of agricultural inputs, and enhancing soil health, nanochitosan contributes to the reduction of GHG emissions from agriculture, helping address the global challenge of climate change. To drive the adoption of nanochitosan in agriculture for its potential to mitigate emissions and enhance sustainability, a coordinated effort is needed involving farmers, researchers, policymakers, and industry stakeholders. Governments and research institutions should allocate resources to support research and development efforts related to nanochitosan applications in agriculture. This includes investigating its efficacy, safety, and environmental impacts. This raises awareness among farmers, agricultural advisors, and extension services about the benefits and best practices of using nanochitosan. This can be achieved through workshops, seminars, and educational materials.

Facilitate on-farm trials and demonstrations to showcase the effectiveness of nanochitosan in different agricultural contexts. These practical examples can encourage adoption among farmers; collaborate with regulatory agencies to develop clear guidelines and safety standards for nanochitosan-based agricultural products; ensure that they meet safety and environmental requirements; and promote collaboration between research institutions, agricultural organizations, and the private sector to share knowledge and data on nanochitosan applications. This can accelerate its adoption and refinement. Governments and agricultural organizations can provide incentives, such as tax breaks or subsidies, to farmers adopting sustainable practices, including the use of nanochitosan. Programs need to be developed to make nanochitosan more accessible and affordable for smallholder farmers, who often lack resources for innovation adoption. Continued research on the long-term environmental and ecological impacts of nanochitosan is essential to ensure that its use aligns with sustainability goals. Establishment of mechanisms for monitoring and reporting on the adoption and impact of nanochitosan in agriculture with transparency and accountability is also essential. Advocation for policies that encourage sustainable agricultural practices and technologies like nanochitosan is required, alongside engagement with policymakers to highlight the benefits of adoption and encouragement of manufacturers and distributors to responsibly market nanochitosan products, providing accurate information on benefits, safety, and usage guidelines. Transparent labeling can build trust, fostering collaborative research networks involving scientists, farmers, and industry representatives to address emerging challenges and refine nanochitosan applications [33].

Encouragement of ongoing innovation and research to expand the range of nanochitosan applications in agriculture, addressing specific regional and crop-related challenges, is imperative. By taking these actions, stakeholders can work together to harness the potential of nanochitosan in agriculture, reduce emissions, improve food security, and promote sustainable farming practices in the face of climate change and environmental challenges.

References

1. Abbas, Q., Yousaf, B., Ullah, H., Ali, M.U., Ok, Y.S., Rinklebe, J., Environmental transformation and nano-toxicity of engineered nano-particles (ENPs) in aquatic and terrestrial organisms. *J. Crit. Rev. Environ. Sci. Technol.*, 50, 23, 2523–81, 2020.
2. Abbaspour, N., Hurrell, R., Kelishadi, R., Review on iron and its importance for human health. *J. Res. Med. Sci.*, 19, 2, 164–74, 2014.

3. Aghdam, M.T.B., Mohammadi, H., Ghorbanpour, M., Effects of nanoparticulate anatase titanium dioxide on physiological and biochemical performance of *Linum usitatissimum* (linaceae) under well-watered and drought stress conditions. *Rev. Bras. Bot.*, 39, 1, 139–46, 2016.
4. Alejandro, S., Höller, S., Meier, B., Peiter, E., Manganese in plants: from acquisition to subcellular allocation. *Front. Plant Sci.*, 11, 1–23, 2020.
5. Alsharef, J. M., Taha, M.R., Firoozi, A.A., Govindasamy, P., Potential of using nanocarbons to stabilize weak soils. *Appl. Environ. Soil Sci.*, 2016, 1, 5060531, 2016. https://doi.org/10.1155/2016/5060531.
6. Choudhary, R.C., Kumaraswamy, R., Kumari, S., Sharma, S., Pal, A., Raliya, R., Biswas, P., Saharan, V., Zinc encapsulated chitosan nanoparticle to promote maize crop yield. *Int. J. Biol. Macromol.*, 127, 126–35, 2019.
7. Ashkavand, P., Tabari, M., Zarafshar, M., Tomášková, I., Struve, D., Effect of SiO2 nanoparticles on drought resistance in hawthorn seedlings. *Res. Pap.*, 76, 4, 350–9, 2015.
8. Elemike, E.E., Uzoh, I.M., Onwudiwe, D.C., Babalola, O.O., The role of nanotechnology in the fortification of plant nutrients and improvement of crop production. *Appl. Sci.*, 9, 3, 1–32, 2019.
9. Das, A. and Das, B., Nanotechnology a potential tool to mitigate abiotic stress in crop plants, in: *Abiotic and biotic stress in plants*, pp. 84–97, IntechOpen, London, 2019.
10. Aziz, H.M.A., Hasaneen, M.N., Omer, A.M., Nano chitosan-NPK fertilizer enhances the growth and productivity of wheat plants grown in sandy soil. *Span. J. Agric. Resour.*, 14, 1, 1–9, 2016.
11. Chhipa, H., Nanofertilizers and nanopesticides for agriculture. *Environ. Chem. Lett.*, 15, 1, 15–22, 2017.
12. Lovecká, P., Macůrková, A., Záruba, K., Hubáček, T., Siegel, J., Valentová, O., Genomic damage induced in Nicotiana tabacum L. plants by colloidal solution with silver and gold nanoparticles. *Plants*, 10, 6, 1260, 2021.
13. Kumari, S., Choudhary, R.C., Kumaraswamy, R., Bhagat, D., Pal, A., Raliya, R., Biswas, P., Saharan, V., Zinc-functionalized thymol nanoemulsion for promoting soybean yield. *Plant Physiol. Biochem.*, 145, 64–74, 2019.
14. Ma, C., White, J.C., Dhankher, O.P., Xing, B., Metal-based nanotoxicity and detoxification pathways in higher plants. *Environ. Sci. Technol.*, 49, 12, 7109–22, 2015.
15. Yuvaraj, M. and Subramanian, K.S., Development of slow release Zn fertilizer using nano-zeolite as carrier. *J. Plant Nutr.*, 41, 3, 311–20, 2018.
16. Chen, M., Zhou, S., Zhu, Y., Sun, Y., Zeng, G., Yang, C., Toxicity of carbon nanomaterials to plants, animals and microbes: recent progress from 2015-present. *Chemosphere*, 206, 255–64, 2018. https://doi.org/10.1016/j.chemosphere.2018.05.020.
17. Zulfiqar, F., Navarro, M., Ashraf, M., Akram, N.A., Munné-Bosch, S., Nanofertilizer use for sustainable agriculture: advantages and limitations. *Plant Sci.*, 289, 1–11, 2019.

18. Choudhary, R.C., Kumaraswamy, R., Kumari, S., Pal, A., Raliya, R., Biswas, P., Saharan, V., Synthesis, characterization, and application of chitosan nanomaterials loaded with zinc and copper for plant growth and protection, in: *Nanotechnology*, pp. 227–47, Springer, Singapore, 2017.
19. Cota-Ruiz, K., Ye, Y., Valdes, C., Deng, C., Wang, Y., Hernández-Viezcas, J.A., Duarte-Gardea, M., Gardea-Torresdey, J.L., Copper nanowires as nanofertilizers for alfalfa plants: understanding nano-bio systems interactions from microbial genomics, plant molecular responses and spectroscopic studies. *Sci. Total Environ.*, 742, 1–9, 2020.
20. Faisal, M., Saquib, Q., Alatar, A.A., Al-Khedhairy, A.A., Ahmed, M., Ansari, S.M., Alwathnani, H.A., Dwivedi, S., Musarrat, J., Praveen, S., Cobalt oxide nanoparticles aggravate DNA damage and cell death in eggplant via mitochondrial swelling and NO signaling pathway. *Biol. Res.*, 49, 1, 1–13, 2016.
21. Yoisungnern, T., Choi, Y.J., Han, J.W., Kang, M.H., Das, J., Gurunathan, S., Internalization of silver nanoparticles into mouse spermatozoa results in poor fertilization and compromised embryo development. *Sci. Rep.*, 5, 11170, 2015.
22. Faisal, M., Saquib, Q., Alatar, A.A., Al-Khedhairy, A.A., Hegazy, A.K., Musarrat, J., Phytotoxic hazards of NiO-nanoparticles in tomato: a study on mechanism of cell death. *J. Hazard. Mater.*, 250, 318–32, 2013.
23. Zhang, G., Zhou, L., Cai, D., Wu, Z., Anion-responsive carbon nanosystem for controlling selenium fertilizer release and improving selenium utilization efficiency in vegetables. *Carbon*, 129, 711–9, 2018.
24. Chhowalla, M., Slow release nanofertilizers for bumper crops. *ACS Cent. Sci.*, 3, 3, 156–7, 2017. https://doi.org/10.1021/acscentsci.7b00091.
25. Dhlamini, B., Paumo, H.K., Katata-Seru, L., Kutu, F.R., Sulphate-supplemented npk nanofertilizer and its effect on maize growth. *Mater. Res. Express*, 7, 9, 1–12, 2020.
26. Ghosh, M., Jana, A., Sinha, S., Jothiramajayam, M., Nag, A., Chakraborty, A., Mukherjee, A., Mukherjee, A., Effects of ZnO nanoparticles in plants: cytotoxicity, genotoxicity, deregulation of antioxidant defenses, and cell-cycle arrest. *Mutat. Res. Genet. Toxicol. Environ. Mutagen.*, 807, 25–32, 2016.
27. Fernández-Cruz, M.L., Lammel, T., Connolly, M., Conde, E., Barrado, A.I., Derick, S., Perez, Y., Fernandez, M., Furger, C., Navas, J.M., Comparative cytotoxicity induced by bulk and nanoparticulated ZnO in the fish and human hepatoma cell lines PLHC-1 and Hep G2. *Nanotoxicology*, 7, 5, 935–52, 2013.
28. Cheng, Y., Yin, L., Lin, S., Wiesner, M., Bernhardt, E., Liu, J., Toxicity reduction of polymer-stabilized silver nanoparticles by sunlight. *J. Phys. Chem. C.*, 115, 11, 4425–32, 2011.
29. Kanjana, D., Preparation and characterization of nanoformulated zinc fertilizer by using biopolymer and their effects on cotton. *Int. J. Chem. Stud.*, 7, 2, 1530–7, 2019.

30. Wang, S.L. and Nguyen, A.D., Effects of Zn/B nanofertilizer on biophysical characteristics and growth of coffee seedlings in a greenhouse. *Res. Chem. Intermed.*, 44, 8, 4889–901, 2018.
31. Jasrotia, P., Kashyap, P.L., Bhardwaj, A.K., Kumar, S., Singh, G.P., Scope and applications of nanotechnology for wheat production: a review of recent advances. *Wheat Barley Res.*, 10, 1, 1–14, 2018.
32. Bhardwaj, A.K., Rajwar, D., Basak, N., Bhardwaj, N., Chaudhari, S.K., Bhaskar, S., Sharma, P.C., Nitrogen mineralization and availability at critical stages of rice (*Oryza sativa*) crop, and its relation to soil biological activity and crop productivity under major nutrient management systems. *J. Soil Sci. Plant Nutr.*, 20, 1238–48, Sep 2020.
33. Xiang, L., Zhao, H.M., Li, Y.W., Huang, X.P., Wu, X.L., Zhai, T., Yuan, Y., Cai, Q.Y., Mo, C.H., Effects of the size and morphology of zinc oxide nanoparticles on the germination of Chinese cabbage seeds. *Environ. Sci. Pollut. Res.*, 22, 14, 10452–62, 2015.
34. Chai, R., Ye, X., Ma, C., Wang, Q., Tu, R., Zhang, L., Gao, H., Greenhouse gas emissions from synthetic nitrogen manufacture and fertilization for main upland crops in china. *Carbon Balance Manag.*, 14, 1, 1–10, 2019.
35. Kizilbash, N., Alruwaili, J., Hai, A., Khachfe, H.M., Ambreen, J., Water soluble carbon nanotubes increase agricultural yields. *J. Biochem. Biotech.*, 3, 1, 1–4, 2020.
36. Iqbal, M.A., Nano-fertilizers for sustainable crop production under changing climate: a global perspective, in: *Sustainable crop production*, M. Hasanuzzaman, M.C.M.T. Filho, M. Fujita, T.A.R. Nogueiraed (Eds.), pp. 292–303, 2019.

4

The Application of Nanochitosan Biopesticides as a Replacement to Synthetic Pesticides

Suleiman Imam Malik[1*], Konjerimam Ishaku Chimbekujwo[2], Abdulmajid Musa Maku[1], Oluwafemi Adebayo Oyewole[1], Isibor Patrick Omoregie[3], Charles Oluwaseun Adetunji[4], Eniola K. I. T.[4], Mohammed Bello Yerima[5, 6] and John Tsado Mathew[7]

[1]*Department of Microbiology, Federal University of Technology, Minna, Nigeria*
[2]*Department of Microbiology, Modibbo Adama University, Yola, Adamawa State, Nigeria*
[3]*Department of Biological Sciences Covenant University, Ota, Ogun State, Nigeria*
[4]*Department of Biological Sciences, Joseph Ayo Babalola University, Ikeji Arakeji, Osun State, Nigeria*
[5]*Department of Microbiology, Edo State University, Uzairue, Edo State, Nigeria*
[6]*Department of Microbiology, Sokoto State University, Sokoto, Sokoto State, Nigeria*
[7]*Department of Chemistry, Ibrahim Badamasi Babangida University Lapai, Niger State, Nigeria*

Abstract

Concerns regarding the detrimental effects of synthetic pesticides on the environment and human health have inspired research into more ecologically friendly alternatives worldwide. Nanochitosan, a modified type of chitosan-derived chitosan, is an interesting candidate for biopesticide applications. Owing to its advantageous properties, such as non-toxicity, biodegradability, low allergenicity, and biocompatibility, nanochitosan can be applied in a variety of fields, including agriculture. The potential of nanochitosan as a biopesticide and its capacity to replace synthetic pesticides in contemporary agriculture are both topics covered in this chapter. This chapter explores the benefits, uses, and potential uses of nanochitosan in pest management and emphasizes how it supports ecologically friendly and sustainable farming methods.

**Corresponding author*: imam.malik@futminna.edu.ng

Charles Oluwaseun Adetunji, Maulin P. Shah, Yerima Mohammed Bello, Daniel Hefft, Jay Singh, Shyam S. Pandey and Ravindra Pratap Singh (eds.) Nanochitosan Applications for Enhanced Crop Production and Food Security, (89–110) © 2025 Scrivener Publishing LLC

Keywords: Nanochitosan, biopesticides, pest, agriculture, plans, animals

4.1 Introduction

The overuse of pesticides is a global problem with serious environmental consequences. Chemicals called pesticides are used in agriculture to eradicate pests and boost crop production. Nevertheless, environmental concerns have been highlighted because of their extensive, and, sometimes, indiscriminate, usage. By harming non-target species, pesticides may cause beneficial insects, such as bees and butterflies, to become less common. According to the United Nations Environment Programme, pesticides are a significant factor in the reduction of pollinators, posing a threat to food security and biodiversity [1]. Water sources can be contaminated by pesticides when they flow into rivers and lakes and leach into groundwater. According to the World Health Organization, pesticide residues have been identified in groundwater and surface water in several areas, endangering aquatic ecosystems and the quality of drinking water [2]. In addition to altering soil ecosystems, pesticides can also negatively affect the health and fertility of agricultural areas. According to the Food and Agriculture Organization, incorrect pesticide application can cause soil deterioration and extinction of beneficial soil species. Pesticide residues have the potential to harm human health by building up in the environment and food chains. To reduce health concerns, the United States Environmental Protection Agency (EPA) focuses on and controls pesticide residue levels in food [3].

For sustainable agriculture and ecosystem health, it is essential to address pesticide misuse and its environmental effects. This has prompted requests for eco-friendly and sustainable solutions to reduce their negative impacts. Nanochitosan, a natural biopolymer made from chitosan, is an alternative. The production of pesticides and pest control are two fields in which nanotechnology is being increasingly employed. Thus, crop output may be increased, and novel pest control methods can be developed using nanotechnology [4]. Recently, significant progress has been made in the application of nanotechnology in agriculture. Pesticides, fungicides, herbicides, and micronutrients may be nano-encapsulated in controlled-release matrices to achieve high penetrability through cuticles and tissues and to allow the release of target chemicals over time [5]. The most significant chitin derivative, chitosan, was obtained when the acetate moiety was removed from chitin. It originates from the cell walls of fungi and crustacean shells, such as shrimp and crabs [6]. An abundant, naturally occurring cationic biopolymer, chitosan, is created when chitin is partially deacetylated. The linear (14) glycosidic connections of the 2-acetamido-d-glucose and 2-amino-d-glucose units made up the biopolymer. Owing to its pKa value (6.5) for its amine functional groups, chitosan is more soluble in neutral and acidic environments. The amine groups also contribute to the physicochemical characteristics of the biopolymer and enable the synthesis of chitosan-based products via artificial processes such as gelation [7].

Chitosan nanoparticles (ChNPs) can operate as growth promoters and powerful antibacterial agents against harmful fungus and bacteria in the agricultural industry. They are also known as chitosan-based agrochemicals because they can serve as nanocarriers for currently used agrochemicals [8] Chitosan could serve as a substitute for the combined use of macronutrients as a potent biocidal agent against agricultural pathogens such as viruses, fungi, and bacteria in plant pathogen management and disease control. Chitosan can easily bind to plant surfaces (such leaves and stems) and is available as nanoparticles (NPs), which prolong the period when agrochemicals are in contact with the target surface [9].

The focus of this chapter is to investigate the potential of nanochitosan as a biopesticide and to replace synthetic pesticides in agricultural practices. A chitosan derivative known as nanochitosan has demonstrated potential as a fungicide, antibacterial agent, and growth promoter. According to some studies, nanochitosan can boost crop performance, boost immunity to diseases, and increase the effectiveness of nutrient utilization [10]. Additionally, nanochitosan has been utilized in conjunction with bioinoculants to increase beneficial microbial populations and enhance soil fertility [11].

4.2 Nanochitosan

Chitin, the primary component of arthropod exoskeletons and fungal cell walls, is used to make chitosan, a biopolymer that is organic, secure, and reasonably priced [12]. Chitin is also the second renewable carbon source after lignocellulosic biomass. Due to its pH sensitivity, biocompatibility, and bioactive properties, this cationic polymer made of (1-4)-2-amino-2-deoxy--glucan has garnered greater interest than its basic polymer, chitin [13]. In contrast to chitosan, ChNPs exhibit more behavior. Compared to chitosan, it improves immune-enhancing, anticancer, and antibacterial effects. Additionally, ChNPs have a more curved surface than larger particles do, which increases the dissolving pressure and, as a result, the solubility in saturation [14].

Several methods, including ionic gelation, emulsion droplet coalescence, emulsion solvent diffusion, reverse micellar approach, polyelectrolyte complexation, and desolvation, can readily transform chitosan into NPs. All of these techniques fall under the category of bottom-up manufacturing procedures, which entail the solution-based assembly of molecules to create predetermined structures (NPs). Because of its advantageous characteristics, including its capacity to interact with organic molecules, sensitivity to enzymatic hydrolysis, and intrinsic physiological activity, ChNPs are widely employed for drug delivery applications. Size, surface potential, molecular weight, and stability of ChNPs are key factors in how well they encapsulate and release drugs [15]. Figure 4.1 showing the deacetylation of chitin into chitosan.

Chitosan has been used in a number of agricultural applications, including the coating of seeds, the addition of nutrients and fertilizers, the improvement of plant growth, the defense against frost, and self-protective mechanisms. Because agrochemicals, like the chitosan matrix, can maintain valuable substances by serving as a reservoir for them, shielding them from their environment, and assessing their qualities, they can be used as efficient delivery systems for plant modification. The release mechanism created for NP-based herbicides significantly aids early weed clearance because it has a large ability to totally erase resistance, preserve the activity of their functional components, and extend their liberation over longer durations. Thus, it was suggested that nanoherbicides help with weed

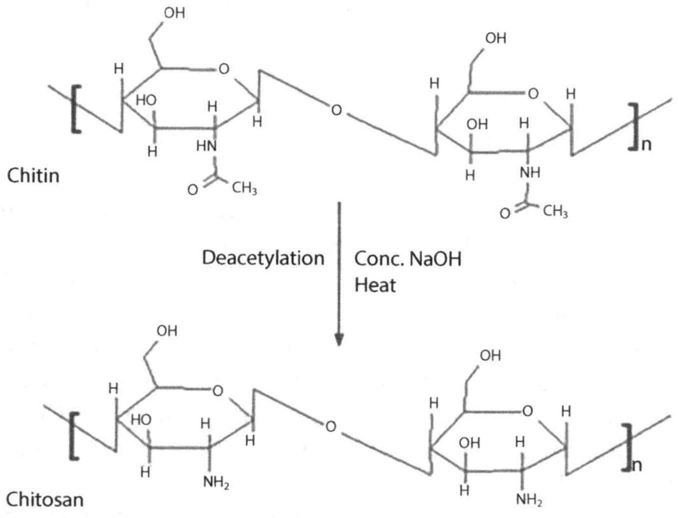

Figure 4.1 Deacetylation of chitin into chitosan [6].

management more successfully than regular herbicides, even at 10-fold lower dosages. The effects of nano-atrazine on maize weeds have also been demonstrated to be equivalent to those of commercial atrazine at a dosage that was 10 times lower [16].

Chitosan NPs are typically created via physical and chemical processes and are utilized all over the world for a variety of purposes, including medicine delivery and agriculture. For their creation and use as organic pest control agents, chitosan-derived NPs of various microbial control agents have been indicated to have biocontrol potential [17]. Environmentally friendly and significantly less associated with worries about toxicity, biodegradability, and physiological stability are known characteristics of polysaccharide-based NPs [18]. The literature makes it abundantly clear that the properties of chitosan NPs can vary greatly based on the techniques used for surface modification and manufacture, which may lead to applications in completely different industries. Over the past few decades, researchers have concentrated on a number of uses, including agriculture, wastewater treatment, and cosmetics, in addition to the tremendous improvements in medical [5].

According to a previous study, Chitosan nanoparticles (CSNPs) protect rice plants from blast fungi. Sandhya *et al.* [19] investigated the efficacy of Harpin-loaded chitosan nanoparticles (H-CSNPs) in tomato against harpin from Pseudomonas syringae pv. syringae (harpinPss) and CSNP controls as a model system for increasing disease resistance and as a fairly cost alternative to conventional antifungal and pesticide treatments. Biodegradability, biocompatibility, nontoxicity, and noteworthy antibacterial capabilities are only a few properties of chitosan [7]. Chitosan, a well-known biostimulant, is used to encourage the development and control stress, such as pests and infections. Low molecular weights, enhanced bioavailability, longer half-lives, and higher surface area–to–volume ratios of NPs make them more effective [20].

4.2.1 Properties of Nanochitosan

1. Size and Morphology: Nanochitosan particles are defined by their nanoscale size. Nanochitosan refers to ChNPs, which are usually between 100 nm and 1000 nm in size. Numerous processes such as ionic gelation, solvent exchange, and emulsion procedures are used to attain these small sizes. Because the size of nanochitosan significantly impacts its quality and uses, it must be precisely controlled. For example, smaller particles often have a larger surface area, which makes them better suited for applications such as medication

administration and wound healing [21]. Nanochitosan particle shape is highly variable and contingent upon the manufacturing process utilized. Particles of nanochitosan may have more intricate geometries that can be spherical or rod-shaped [22]. The behavior of a particle in various applications is significantly influenced by its morphology. For example, rod-shaped nanochitosan particles have demonstrated potential as reinforcing agents in composite materials owing to their high aspect ratio and capacity to increase the strength of the material [23].

2. Surface Charge: When chitosan is converted into nanoscale particles, it transforms into nanochitosan, which has better properties such as a higher surface area and reactivity. Nanochitosan colloidal behavior, stability, and interactions in aqueous environments are largely determined by their surface charges. Zeta potential measurements are commonly used to quantify the surface charge of nanochitosan. The electrical potential difference between the dispersion medium and fluid layer that is stationary and adheres to the surface of the nanochitosan particles is known as the zeta potential. This is a crucial factor in understanding the behavior and colloidal stability of NPs. The surface charges can be indicated by either positive or negative zeta potential values [24].

   The zeta potential was negative for the nanochitosan. The main reason for this negative charge is that its surface contains amino groups ($-NH_2$). These amino groups can ionize in an aqueous solution, forming negatively charged amino groups (NH_2-) and positively charged amino groups (NH_3+) on the surface of the NPs [22]. The stability of the nanochitosan dispersions depends on the net negative zeta potential because it creates electrostatic repulsion between the particles, which prevents aggregation. The ability to release loaded medications under regulated conditions makes this feature extremely useful in drug delivery systems. Nanochitosan can be altered such that it has a positive zeta potential under specific circumstances. Chemical reactions and the attachment of cationic molecules to the surface of nanochitosan are the two methods used to create this alteration. For gene delivery and antibacterial applications, nanochitosan can interact more effectively with negatively charged surfaces such as cell membranes when it has a positive zeta potential [25].

3. Enhanced Physicochemical Properties: Nanochitosan has distinct physicochemical features owing to its nanostructure, including better bioavailability, greater reactivity, and improved solubility [19]. Owing to the interactions between amino and hydroxyl groups, chitosan in its natural state has low solubility in water, which leads to poor dispersion and colloidal stability. Nanochitosan, on the other hand, shows enhanced solubility, which is very advantageous for a variety of applications. Better dispersion in water and other solvents is made possible by the nanoscale size and surface changes of nanochitosan, which break the hydrogen bonds between the chitosan chains [26]. The use of nanochitosan in medical delivery systems, agricultural formulations, and wastewater treatment is made possible by its improved solubility.

   Although nanochitosan particles are smaller and can interact with biological systems more effectively, they have higher bioavailability. Nanochitosan can encapsulate and safeguard active medicinal components in drug delivery systems, thereby improving their bioavailability and facilitating targeted distribution to certain cells or tissues [27]. As a result of its high surface area-to-volume ratio, nanochitosan is a very effective adsorbent for a wide range of pollutants. The improved physicochemical characteristics of this material allow effective adsorption of organic contaminants, dyes, and heavy metals from wastewater and water. Functional groups can be added to the surface of nanochitosan to further customize its adsorption ability for certain pollutants [28]. This feature is essential for water purification and environmental clean-up. Nanochitosan-based carriers have demonstrated significant promise in enhancing crop protection, lowering the environmental impact, and enhancing the controlled release of pesticides in agriculture.

4. Nanochitosan is biocompatible; living organisms including humans can easily tolerate it. It is derived from chitosan, a naturally occurring polymer formed from chitin that is used by crustaceans to make their exoskeletons. Furthermore, nanochitosan is biodegradable and gradually degrades through natural processes [22]. The fact that nanochitosan is nontoxic is one of its biggest benefits. Several studies have

shown that nanochitosan displays low cytotoxicity in different cell lines. Owing to its excellent biodegradability, nanochitosan is a perfect material for tissue engineering and drug delivery applications. It releases nontoxic byproducts during degradation, which lowers the possibility of negative physiological responses [29]. Nanochitosan has many potential applications in agriculture owing to its remarkable biocompatibility and adaptability. It is an extremely appealing biomaterial for a variety of applications owing to its nontoxic nature, biodegradability, immunomodulatory properties, and effective biological barrier permeability.

5. Antimicrobial Activity: Chitosan (CS) is a linear polysaccharide with antibacterial activity against viruses, fungi, and bacteria. Nanochitosan, a nanoscale version of chitosan, is considerably more effective because of its larger surface area and better dispersibility. Considering its cationic nature, nanochitosan can interact with negatively charged microbial cell membranes, causing disruption and growth inhibition. Through disruption of cell membranes, interference with metabolic processes, and disruption of DNA replication, nanochitosan efficiently limits the development of harmful bacteria. Their usefulness in pharmacological and agricultural applications stems from their efficacy against fungi. The possibility of using nanochitosan as an antiviral agent to prevent the reproduction of enveloped viruses has also been investigated. It uses oxidative stress, enzyme inhibition, membrane disruption, and electrostatic interactions to exert antibacterial effects. Its broad-spectrum and adaptable properties render it an effective antibacterial agent [30].
6. Encapsulation and Delivery: Drugs, pesticides, and essential oils are examples of bioactive chemicals that can be packaged and delivered by using nanochitosan as a carrier. These compounds can be effectively encapsulated and released under regulated conditions owing to their large surface area and compact size of nanochitosan particles [31].

Encapsulation of many bioactive molecules, including medicines, proteins, genes, and imaging agents, is the principal use of nanochitosan. Owing to its large surface area and surface charge, nanochitosan is a perfect carrier for these substances. Therapeutic chemicals can be delivered continuously and precisely owing to the controlled release feature of the encapsulation method, which also shields the agents from degradation [32].

7. Barrier Properties: The oxygen permeability and moisture resistance of nanochitosan are barrier properties that may be helpful for the packaging and preservation of food. It can help increase the shelf life of perishable foods by acting as a barrier between these items and external effects [30]. Excellent gas barrier qualities, water vapor barrier, UV barrier, and oil and grease resistance are just a few of the many qualities of nanochitosan. Its nanostructure provides gas molecules with a winding route that lowers the rates at which oxygen and moisture are transmitted, which makes it appropriate for use in food packaging and medicine. In addition to lowering the water vapor transmission, it is more environmentally responsible. Furthermore, nanochitosan improves oil and grease resistance, making it useful for applications in food services and packaging [33].
8. Environmental Friendliness: Nanochitosan is considered environmentally advantageous because of its natural origin, biodegradability, and low toxicity compared to synthetic alternatives. This upholds the principles of robust materials and environmental friendliness. Crustacean shells, such as those of shrimp and crab, are a source of chitosan, the predecessor of nanochitosan. Because the seafood business generates millions of tons of waste annually, chitosan is a sustainable resource that may be properly utilized. As chitosan degrades quickly, it is a more ecologically friendly option than synthetic polymers. The conversion of chitosan into NPs can further improve its biodegradability, as smaller particles break down more quickly and reduce the environmental effects of waste disposal [30].

Given that nanochitosan is composed of natural substances, it is regarded as non-toxic and safe for use. Its biocompatibility is beneficial for many applications, particularly in agriculture and medicine. The numerous environmental advantages of nanochitosan include decreased chemical use in agriculture, less plastic waste in biodegradable packaging, lower production energy consumption, and mitigation of water pollution during wastewater treatment. Industries may decrease their environmental impact and preserve aquatic ecosystems by substituting nanochitosan-based goods and chemicals with conventional materials and chemicals [33].

Overall, nanochitosan demonstrates special qualities that make it a potential substance for a range of uses including biopesticides. It has the potential to be an efficient and healthy substitute for synthetic pesticides owing to its size, surface charge, biocompatibility, antibacterial activity, and encapsulation characteristics.

4.2.2 Synthesis of Nanochitosan

Several techniques have been devised to create CS particles as delivery systems. The nature of the bioactive chemicals, particle size, release kinetics, stability, residual toxicity, and mechanism of ultimate product delivery are the primary factors influencing these techniques. The manufacture of CSNPs has frequently made use of ionic gelation, emulsion cross-linking, and coacervation or precipitation techniques. Although currently not commonly used, reverse micellar, spray drying, emulsion-droplet coalescence, screening, and other procedures are also used to produce NPs [32]. Chitosan NPs were first introduced by Ohya *et al.* [34] for the systemic administration of the chemotherapy medication 5-fluorouracil. Chitosan NPs have been thoroughly studied, and numerous approaches have been developed while taking a range of factors into account, including size, stability, drug loading capacity, and retention period.

4.2.2.1 Emulsion Cross-Linking

Using the amino and aldehyde groups of chitosan, emulsification and cross-linking is the first method for creating ChNPs. This required mixing an oil phase with a chitosan solution, stabilizing it with Span 80, and then cross-linking it with toluene. Small NPs with a restricted distribution were generated using this approach [8].

4.2.2.2 Reverse Micellar Method

To create ChNPs, a procedure known as the reversed micellar method, which is akin to emulsification and cross-linking, was employed. This includes combining an organic phase made up of a lipophilic surfactant and an organic solvent with an aqueous phase. Through cross-linking, the micelle core, which includes chitosan, produces NPs. The surfactant is precipitated during this process, which also includes dialysis and freeze-drying. Utilizing safe solvents and cross-linkers, this technique has been modified [18].

4.2.2.3 *Precipitation Method*

Precipitation-based techniques can be used to create ChNPs. The phase-inversion precipitation technique produces nanometer-sized droplets by emulsification and precipitation. The desolvation procedure, also known as the emulsion-droplet coalescence method, involves the coalescence of two water-in-oil emulsions. The final NPs were made using centrifugation, washing with solvents like toluene, ethanol, and water, and freeze-drying. These NPs are typically larger than 600–800 nm. They are not commonly used due to the usage of organic solvents and high energy requirements, although they have a high encapsulation efficiency for hydrophobic drugs like cyclosporin A [35].

4.2.2.4 *Ionic Gelation*

Over the past two decades of research, the most widely used technique for producing ChNPs has been ionic gelation. In order to create NPs using this technique, chitosan is complexed with a counter ion, such as sodium tripolyphosphate (TPP), in an aqueous solution. As a result, spherical NPs are created, which can be acquired through oven- or freeze-drying. By altering the chitosan/TPP ratio, which affects the efficiency and delivery of drug encapsulation, the process is straightforward, non-toxic, and flexible [18].

4.2.3 Preparation of Nanochitosan

Methacrylic acid (MAA) can be polymerized in a chitosan (CS) solution to create nanochitosan by this method [30]. The first step of the procedure is to dissolve chitosan for 12 h in a 0.5% solution of MAA. Typically, 0.8% weight/weight of chitosan is utilized in the synthesis. Nanochitosan particles are created when MAA polymerizes in the chitosan solution. The precise characteristics of the polymerization process, such as the reaction conditions and catalysts used, may change depending on the technique used. The main objective of this study was to incorporate MAA into the chitosan matrix and produce nanoscale particles. The resulting nanochitosan particles have special qualities that make them suitable for a variety of applications. These properties include the capacity to encapsulate and distribute bioactive chemicals, antibacterial activity, and biocompatibility. It is crucial to remember that the process mentioned above is simply an illustration of the preparation of nanochitosan. Depending on the intended characteristics and uses of nanochitosan, additional techniques and modifications are available. To improve the characteristics of nanochitosan and broaden its possible applications, more work is being done in the field of nanotechnology to investigate and improve its synthesis processes.

4.2.4 Application of Nanochitosan in Agriculture

The number of studies on the use of ChNPs in agricultural applications has significantly increased in response to the demand for sustainable and ecologically friendly agrochemicals, such as fertilizers and insecticides. Chitosan NPs are mostly used as controlled release nanocarriers to increase the stability of active compounds, much like how they are used in medicine. These results suggest that agrochemicals can be used at lower dosages and require fewer treatments, reducing the risk of environmental contamination and negative impacts on unintended species. Medium and highly positively charged NPs are often used to facilitate the slow and steady release of active chemicals into the soil, because they offer greater stability in aquatic conditions. Agrochemicals or active substances, such as essential oils, have been encapsulated in ChNPs through nanoencapsulation, which occurs spontaneously during ionic gelation [36]. Chitosan NPs are frequently employed in the delivery of herbicides for the control of weeds [37], the treatment of insects and fungus [9], and a variety of deliveries, including plant growth regulators and fertilizers for healthy plant nutrition. Alginate/chitosan (ALG/CS) NPs, chitosan-poly (acrylic acid) (CS-PAA), and chitosan-poly (MAA) (CS-PMAA) are examples of chitosan derivatives that have also been employed [18].

4.3 Efficacy of Nanochitosan Compared to Synthetic Pesticides

A key factor in assessing the potential of nanochitosan as a biopesticide is how effective it is in comparison to synthetic pesticides. Although there is little comparable data available, research indicates that nanochitosan and other biopesticides may be preferable to synthetic pesticides in terms of their effects on the environment and hazards compared to human health. According to Mkindi *et al.* [38], synthetic pesticides have been linked to dangerous environmental consequences and detrimental effects on natural adversaries, such as ladybugs and spiders. In contrast, it has been discovered that nanochitosan-based biopesticides offer effective control of important pest species while having fewer adverse effects on natural enemies.

Additionally, studies have shown that the use of plant pesticides, such as nanochitosan, is more economical than synthetic pesticides.

In the context of sustainable agriculture, the use of nanochitosan as a biopesticide is particularly crucial. To achieve sustainable and environmentally friendly pest control methods, biopesticides like nanochitosan are being developed and used as substitutes for synthetic pesticides. According to Yan and Zhang [39], biopesticides—including botanical pesticides like nanochitosan—are viewed as less dangerous to both human health and the environment. Due to their low risk to the environment and people, they have been utilized as substitutes for synthetic pesticides for a long time [40]. The usage of natural and organic pesticides is becoming more and more popular as a more environmentally friendly alternative to synthetic pesticides. Compared to synthetic chemicals, these biopesticides offer superior selectivity, lessen insect resistance, and have lower environmental persistence [41].

The potential advantages of biopesticides, such as nanochitosan, are highlighted by Mkindi *et al.* [38] in terms of their short persistence, naturally low active ingredient concentrations, and anti-feedant/repellent modes of action. Compared to concentrated synthetic treatments, biopesticides are more environmentally friendly because to these qualities. However, more field research is required to compare the effects of synthetic and biological pesticides on ecosystems. The benefits of biopesticides, especially nanobiopesticides like nanochitosan, over synthetic pesticides are emphasized by Ayilara *et al.* [42]. In comparison to synthetic pesticides, phytopesticides, which include biopesticides generated from plants, have a variety of modes of action and pose fewer dangers to human health and the environment. Higher pesticidal action, targeted or controlled release, and excellent biocompatibility and biodegradability are all features of nanobiopesticides like nanochitosan [42]. According to these findings, nanochitosan and other biopesticides have the potential to be safer for human health and the environment than synthetic pesticides. It is crucial to remember that additional study is required to fully compare the effectiveness of nanochitosan and synthetic insecticides in actual field situations.

4.3.1 Nanochitosan's Mechanism of Action Against the Pathogens

Chitosan exerts antibacterial effects on microorganisms. Due to electrostatic interactions with the pathogen's surface, made possible by its positive charge, cells are damaged and eventually die. Chitosan produces a barrier to the extracellular transit of nutrients, inhibits DNA/RNA interaction, and chelates critical components for pathogen growth. Due to their high surface area and compact size, chitosan nanoformulations can increase the effectiveness of this process. Additionally, chitosan stimulates the production of defense-related enzymes as catalase, peroxidase, polyphenol-oxidase, and phenylalanine ammonia-lyase (PAL). L-phenylalanine is converted by PAL into trans-cinnamic acid and ammonium, which lowers the occurrence of gray mold. In order to create lignin, polyphenol oxidase catalyzes the conversion of phenolic compounds, creating a barrier to disease penetration. Reactive oxygen species (ROS) can cause oxidative damage to plant cells, although catalase can counteract this damage while peroxidases help create harmful ROS. Peaches treated with chitosan exhibit significant mycelial and spore germination suppression [43].

In plants exposed to chitosan, it has been discovered that defense-related secondary metabolites such as phytoalexins, phenolic compounds, lignin, and callose increase. These metabolites are created when a disease infects a host and has antibacterial and antioxidant effects. According to studies, chitosan-infused plants can produce phytoalexins that can protect them against infections. By creating lignin, signaling molecules, and defensive response chemicals, plant phenolic compounds play a significant part in fostering plant resistance. The pathogenesis-related proteins chitinase and 1,3-glucanase, which function as catalytic converters in the hydrolysis of chitin and -D-glucans, degrade the fungal cell wall, and prevent fungal development on the host plant, are also present in chitosan-treated plants [44].

4.3.2 Bioactivity of Nanochitosan as a Biopesticide

Owing to its distinct characteristics and the potential for tailored pesticide administration, nanochitosan has attracted interest as a potential biopesticide. Numerous insecticides, including thiamethoxam, iprofloxacin-HCl, and geraniol, have been loaded onto polysaccharide nanocarriers, such as nanochitosan. The controlled release capabilities of these nanocarriers enable the targeted administration of insecticides to species [45]. Oliveira *et al.* [46] investigated the use of chitosan/gum Arabic NPs containing geraniol as biopesticides. Geraniol was nanoencapsulated in chitosan/gum Arabic NPs, which improved the compound's stability and efficiency and increased its effectiveness in integrated pest management. These NPs' biological activity against the whitefly was assessed. Geraniol was incorporated in them. Furthermore, ciprofloxacin-HCl's antibacterial activity has been improved by the introduction of NPs made of chitosan. The enhanced antibacterial activity of these NPs points to their potential for use as a biopesticide against bacterial pests [47].

Nanopesticides, particularly formulations based on nanochitosan, are being used, although this raises questions regarding their potential ecotoxicological consequences and environmental sustainability. To guarantee that nanopesticides are used safely and sustainably, it is crucial to take into account the regulatory implications and assess their environmental impact [48]. Taking into account their biocompatibility, biodegradability, and low toxicity, natural polysaccharides, such as chitosan, are frequently employed in the design of drug delivery systems. Chitosan-based nano-carriers have been used to transport small interfering RNA (siRNAs), which can be used to target genes in pests to control them [49].

Considering its capacity to operate as a carrier for the precise delivery of insecticides and to increase their bioactivity, nanochitosan has potential as a biopesticide. Botanical chemicals, such as geraniol, can be encapsulated in chitosan-based NPs to boost their stability and efficacy against pests. Further research is required to assess the environmental sustainability and potential ecotoxicological impacts of nanopesticides. Additionally, siRNA delivery using chitosan-based nanocarriers has the potential to be tailored for pest management.

4.3.3 Environmental Benefits of Nanochitosan

Nanochitosan provides a number of environmental advantages because of its unique characteristics and uses. One of the main benefits is its ability to lower pesticide dosages and reduce human exposure to dangerous chemicals [50]. Nanochitosan-based formulations of nanoencapsulated pesticides have been demonstrated to protect crops effectively while being environmentally beneficial [50]. Nanochitosan can help with sustainable agricultural practices by lowering the amount of pesticides required. In addition, nanochitosan has shown to possess antibacterial qualities, such as quick bactericidal activity and ongoing antimicrobial agent release [50]. By eliminating the need for conventional antimicrobial treatments that might harm ecosystems, this can have a positive impact on environmental health.

Nanochitosan has demonstrated promise in the biomedical sector as a cancer treatment. Tumor growth has been reported to be inhibited by the encapsulation of anticancer medications, such as paclitaxel, in nanochitosan NPs [51]. This customized drug delivery system may be able to lessen chemotherapy's adverse effects as well as the negative environmental effects of traditional cancer therapies. Additionally, nanochitosan has uses for enhancing soil health. According to studies, the addition of beneficial bacteria and nanochitosan can improve soil quality and boost nutrient status [52]. By encouraging healthy soil ecosystems and lowering the demand for artificial fertilizers, this can support sustainable agricultural practices. Furthermore, nanochitosan synthesis itself may be environmentally favorable. Exoskeletons of crustaceans are one source of chitin, the building block of chitosan [22]. Nanochitosan manufacture supports the circular economy and lessens waste by using these waste products.

In conclusion, nanochitosan has a variety of advantages for the environment. Its inclusion in pesticide formulations helps minimize human exposure to chemicals and their use. Its antibacterial qualities can offer substitutes for traditional antimicrobial substances. Nanochitosan-based medication delivery systems for cancer therapy can target tumors and reduce adverse effects. Additionally, nanochitosan can support sustainable farming methods and the improvement of soil health. The creation of nanochitosan from waste materials also adheres to the circular economy's guiding principles. All things considered, nanochitosan has enormous potential for addressing environmental issues and supporting sustainable practices.

4.4 Challenges and Future Prospects for the Use of Chitosan Nanoparticles as Biopesticides

The use of nanochitosan as a biopesticide presents a number of difficulties for this innovative and developing technology. To begin with, assuring its effectiveness depends on elements such as the target pest, administration technique, dosage, and environmental circumstances. Nanochitosan may need to be formed into a variety of delivery systems; therefore, formulation and stability are also essential. The regulatory environment presents another difficulty because nanochitosan may need more investigation to verify its efficacy and safety. Nanochitosan's potential negative effects on the environment and non-target creatures are another issue. Another difficulty is cost-effectiveness, as the production procedure must be competitive with other pest control strategies. Additionally, crucial are knowledge and understanding of the advantages of nanochitosan, its methods of application, and its safety considerations. By overcoming these obstacles, nanochitosan's potential as a biopesticide can be unlocked, which will aid in the development of sustainable pest control methods.

A breakthrough biopesticide called nanochitosan has a lot of potential across the board. By targeting particular plant tissues, enhancing formulation and delivery methods, combining with other biopesticides, or using conventional pesticides, it can improve insect control. It adheres to the principles of sustainable agriculture and considers the long-term consequences on the health of the soil, beneficial creatures, and non-target organisms. Being a relatively new technology, nanochitosan requires careful regulatory scrutiny. Commercialization and production scaling up are equally crucial. Research can concentrate on creating affordable procedures for synthesizing and formulating nanochitosan, spreading knowledge of its advantages, and encouraging its implementation in pest management techniques. In conclusion, nanochitosan has a bright future ahead of it thanks to ongoing research and development that will result in better formulations, more precise delivery methods, and increased pest control effectiveness.

4.5 Conclusion

A notable development in agricultural technology is nanochitosan, which performs a number of tasks that are advantageous to both crop yield and environmental sustainability. Its potential to change current agriculture is highlighted by its capacity to increase nutrient delivery, control pests and diseases, boost soil quality, and limit environmental impacts. They can also be used as controlled fertilizers, pesticides, and plant growth stimulants in agriculture. Greenly synthesized ChNPs may be affordable substitutes for harmful pesticides, which would help the scientific community in crop protection and enhancement. Thus, it can be said that nanochitosan has the potential to replace synthetic pesticides in agriculture in a way that is both efficient and safe for the environment. Utilizing nanochitosan can result in higher agricultural yields and less environmental harm because of its wide range of activities against different pests. Nevertheless, there are still issues that must be resolved, such as guaranteeing its efficacy, stability, and safety, as well as lowering its cost compared to other pest management techniques. The potential of nanochitosan as a biopesticide needs to be fully realized through further studies and development.

4.6 Recommendations

1. Synthesis and formulation of nanochitosan should be improved through further research and development, with an emphasis on developing accessible methods for doing so, disseminating information about its benefits, and promoting the use of nanochitosan in pest control strategies.
2. It is important to monitor the regulations to ensure that nanochitosan is safe and efficient when used as a biopesticide.
3. It is important to raise the level of awareness and understanding of nanochitosan among farmers, researchers, and policymakers to support its use in pest-management plans.
4. Further research is required to assess the long-term effects of nanochitosan on the health of soil, beneficial species, and non-target organisms.
5. Making nanochitosan more affordable than alternative pest management techniques, such as the use of synthetic pesticides, should be a priority.

References

1. UNEP, Global Bee Colony Disorders and Other Threats to Insect Pollinators [Internet], 2020, [cited 2023 Nov 25]. Available from: https://www.unep.org/resources/report/global-bee-colony-disorders-and-other-threats-insect-pollinators.
2. WHO, Guidelines on Protecting Workers from Occupational Pesticide Exposure [Internet], 2020, [cited 2023 Nov 25]. Available from: https://www.who.int/publications/i/item/9789240001440.
3. EPA, Pesticide Residues in Food [Internet], 2020, [cited 2023 Nov 25]. Available from: https://www.epa.gov/pesticide-tolerances/pesticide-residues-food.
4. Wang, X., Xu, J., Wang, X., Qiu, B., Cuthbertson, A.G., Du, C., *et al.*, Isaria fumosorosea-based zero-valent iron nanoparticles affect the growth and survival of sweet potato whitefly, Bemisia tabaci (Gennadius). *Pest. Manage. Sci.*, 75, 8, 2174–2181, 2019.
5. Hijazi, N., Le Moigne, N., Rodier, E., Sauceau, M., Vincent, T., Benezet, J.C., *et al.*, Biocomposite films based on poly (lactic acid) and chitosan nanoparticles: Elaboration, microstructural and thermal characterization. *Polym. Eng. Sci.*, 59, S1, E350–E360, 2019.
6. Mohammed, M.A., Syeda, J.T., Wasan, K.M., Wasan, E.K., An overview of chitosan nanoparticles and its application in non-parenteral drug delivery. *Pharmaceutics*, 9, 4, 53, 2017.

7. Van Bavel, N., Issler, T., Pang, L., Anikovskiy, M., Prenner, E.J., A Simple Method for Synthesis of Chitosan Nanoparticles with Ionic Gelation and Homogenization. *Molecules*, 28, 11, 4328, 2023.
8. Ingle, P.U., Shende, S.S., Shingote, P.R., Mishra, S.S., Sarda, V., Wasule, D.L., *et al.*, Chitosan nanoparticles (ChNPs): A versatile growth promoter in modern agricultural production. *Heliyon*, 8, 11, e11893, 2022.
9. Sahab, A.F., Waly, A.I., Sabbour, M.M., Nawar, L.S., Synthesis, antifungal and insecticidal potential of Chitosan (CS)-g-poly (acrylic acid)(PAA) nanoparticles against some seed borne fungi and insects of soybean. *Int. J. Chem. Tech. Res.*, 8, 2, 589–98, 2015.
10. Shahrajabian, M., Chaski, C., Polyzos, N., Tzortzakis, N., Petropoulos, S., Sustainable agriculture systems in vegetable production using chitin and chitosan as plant biostimulants. *Biomolecules*, 11, 6, 819, 2021. https://doi.org/10.3390/biom11060819.
11. Sreelatha, S., Kumar, N., Sarojam, R., Biological effects of thymol loaded chitosan nanoparticles (tcnps) on bacterial plant pathogen xanthomonas campestris pv. campestris. *Front. Microbiol.*, 13, 1085113, 2022. https://doi.org/10.3389/fmicb.2022.1085113.
12. Malerba, M. and Cerana, R., Chitosan effects on plant systems. *Int. J. Mol. Sci.*, 17, 7, 996, 2016.
13. Sivashankari, P.R. and Prabaharan, M., Deacetylation modification techniques of chitin and chitosan, in: *Chitosan Based Biomaterials Volume 1*, pp. 117–133, Woodhead Publishing, Sawston, Cambridge UK, 2017.
14. Dawoud, S.F., Al-Akra, T.M., Zedan, A.M., Hepatoprotective effects of chitosan and chitosan nanoparticles against biochemical, genetic, and histological disorders induced by the toxicity of emamectin benzoate. *Rep. Biochem. Mol. Biol.*, 10, 3, 506, 2021.
15. Abdelhamid, A.E., Ahmed, E.H., Awad, H.M., Ayoub, M.M., Synthesis and cytotoxic activities of selenium nanoparticles incorporated nano-chitosan. *Polym. Bull.*, 81, 2, 1–17, 2023.
16. Khan, B.A., Nadeem, M.A., Iqbal, M., Yaqoob, N., Javaid, M.M., Maqbool, R., *et al.*, Chitosan nanoparticles loaded with mesosulfuron methyl and mesosulfuron methyl+ florasulam+ MCPA isooctyl to manage weeds of wheat (Triticum aestivum L.). *Green Process. Synth*, 12, 1, 20228152, 2023.
17. Wu, J., Du, C., Zhang, J., Yang, B., Cuthbertson, A.G., Ali, S., Synthesis of Metarhizium anisopliae–chitosan nanoparticles and their pathogenicity against Plutella xylostella (Linnaeus). *Microorganisms*, 10, 1, 1, 2021.
18. Yanat, M. and Schroën, K., Preparation methods and applications of chitosan nanoparticles; with an outlook toward reinforcement of biodegradable packaging. *React. Funct. Polym.*, 161, 104849, 2021.
19. Siddaiah, C., Prasanth, K., Satyanarayana, N., Mudili, V., Gupta, V., Kalagatur, N., *et al.*, Chitosan nanoparticles having higher degree of acetylation induce resistance against pearl millet downy mildew through nitric oxide generation. *Sci. Rep.*, 8, 1, 2485, 2018. https://doi.org/10.1038/s41598-017-19016-z.

20. Nadendla, S.R., Rani, T.S., Vaikuntapu, P.R., Maddu, R.R., Podile, A.R., HarpinPss encapsulation in chitosan nanoparticles for improved bioavailability and disease resistance in tomato. *Carbohydr. Polym.*, 199, 11–19, 2018.
21. Bashir, S.M., Ahmed Rather, G., Patrício, A., Haq, Z., Sheikh, A.A., Shah, M.Z.U.H., *et al.*, Chitosan nanoparticles: a versatile platform for biomedical applications. *Materials*, 15, 19, 6521, 2022.
22. Lee, C., Hao, L., Park, J., Oh, D., Hwang, D., Nanochitin and nanochitosan: chitin nanostructure engineering with multiscale properties for biomedical and environmental applications. *Adv. Mater.*, 35, 4, 2203325, 2022. https://doi.org/10.1002/adma.202203325.
23. Jeyaraj, M., Gurunathan, S., Qasim, M., Kang, M.H., Kim, J.H., A comprehensive review on the synthesis, characterization, and biomedical application of platinum nanoparticles. *Nanomaterials*, 9, 12, 1719, 2019.
24. Gopmandal, P.P. and Duval, J.F., Electrostatics and electrophoresis of engineered nanoparticles and particulate environmental contaminants: beyond zeta potential-based formulation. *Curr. Opin. Colloid Interface Sci.*, 101605, 2022.
25. Inam, W., Bhadane, R., Akpolat, R.N., Taiseer, R.A., Filippov, S.K., Salo-Ahen, O.M., *et al.*, Interactions between polymeric nanoparticles and different buffers as investigated by zeta potential measurements and molecular dynamics simulations. *View*, 3, 4, 20210009, 2022.
26. Shukla, S.K., Mishra, A.K., Arotiba, O.A., Mamba, B.B., Chitosan-based nanomaterials: A state-of-the-art review. *Int. J. Biol. Macromol.*, 59, 46–58, 2013.
27. Wilson, B., Samanta, M.K., Santhi, K., Kumar, K.S., Ramasamy, M., Suresh, B., Chitosan nanoparticles as a new delivery system for the anti-Alzheimer drug tacrine. *Nanomed. Nanotechnol. Biol. Med.*, 6, 1, 144–152, 2010.
28. Sanchez-Salvador, J.L., Balea, A., Monte, M.C., Negro, C., Blanco, A., Chitosan grafted/cross-linked with biodegradable polymers: A review. *Int. J. Biol. Macromol.*, 178, 325–343, 2021.
29. Anitha, A., Sowmya, S., Kumar, P.S., Deepthi, S., Chennazhi, K.P., Ehrlich, H., *et al.*, Chitin and chitosan in selected biomedical applications. *Prog. Polym. Sci.*, 39, 9, 1644–1667, 2014.
30. Ngo, T., Nguyen, T., Dang, T., Tran, T., Rachtanapun, P., Characteristics and antimicrobial properties of active edible films based on pectin and nanochitosan. *Int. J. Mol. Sci.*, 21, 6, 2224, 2020. https://doi.org/10.3390/ijms21062224.
31. Loiseau, P., Pomel, S., Croft, S., Chitosan contribution to therapeutic and vaccinal approaches for the control of leishmaniasis. *Molecules*, 25, 18, 4123, 2020. https://doi.org/10.3390/molecules25184123.
32. Singh, A., Mittal, A., Benjakul, S., Chitosan nanoparticles: Preparation, food applications and health benefits. *Sci. Asia.*, 47, 2021, 1–10, 2021.
33. Salim, S., Rihayat, T., Safitri, A., Compatibilization study of blending polymers pla-pcl with fillers catechin-chitosan as new materials for

bioplastic manufacture. *Int. J. Res. Granthaalayah*, 10, 10, 241–250, 2022. https://doi.org/10.29121/granthaalayah.v10.i10.2022.4834.
34. Ohya, Y., Shiratani, M., Kobayashi, H., Ouchi, T., Release behavior of 5-fluorouracil from chitosan-gel nanospheres immobilizing 5-fluorouracil coated with polysaccharides and their cell specific cytotoxicity. *J. Macromol. Sci., Pure Appl. Chem.*, 31, 5, 629–642, 1994.
35. Faqir, Y., Ma, J., Chai, Y., Chitosan in modern agriculture production. *Plant Soil Environ.*, 67, 12, 679–699, 2017.
36. Grillo, R., Pereira, A.E., Nishisaka, C.S., De Lima, R., Oehlke, K., Greiner, R., *et al.*, Chitosan/tripolyphosphate nanoparticles loaded with paraquat herbicide: an environmentally safer alternative for weed control. *J. Hazard. Mater.*, 278, 163–171, 2014.
37. Maruyama, C.R., Guilger, M., Pascoli, M., Bileshy-José, N., Abhilash, P.C., Fraceto, L.F., *et al.*, Nanoparticles based on chitosan as carriers for the combined herbicides imazapic and imazapyr. *Sci. Rep*, 6, 1, 19768, 2016.
38. Mkindi, A., Mpumi, N., Tembo, Y., Stevenson, P., Ndakidemi, P., Mtei, K., *et al.*, Invasive weeds with pesticidal properties as potential new crops. *Ind. Crop Prod*, 110, 113–122, 2017. https://doi.org/10.1016/j.indcrop.2017.06.002.
39. Yan, F. and Zhang, A., A floral fragrance, methyl benzoate, is an efficient green pesticide. *Sci. Rep.*, 7, 1, 42168, 2017. https://doi.org/10.1038/srep42168.
40. Acheuk, F., Basiouni, S., Shehata, A., Dick, K., Hajri, H., Lasram, S., *et al.*, Status and prospects of botanical biopesticides in europe and mediterranean countries. *Biomolecules*, 12, 2, 311, 2022. https://doi.org/10.3390/biom12020311.
41. Catania, R., Lima, M., Potrich, M., Sgolastra, F., Zappalà, L., Mazzeo, G., Are botanical biopesticides safe for bees (hymenoptera, apoidea)? *Insects*, 14, 3, 247, 2023. https://doi.org/10.3390/insects14030247.
42. Ayilara, M., Adeleke, B., Akinola, S., Fayose, C., Adeyemi, U., Gbadegesin, L., *et al.*, Biopesticides as a promising alternative to synthetic pesticides: a case for microbial pesticides, phytopesticides, and nanobiopesticides. *Front. Microbiol.*, 14, 1040901, 2023. https://doi.org/10.3389/fmicb.2023.1040901.
43. Maluin, F.N. and Hussein, M.Z., Chitosan-based agronanochemicals as a sustainable alternative in crop protection. *Molecules*, 25, 7, 1611, 2020.
44. Xing, K., Zhu, X., Peng, X., Qin, S., Chitosan antimicrobial and eliciting properties for pest control in agriculture: a review. *Agron. Sustain. Dev.*, 35, 569–588, 2015.
45. Scott-Fordsmand, J., Fraceto, L., Amorim, M., Nano-pesticides: the lunchbox principle—deadly goodies (semio-chemical functionalised nanoparticles that deliver pesticide only to target species). *J. Nanobiotechnol.*, 20, 1, 13, 2022. https://doi.org/10.1186/s12951-021-01216-5.
46. Oliveira, J., Campos, E., Pereira, A., Nunes, L., Silva, C., Pasquoto, T., *et al.*, Geraniol encapsulated in chitosan/gum arabic nanoparticles: a promising system for pest management in sustainable agriculture. *J. Agric. Food. Chem.*, 66, 21, 5325–5334, 2018. https://doi.org/10.1021/acs.jafc.8b00331.

47. Marei, N., Elwahy, A., Salah, T., Sherif, Y., El-Samie, E., Enhanced antibacterial activity of egyptian local insects' chitosan-based nanoparticles loaded with ciprofloxacin-hcl. *Int. J. Biol. Macromol.*, 126, 262–272, 2019. https://doi.org/10.1016/j.ijbiomac.2018.12.204.
48. Grillo, R., Fraceto, L., Amorim, M., Scott-Fordsmand, J., Schoonjans, R., Chaudhry, Q., Ecotoxicological and regulatory aspects of environmental sustainability of nanopesticides. *J. Hazard. Mater.*, 404, 124148, 2021. https://doi.org/10.1016/j.jhazmat.2020.124148.
49. Serrano-Sevilla, I., Artiga, A., Mitchell, S., Matteis, L., Fuente, J., Natural polysaccharides for sirna delivery: nanocarriers based on chitosan, hyaluronic acid, and their derivatives. *Molecules*, 24, 14, 2570, 2019. https://doi.org/10.3390/molecules24142570.
50. Sivanesan, I., Manikandan, M., Gopal, J., Hasan, N., Ali, S., Shin, J., *et al.*, Nanochitosan: commemorating the metamorphosis of an exoskeletal waste to a versatile nutraceutical. *Nanomater.*, 11, 3, 821, 2021. https://doi.org/10.3390/nano11030821.
51. Adhikari, H. and Yadav, P., Anticancer activity of chitosan, chitosan derivatives, and their mechanism of action. *Int. J. Biomater.*, 2018, 1–29, 2018. https://doi.org/10.1155/2018/2952085.
52. Chaudhary, P., Chaudhary, A., Bhatt, P., Kumar, G., Khatoon, H., Rani, A., *et al.*, Assessment of soil health indicators under the influence of nanocompounds and bacillus spp. in field condition. *Front. Environ. Sci.*, 9, 769871, 2022. https://doi.org/10.3389/fenvs.2021.769871.

5

The Use of Nanochitosan for Enhancement in the Quality and Yield of Fruit Crops

Ruth Ebunoluwa Bodunrinde[1], Konjerimam Ishaku Chimbekujwo[2], Oluwafemi Adebayo Oyewole[3], Charles Oluwaseun Adetunji[4*], Abel Inobeme[5], Eniola K. I. T.[6], Mohammed Bello Yerima[7] and John Tsado Mathew[8]

[1]*Department of Microbiology, Federal University of Technology, Akure, Nigeria*
[2]*Department of Microbiology, Modibbo Adama University, Yola, Nigeria*
[3]*Department of Microbiology, Federal University of Technology, Minna, Nigeria*
[4]*Department of Microbiology, Edo State University, Uzairue, Nigeria*
[5]*Department of Chemistry, Edo State University, Uzairue, Nigeria*
[6]*Department of Biological Sciences, Joseph Ayo Babalola University, Ikeji Arakeji, Osun State, Nigeria*
[7]*Department of Microbiology, Sokoto State University, Sokoto, Sokoto State, Nigeria*
[8]*Department of Chemistry, Ibrahim Badamasi Babangida University, Lapai, Niger State, Nigeria*

Abstract

The global economy is largely sustained by agriculture, which accounts for the majority of its inhabitants. Chitosan, normally occurring substance made from crab shells, is produced in industry. The material has been utilized to enhance vegetable growth and output, as well as to safeguard against fungi, bacteria, and viruses. The biopolymer "nanochitosan" has generated considerable interest due to its high biological compatibility, biodegradability, and biological activity, which make it a potential agricultural polymer. Many plant species use it as an inducer, which not only stimulates the immune system but also raises crop yields. Chitosan enhanced plant effectiveness by reducing the adverse effects of unfavorable circumstances and plant growth. A new approach to improve crop yield through Chitosan NPs is being developed. This review examines the applications of ChNPs in improving fruit crops' standard and yield.

**Corresponding author*: adetunji.charles@edouniversity.edu.ng

Charles Oluwaseun Adetunji, Maulin P. Shah, Yerima Mohammed Bello, Daniel Hefft, Jay Singh, Shyam S. Pandey and Ravindra Pratap Singh (eds.) Nanochitosan Applications for Enhanced Crop Production and Food Security, (111–122) © 2025 Scrivener Publishing LLC

***Keywords*:** Chitosan nanoparticles (ChNPs), growth, fruits, yield, crops

5.1 Introduction

Recent breakthrough in biotechnology and science has shown that nanotechnology and nanosynthetic processes can increase the viability of crop and fruit yield through the use of nanochitosan. Nanochitosan, a nanoscale derivative of chitosan, has shown great promise in agriculture, particularly in enhancing the quality and yield of fruit crops. Chitosan itself, derived from chitin found in the shells of crustaceans, has long been recognized for its antifungal and antimicrobial properties. When processed into nanoscale particles, it gains several advantages such as increased surface area and reactivity, making it highly effective in various applications. In the context of fruit crops, nanochitosan has been studied for its ability to improve plant growth, disease resistance, and overall yield [1, 2]. Besides its low production costs, chitosan is also biologically broken down, biocompatible, and non-allergenic while being destroyed by both specific and insensitive enzymes, making it mostly non harmful for human beings. The properties of chitosan include destroying cell walls, hindering protein and ribonucleic acid synthesis, altering growth of fungal and poison production, and stimulating the defense mechanisms [3] of the host plant. Chitosan can significantly increase bacterial growth in agriculture, even without chemical fertilizer [4]. The advantageous impact of chitosan is typically determined by its accumulation, preparation technique, surrounding conditions, and growth state [5].

5.2 Chitosan

Biopolymers such as chitin and chitosan rank as the world's second most prevailing polymer. These biopolymers have a long linear chain that is linked to the surfactant molecules through the -d-glucosides linkage. The deacetylation process that produces chitosan, a linear electrostatic polysaccharide, is found in insects, mushrooms, and algae [6]. While insects are being used as a potential source of chitosan in current research, commercial tinfoil is primarily obtained from fungi, crustaceans, and squids. Protein is used to make food, and the exoskeleton of insects provides a plentiful byproduct [7]. By means of chemical or enzymatic deacetylation, chitin produces up to 44% of the external skeleton in arthropods (crabs), 40% in cephalons' internal skeleton, and 42% in fungus' chassis.

The transformation of chitin into chitosan entails several early stages, including removal of mineralize with HCl, microbe degradation, and removal of pigments [7]. Shrimp coverings are high in protein and chitin but have little economic value because they are used as substrate for animals. The waste can be transformed into chitin and its products, chitosan, which have significant financial value at low cost. The production of chitosan from shrimp outer shell is an economically viable option, as it converts them from useless garbage to valuable resourcefulness [8].

5.3 The Impact of Chitosan NPs on the Growth and Yields of Some Fruit Crops

5.3.1 Cucumber

The impact of foliar spraying of both chitosan and yeast on cucumber was investigated. To ensure the growth and efficacy of cucumber crops in biological conditions, different amounts of chitosan (1, 2, 3, and 4 milliliters per liter), as well as yeast at 1, 2, 3, and 4 grams per liter were used during two different seasons 2010 and 2011. As a result, it was discovered that the 4 milliliters per liter chitosan concentration enhanced both the K and P compounds percent over two seasons. He further demonstrated that, at foliar level 4 milliliters per liter, the variable for both vegetative and reproductive parameters of cucumber are more relevant. The study found that cucumber plants exhibited higher standard and higher overall output when administered with active chitosan on the foliar. In terms of cucumber quality and yield, chitosan is the preferred bio-stimulant over yeast in laboratory experiments. Chitosan treatment may have a stimulating effect on vegetative growth by altering cell osmotic pressure to increase the availability and absorption of water and essential minerals [9]. *Colletotrichum* spp. are responsible in anthracnose disease in cucumbers. The utilization of foliar at 0.05% and 0.1% chitosan is present [10].

5.3.2 Tomato

The NPs present in chitosan may enhance the productivity of tomato (Lycopersicon esculentum L.) plant under H_2O stress state. The use of chitosan at different concentrations, such as 0, 50, 100, and 150 milligrams per liter, was necessary to maintain glasshouse condition using a completely randomized fashion in three replicas. They found that the effects on plant characteristics, plant height, leaf area, chlorophyll content, and tomato fruit weight are positively correlated with exposure to H_2O stress for 6 days at 100 milligrams per liter. Chitosan foliar spray has been shown to reduce the effects of H_2O stress on photosynthesis and potentially increase endogenous cytokinins, which are responsible for the high pigment content in leaves. The use of chitosan in different amount enhanced chlorophyll percentage in tomato leaves, whereas applying it on the foliar side decreased loss of water and gases exchange, reducing water requirements, and improving plant photosynthetic processes. A 6-day treatment plan that includes a foliar spray of chitosan at 10 times every day with water stress is believed to improve tomato performance [11]. Another study examined the impact of chitosan and copper nanoparticle on tomato plant development, antioxidant capability in plants, contents of mineral, and salt stress status [12]. According to the current study, chitosan and Cu-NPs were found to promote both vegetative and reproductive growth in tomato without altering salt stress.

5.3.3 Mango

Micro-chitosan and micro-K silicate were foliar spraying the Ewais mango cultivar to improve flowering, fertility, and the standard of fruit for this study. The trial involved seven treatments, with ChNPs at 50, 100, or 150 ppm, and K-NPs silicate at 0.4, 0.6, and 0.8 grams per liter and control dose. In this experiment, the Ewais mango cultivar exhibited better results with NPs treatments for chitosan or K-silicate, than the not treated trees, which had lower effects on flowering and fruit set, yield, fruiting attribute, and leaf constituents of green pigments, Zn, and K [13]. The process of spraying nano-K silicate with 0.6 and 0.8 grams per liter achieved greatest yield and the best fruit property, was more efficient in the growth of panicle length, and reduced flora break down than not treated trees. K-NPs silicate with a value of 0.8 grams per liter improved the final fruit category compared to other treatments. Moreover, they also sprayed the trees with chitosan at 50 or 100 ppm, which yielded similar results to K-NPs silicate. The addition of chitosan NPs at 50 or 100 ppm can enhance the green pigmentation, Zn, and K constituent in leaves [13].

5.3.4 Orange

Micromaterials are now widely used in after harvest. Valencia oranges were studied for their standard and storage ability using ChNPs. High-energy ball milling technique was utilized to prepare a ChNPs. The identification of ChNPs was achieved through vital light scattering, fourier transform infrared spectoscopy (FTIR) spectroscopy, and surface morphology via transmission electron microscopy. Valencia oranges that were fully grown were sprayed with 0.2%, 0.4%, and 0.8% concentrate of ChNPs. For 75 days, the fruits were kept in room storage. At 0, 15, 30, 45, 60, and 75 days per week, the standard variables such as fruit weight loss, level of decay, durability, total acidity, level of dissolvable solids, and concentrate of ascorbic acid were assessed [14] apart from the *in vitro* evaluation of ChNPs, based on the results of the next two seasons, the ChNPs at 0.8% treatment had the most favorable outcomes with regard to weight loss of fruit, deterioration, T. S. S. and acid ratio proportion of the treatments to the total acid in each season. Moreover, fruit pulp and firmness were noted to be at their highest levels with 0.8% of ChNPs. Pre-packaged fruits experienced a steady increase in weight loss and spoilage. Valencia oranges fruits were most successful when 0.8% of ChNPs was employ for 75 days at normal temperature [14].

5.4 Application of Chitosan Nanoparticles (ChNPs)

5.4.1 In Agriculture

Chitosan nanoparticles can act as an anti-bacterial or anti-fungal agent against crop pathogens like fungi in agriculture. *Pyricularia grisea, Alternaria solani,* and *Fusarium oxysporum* are among the bacteria that serve as growth proponent for plants [15]. By using chitosan nanoparticle constituents, the plant protective machineries can be strengthened by recognizing the protective enzyme's mechanisms. Xing *et al.* [16] stated that oleoyl-ChNP constituent was found to have a significant antifungal effect against various pathogens. Maize seedlings were subjected to Copper nanochitosan for 4 h at 0.04%–0.16%, Chitosan 50–190 kDa, deacetylation rate at 80%. Sprinkling plants daily for 35 days, with chitosan of 16% and deacetylation of 50–190 kDa, results in increased plant height, root dimension and number, green pigment content, ear diameter, and grain output per plot [17]. Recent studies on nanochitosan have revealed several effective methods of utilizing them in crops and fields. By promoting antifungal and antibacterial activity in multiple modes, it can also combat bacterial and fungal pathogens by stimulating the immune response in plants [18]. Direct application of the plant as a soil candidate or foliar candidate can reveal the structural impact on the entire plant body [19]. To prevent damage from microbes and prolong shelf life, ChNP formulations are used as post-harvest coverings on fruits [16].

5.4.2 Plant Growth Enhancement and Increased Productivity

Extensive research has been done on chitosan nanoparticles and its nano-formulations as a potential plant growth-enhancing agent. Protonated chitosan with a high charge shows increased attraction toward cell membrane, leading to high plant activity. Chitosan is a plant-derived macronutrient with dietary values of approximately 9%–10% [20]. ChNP constituents have been widely utilized as an alternative method for increasing the rate sprouting and building organic matter. In addition, ChNP formulas have been utilized as a growth-promoting agent by increasing the levels of nutrients uptake, pigment content, and photosynthetic rate. The treatment of chitosan nanoparticles has resulted in an increase of 61% and 29% in small-sized and of 82% and 55% in medium-sized nanoparticles [20]. Chitosan nanoparticles have been shown to enhance the vegetative growth of seedling, in contrast to their chitosan oligomer and not treated control plants. Saharan *et al.* [21] examined the effects of $CuSO_4$, bulk chitosan, and copper-chitosan nanoparticles on the growth of maize seedlings. The nano-formulations had a significant impact on the growth of maize seedlings, as well as their activity in enzymes (α-amylase and protease) and overall protein content. The bioavailability of hormones could be improved by using chitosan nano-conveyor to efficiently release plant growth regulator-containing hormone ingredients through an efficient nano delivery system. Plant hormones, including their class II and class III classes I, are the only known plant growth controller. French beans exhibited a 37% increase in root growth and 82% in leaf area when using Gibberellic acid-ChNPs [22].

5.4.3 Biocides Against Plant Pathogens and Pests

The use of chitosan, either as single or in combination with other energetic agents, demonstrated a hopeful approach of fungicides in treating such as blast disease from wheat, rice, maize, and potato, Fusarium wilt, head blight disease. A multitude of research studies have established the chitosan nanoparticle formula to improve plant protective mechanisms by obtaining the protective catalytic functions. Besides, Xing *et al.* [16] use oleoyl-ChNP formula *in vitro*, and it has been shown to be effective against crop pathogenic fungi, including *Alternaria tenuissima*, and has resulted in a notable antifungal impact. *G. zeae* and *F. culmorum* Ch-resistant fungi are comparable to culmorum. The antifungal activity of chitosan nanoparticles and mass chitosans on *A. solani*, *P. grisea*, and *F. oxysporum* was reported Sathiyabama and Parthasarathy [23]. Mass chitosan failed to hinder mycelia as effectively as the nanochitosan. Nanochitosan' reduced dimension, permeability, and zeta potential make them very stable, and it has been shown to have an impact on pathogenic fungi that have been tested. Furthermore, the nanochitosan conveyor system was equipped with agrochemicals as the active agent, which possesses proscribed release properties and faster efficiency in reaching the target cell or parts of the plant [24]. The loading, encapsulation capability, discharge profile, particle dimension, and structure are crucial factors in the plan and formation of nanochitosan-agrochemicals NPs. Multiple studies examined and reported all the parameters mentioned while creating and developing nano-formula. Maluin and Hussein [15] examined the sustainability of chitosan-based agro-nanochemicals for crop protection, detailing their use in controlled release formulations, plant growth enhancers or biocidal activities.

5.5 Other Potential Use of Nanochitosan for Enhancing Fruit Crops

5.5.1 Enhanced Disease Resistance

Nanochitosan, with its increased surface area, can form a protective barrier on plant surfaces, making it difficult for pathogens to penetrate. This can help in reducing diseases and pests, leading to healthier fruit crops [25]. Nanochitosan has the ability to elicit resistance against numerous plant diseases, which has been confirmed in many host-pathogen interfaces [21, 26]. Studies revealed that Cu-chitosan nanoparticles are effective in the control of tomato early blight and *Fusarium* wilt [26]. Chitosan nanoparticles effectively suppressed rice and finger millet blast fungus *Pyricularia grisea* [27]. Chitosan nanoparticles when treated to pearl finger millet seeds and assayed against downy mildew disease under greenhouse conditions provided a significant downy mildew management over the untreated experimental control [28, 29].

5.5.2 Improved Nutrient Absorption

Nanochitosan particles has been designed to encapsulate nutrients, enabling targeted delivery to plants. This targeted approach ensures efficient nutrient absorption, promoting better growth and development [30].

5.5.3 Stress Tolerance

Nanochitosan has been shown to enhance plant tolerance to various environmental stresses such as drought and salinity. This property is crucial for fruit crops, as they often face adverse growing condition [31, 32].

5.5.4 Post-Harvest Preservation

Nanochitosan coatings on fruits can extend their shelf life by inhibiting microbial growth and reducing spoilage. This is essential for minimizing food wastage and ensuring a longer availability of fresh produce in the market [23, 33]. Nanochitosan improved post-harvest antioxidant ability, lengthens shelf life, improved nutritional value, and promoted tolerance to disturbing stress. Additionally, the CHNPs reduced the incidence of post-harvest phytopathogens. In most instances, smaller nanochitosan (<150 nm) showed higher potentials than larger ones (>150 nm). This was likely a result of the greater plant tissue easy penetrability and surface area of the smaller nanochitosan [34, 35].

5.5.5 Environmental Impact

Nanochitosan being a natural derivative has minimal environmental impact compared to synthetic chemicals. Its biodegradability and eco-friendly nature make it a sustainable choice for agriculture [36–39].

5.6 Conclusion

Chitosan is known as an efficient biomedical material due to its several functional use. It is one of the stable and practically supple adsorbents because it is made up of hydroxyl (–OH) and amine ($-NH_2$) groups, while these points highlight the potential benefits of nanochitosan. A study has been conducted to look into the impact of foliar sprinkling of particulate matter such as chitosan on fruit crops' development and output. By utilizing chemical additives like potassium bicarbonate, yeast, or other compounds alone, these products can help to lower disease rates and inhibit microbes from spreading, thereby protecting both standard of the crop and output. Foliar sprinkling of ChNPs is more efficient than other biological stimulants and chemical additives in terms of crop development and output, with a promising future of continual and ecofriendly cultivation.

References

1. Saharan, V., Kumaraswamy, R., Choudhary, R.C., Kumari, S., Pal, A., Raliya, R., Biswas, P., Cu-chitosan nanoparticle mediated sustainable approach to enhance seedling growth in maize by mobilizing reserved food. *J. Agric. Food. Chem.*, 64, 6148–6155, 2016.
2. Benettayeb, A., Seihoub, A.F.Z., Pal, P., Ghosh, S., Usman, M., Chia, C.H., Usman, M., Sillanpää, M., Chitosan Nanoparticles as Potential Nano-Sorbent for Removal of Toxic Environmental Pollutants. *Nanomaterials (Basel)*, 13, 3, 447, 2023 Feb. Published online 2023 Jan 21, doi: 10.3390/nano13030447.
3. Kramchote, S. and Suwor, P., Preharvest chitosan effects on growth, yield and quality of 'SuperHot' and 'Num Khao' chili pepper (Capsicum annuum L.). *Hortic. J.*, 91, 4, 522–530, 2022, https:// doi.org/10.2503/hortj.UTD-360.
4. Parvin, M.A., Zakir, H.M., Naznin, S., Kafi, A., Seal, H.P., Author's Complementary Copy Effects of different application methods of chitosan on growth, yield and quality of tomato (Lycopersicon esculentum Mill.). *Arch. Agric. Environ. Sci.*, 9, 1, 261–267, September, 2019, https://doi.org/10.26832/24566632.2019.040301.
5. Pirbalouti, A.G., Malekpoor, F., Salimi, A., Golparvar, A., Scientia Horticulturae Exogenous application of chitosan on biochemical and physiological characteristics, phenolic content and antioxidant activity of two species of basil (Ocimum ciliatum and Ocimum basilicum) under reduced irrigation. *Sci. Hortic. (Amsterdam)* [Internet], 217, 114–22, 2017, Available from: http:// dx.doi.org/10.1016/j.scienta.2017.01.031.
6. Confederat, L.G., Tuchilus, C.G., Dragan, M., Sha, M., Dragostin, O.M., Preparation and antimicrobial activity of chitosan and its derivatives : A concise review. *Molecules*, 17, 26, 12, 3694, 2021, https://doi.org/10.3390/molecules26123694.
7. Rkhaila, A., Chtouki, T., Erguig, H., Haloui, N.E., Ounine, K., Their synergic effects with endophytic Bacillus species: Unlimited applications in agriculture. *Molecules*, 26, 1117, 2021, https://doi.org/10.3390/molecules26041117.
8. Hosney, A., A review of the chemical extraction of chitosan from shrimp wastes and prediction of factors affecting chitosan yield by using an artificial neural network. *Mar. Drugs*, 28, 20, 11, 675, 2022, https://doi.org/10.3390/md20110675.
9. Shehata, S.A. and El-ramady, H.R., Response of cucumber plants to foliar application of chitosan and yeast under greenhouse conditions. *Aust. J. Basic Appl. Sci.*, 6, 4, 63–71, 2012.
10. Dodgson, J.L. and Dodgson, W., Comparison of effects of chitin and chitosan for control of Colletotrichum sp. on cucumbers. *J. Pure Appl. Microbiol.*, 11, 1, 87–94, 2017.
11. Alam, A.M., Ullah, I., Tanveer, S., Zuhair, S.A., Ullah, I., Quality indices of tomato plant as affected by water stress conditions and chitosan

application. *Pure Appl. Biol.*, 9, 2, 1364–75, 2020, http://dx.doi.org/10.19045/bspab.2020.90143.
12. Benavides-mendoza, A., Effects of chitosan–PVA and Cu nanoparticles on the growth and antioxidant capacity of tomato under saline stress. *Molecules*, 23, 178, 2018, https://doi.org/10.3390/ molecules23010178.
13. Gad, M.M., Abdel-Mohsen, M.A., Zagzog, O.A., Improving the yield and fruiting characteristics of Ewais Mango Cultivar by spraying with Nano-chitosan and Nano-potassium silicate. *Sci. J. Agric. Sci.*, 3, 2, 68–77, 2021, DOI: 10.21608/sjas.2021.102597.1161.
14. Alshallash, K.S., Sharaf, M., Abdel-Aziz, H.F., Arif, M., Hamdy, A.E., Khalifa, S.M., Hassan, M.F., ghazala MM, A., Bondok, A., Ibrahim, M.T.S., Alharbi, K., Elkelish, A., Postharvest physiology and biochemistry of Valencia orange after coatings with chitosan nanoparticles as edible for green mold protection under room storage conditions. *Front. Plant Sci.*, 13, 1034535, 2022, doi: 10.3389/fpls.2022.1034535.
15. Maluin, F.N. and Hussein, M.Z., Chitosan-based agronanochemicals as a sustainable alternative in crop protection. *Molecules*, 25, 7, 1611, 2020.
16. Xing, K., Shen, X., Zhu, X., Ju, X., Miao, X., Tian, J., Feng, Z., Peng, X., Jiang, J., Qin, S., Synthesis and *in vitro* antifungal efficacy of oleoyl-chitosan nanoparticles against plant pathogenic fungi. *Int. J. Biol. Macromol.*, 82, 830-836, 2016.
17. Siddaiah, C.N., Prasanth, K.V.H., Satyanarayana, N.R., *et al.*, Chitosan nanoparticles having higher degree of acetylation induce resistance against pearl millet downy mildew through nitric oxide generation. *Sci. Rep.*, 8, 2485, 2018.
18. Iriti, M. and Varoni, E.M., Chitosan-induced antiviral activity and innate immunity in plants. *Environ. Sci. Pollut. Res.*, 22, 4, 2935–2944, 2015.
19. Bandara, S., Du, H., Carson, L., Bradford, D., Kommalapati, R., Agricultural and biomedical applications of chitosan-based nanomaterials. *Nanomater. (Basel)*, 10, 10, 1903, 2020.
20. Agarwal, M., Nagar, D., Srivastava, N., Agarwal, M., Chitosan nanoparticles based drug delivery: an update. *Int. J. Adv. Multidiscip. Res.*, 2, 1–13, 2015.
21. Saharan, V., *et al.*, Synthesis and *in vitro* antifungal efficacy of Cu–chitosan nanoparticles against pathogenic fungi of tomato. *Int. J. Biol. Macromol.*, 75, 346–353, 2015, https://doi.org/10.1016/j.ijbiomac.
22. Pereira, A.E.S., Sandoval-Herrera, I., Zavala-Betancourt, S., Oliveira, H., Ledezma- P erez, A., Romero, J., Fraceto, L.F., γ-Polyglutamic acid/chitosan nanoparticles for the plant growth regulator gibberellic acid: characterization and evaluation of biological activity. *Carbohydr. Polym.*, 157, 1862–1873, 2017.
23. Sathiyabama, M. and Parthasarathy, R., Biological preparation of chitosan nanoparticles and it's *in vitro* antifungal efficacy against some phytopathogenic fungi. *Carbohydr. Polym.*, 151, 321–325, 2016.

24. Duhan, J.S., Kumar, R., Kumar, N., Kaur, P., Nehra, K., Duhan, S., Nanotechnology: the new perspective in precision agriculture. *Biotechnol. Rep.*, 1, 11–23, 2017.
25. Zhou, Y., Zhang, M., Wang, X., Huang, Q., Min, Y., Ma, T., Niu, J., Removal of crystal violet by a novel cellulose-based adsorbent: Comparison with native cellulose. *Ind. Eng. Chem. Res.*, 53, 5498–5506, 2014, doi: 10.1021/ie404135y.
26. Saad, E.M., Elshaarawy, R.F., Mahmoud, S.A., El-Moselhy, K.M., New Ulva lactuca Algae Based Chitosan Bio-composites for Bioremediation of Cd (II) Ions. *J. Bioresour. Bioprod.*, 6, 223–242, 2021, doi: 10.1016/j.jobab.2021.04.002.
27. Manikandan, A. and Sathiyabama, M., Preparation of chitosan nanoparticles and its effect on detached rice leaves infected with Pyricularia grisea. *Int. J. Biol. Macromol.*, 84, 58–61, 2016, https://doi.org/10.1016/j.ijbiomac.2015.11.083.
28. Zeng, D., Luo, X., Tu, R., Application of bioactive coatings based on chitosan for soybean seed protection. *Int. J. Carbohydr. Chem.* 2012, 1, ID 104565, 2012, https://doi.org/10.1155/2012/104565.
29. Jannoey, P. and Channei, D., Synthesized nanochitosan induced rice chitinase isozyme expression; application in brown planthopper (BPH)control. *NU. Int. J. Sci.*, 12, 25–37, 2016.
30. Abdel-Aziz, H.M., Hasaneen, M.N., Omer, A.M., Nano chitosan-NPK fertilizer enhances the growth and productivity of wheat plants grown in sandy soil. *Span. Jour. Agric. Res.*, 14, e0902, 2016.
31. Elshamy, M.T., ELKhallal, S.M., Husseiny, S.M., Farroh, K.Y., Application of nano-chitosan NPK fertilizer on growth and productivity of potato plant. *J. Sci. Res. Sci.*, 36, 424–441, 2019.
32. Attaran Dowom, S., Karimian, Z., Mostafaei Dehnavi, M., Samiei, L., Chitosan nanoparticles improve physiological and biochemical responses of Salvia abrotanoides (Kar.) under drought stress. *BMC Plant Biol.*, 22, 364, 2022.
33. Ali, M.E.A., Aboelfadl, M.M.S., Selim, A.M., Khalil, H.F., Elkady, G.M., Chitosan nanoparticles extracted from shrimp shells, application for removal of Fe(II) and Mn(II) from aqueous phases. *Sep. Sci. Technol.*, 53, 2870–2881, 2018, doi: 10.1080/01496395.2018.1489845.
34. Al-Sheikh, H. and Yehia, R.S., *In vitro* antifungal efficacy of *Aspergillus niger* ATCC 9642 chitosan-AgNPs composite against post-harvest disease of citrus fruits. *Appl. Biochem. Microbiol.*, 52, 4, 413–420, 2016.
35. Abdel-Rahman, F.A., Monir, G.A., Hassan, M.S.S., Ahmed, Y., Refaat, M.H., Ismail, I.A., El-Garhy, H.A.S., Exogenously applied chitosan and chitosan nanoparticles improved apple fruit resistance to blue mold, upregulated defense-related genes expression, and maintained fruit quality. *Horticulturae*, 7, 8, 224, 2021, https://doi.org/10.3390/horticulturae7080224.
36. Fan, L., Zhang, Y., Luo, C., Lu, F., Qiu, H., Sun, M., Synthesis and characterization of magnetic β-cyclodextrin-chitosan nanoparticles as nano-adsorbents

for removal of methyl blue. *Int. J. Biol. Macromol.*, 50, 444–450, 2012, doi: 10.1016/j.ijbiomac.2011.12.016.

37. Vakili, M., Rafatullah, M., Salamatinia, B., Abdullah, A.Z., Ibrahim, M.H., Tan, K.B., Gholami, Z., Amouzgar, P., Application of chitosan and its derivatives as adsorbents for dye removal from water and wastewater: A review. *Carbohydr. Polym.*, 113, 115–130, 2014, doi: 10.1016/j.carbpol.2014.07.007.
38. Pap, S., Kirk, C., Bremner, B., Turk Sekulic, M., Gibb, S.W., Maletic, S., Taggart, M.A., Synthesis optimisation and characterisation of chitosan-calcite adsorbent from fishery-food waste for phosphorus removal. *Environ. Sci. Pollut. Res.*, 27, 9790–9802, 2020, doi: 10.1007/s11356-019-07570-0.
39. Igwegbe, C.A., Ighalo, J.O., Ghosh, S., Ahmadi, S., Ugonabo, V.I., Pistachio (*Pistacia vera*) waste as adsorbent for wastewater treatment: A review. *Biomass Convers. Biorefin.*, 10, 8793–8811, 2012. doi: 10.1007/s13399-021-01739-9.

6

Application of Nanochitosan for Effective Fruit Production

Ibrahim Dawud[1], Ruth Ebunoluwa Bodunrinde[2], Oluwafemi Adebayo Oyewole[3], Charles Oluwaseun Adetunji[4*], Abel Inobeme[5], Eniola K. I. T.[6], Mohammed Bello Yerima[7] and John Tsado Mathew[8]

[1]*Department of Applied Biology, Federal University of Technology Babura, Jigawa State, Nigeria*
[2]*Department of Microbiology, Federal University of Technology, Minna, Nigeria*
[3]*Department of Microbiology, Edo State University, Uzairue, Nigeria*
[4]*Department of Chemistry, Edo State University, Uzairue, Nigeria*
[5]*Department of Biological Sciences, Joseph Ayo Babalola University, Ikeji Arakeji, Osun State, Nigeria*
[6]*Department of Microbiology, Sokoto State University, Sokoto, Sokoto State, Nigeria*
[7]*Department of Chemistry, Ibrahim Badamasi Babangida University, Lapai, Niger State, Nigeria*
[8]*Department of Microbiology, Federal University of Technology, Akure, Nigeria*

Abstract

Agricultural practices are one of the supportive pillars of global economy with a past, present, and future prospects. Orchid farming is a lucrative farming that supports the growth of numerous varieties of fruits trees with economic benefits. Fruits are seeds bearing fleshy or dry ripened structures in a flowering plant formed from the ovary that are typically sweet, sour, or unique in taste, which are incredibly edible at a raw or processed state. Fruits contain a wide range of micronutrients and non-nutrient bioactive compounds such as dietary fiber, minerals (potassium, calcium, and magnesium), vitamins (A, C, and E), and phytochemicals. An increase in the global fruit farming and consumption has been observed in recent years due to the increase awareness on the health and economic benefits attached to many fruits. Some of the major challenges faced by orchid farmers/

**Corresponding author*: adetunji.charles@edouniversity.edu.ng

Charles Oluwaseun Adetunji, Maulin P. Shah, Yerima Mohammed Bello, Daniel Hefft, Jay Singh, Shyam S. Pandey and Ravindra Pratap Singh (eds.) Nanochitosan Applications for Enhanced Crop Production and Food Security, (123–136) © 2025 Scrivener Publishing LLC

horticulturist include the adulteration of the nutritional composition of the soil and post-harvest disease havoc caused by pests and other environmental factors, these problems cause a debilitating effect in quantity and quality of the fruits. Chitosan is natural occurring biodegradable polymers with appealing properties such as biodegradability, biocompatibility, low allergenic, and non-toxicity. Their unique properties proffer them a good candidate for various agricultural activities. As such, chitosan nanoparticles are extensively studied and applied as a bioagent/coating agent in the agricultural setup, ranging from fungicidal agent, antimicrobial agent, growth-promoting agents, wastewater treatment, and shelf live enhancement in fruits and other agricultural products.

***Keywords*:** Nanochitosan, fruits, agriculture, chitosan, antimicrobial, nanoparticles

6.1 Introduction

Reduction in the volume of potassium content in the soil has led to the quest for alternative to this problem via the used of nanotechnology; this technology was reported to help in the increase in the effectiveness of potassium. The work of Abd-Elrahman *et al.* [1] showed that the usage of common potassium fertilizer on an acre of land resulted in great increase in soluble and transferable potassium in the soil, when compared to those with those that were not nanosynthesized. The application of chemical fertilizer during the cultivation of strawberry has caused pollution in the environment and distorting the level of vitamins that are inherent in the plant. Chemical fertilizers have produced several hazardous deposits in the soil; they have impacted the ecosystem negatively and impaired human health and lack of stable water due to water contamination. Nanotechnology has been extensively utilized, because they are less toxic in the promotion of plant growth and in the increase of plant yield [2, 3].

Nanochitosan is a chitin derivative from biopolymer that enhances the growth of plant root, soil absorbency, and water-use effectiveness [4]. Research has shown that the derivatives of chitosan increase chlorophyll fluorescence and the rate of photosynthesis [5]. Chitosan can also control gene expression in the defense pathways of plants, and this feature has assisted in keeping fruit for a very long period of time, particularly after fruit harvest and during storage [4]. Additionally, nanochitosan is biologically compatible and biodegradable, and complexes comprising chitosan have done well without causing any harm on beneficial soil microorganisms and also act as a stabilizer to the soil [5, 6]. Zulfiqar *et al.* [7] and Elgala *et al.* [8] reported that nanofertilizer help to reduce to the problem of fertilizer wastage and increase the chance of available nutrient for plants. According to Abdel-Aziz *et al.* [9], nanochitosan-NPK fertilizer augmented the rate of growth and productivity of wheat that was planted in sandy soil.

6.2 Mechanism of Action

Nanosized nature of nanofertilizer made it easy for the penetration of the materials into the cell wall and for easy migration to the plasma membrane. The ability of the nanofertilizer to possess a size smaller than the pore size of the cell wall helps to move via apoplastic and symplastic paths, making it easier to penetrate plants for metabolic and physiological actions [10, 11].

6.3 Fruits

Fruits are derived from the Latin word "fruor," which means "I delight in." Fruits are seeds bearing fleshy or dry ripened structure in a flowering plant formed from the ovary that are typically sweet, sour, or unique in taste, which are incredibly edible at a raw or processed state. Some common examples of fruits include bananas, strawberries, mango, watermelon, apple, oranges, pineapple, nuts, tomatoes, and corn kernels [12]. Fruit propagations and dispersion have a long-standing history existing between the plants, human, animals, and some abiotic (such as wind) interactions within the ecological community in a symbiotic manner. Fruits key components of the human diet as they contain compounds of nutritional importance, including vitamins that are not synthesized by the human body.

6.3.1 Economic Importance of Fruits

The economic importance of fruits is enormous and cannot be cannot be over emphasized because of the fact that it provides food for human and animals and raw materials for industries and increases the nation's growth domestic product. Furthermore, fruit cultivation is critical to the economic well-being of any nation's population and is used to determine the overall standard of living of a population. Below are some of the economic importance of fruit and fruit farming:

i. Excellent source of vitamin and minerals: Fruits are excellent source of numerous vitamins such as vitamin C (crucial in free radical neutralization), vitamin A (good in eyes and hair development), magnesium (heart protections against some disease and muscles relaxation) diseases), and potassium (blood pressure regulation). Without some of these, maintaining an optimal health conditions in the living system will be very difficult [12, 13].

ii. Good source of revenue generation: Fruit production helps in improving the nation's economic activities through increase in revenue generation and employment provision; varieties of the fruits stand a chance of been an export commodity from one country to another, which provides a significant amount of foreign exchange and also provides job opportunities to the ever-growing global population. It also enhances the profit nets of orchard farmers [14, 15].

iii. Food security: Fruits can serve as a food security to a nation and the world at large; this is due to the fact that numerous fruits contain the essential elements of food such as carbohydrate, proteins, lipids, and other essential nutrient for human survival in a reasonable amount that can be used to feed or augment the nutritional requirement of the population. As a result, majority of people are likely to increase their calories intake at a higher pace through fruit consumptions; fruit consumption is crucial for a well-balanced diet and for maintaining excellent health maintenance [16].

iv. Excellent source of raw materials for agro-based industries: Fruit farming provides some essential raw materials for varieties of agro-based industrial production process; for instance, utilization of cocoanut husk in canning and preservation, papaya, Aonla, and jamun seeds for pharmaceutical products and other essential services for packaging and transportation industries [17].

v. Creating avenue for efficient utilization of resource: Fruit farming is mostly perennial, which allows farmers to remain engaged in farming operation throughout the year and ensuring maximum utilization of farm resources and assets, which include the land, water, labor, and farm machinery for optimum output [17].

vi. Enhances optimum utilization of waste and barren lands: Some land can be declared barren due to poor shallow and undulated soil that makes it unfit for biennial agricultural activities. Most fruits are perennial such as mango, cashew, and custard apple. These fruits can be grown in the barren and undulated soils that may not support numerous agricultural activities. Growing of fruits in agricultural unfits land can help in avoiding wastage and regaining the soil fertility as it enhances microbial activities within the soil [15].

6.3.2 Nanochitosan in Fruits and Fruit Production

For many centuries, humans have been trying to improve the nutritional quality of their diet; fruits as one of the key components of human and animal diets have gained an increasing attention toward production and quality enhancement so as to satisfy the nutritional needs of human race. This makes the horticultural of sector to be growing at a phenomenal pace, outstripping the crop agricultural sectors with a production index of about 100 million tons [17]. Fruits crops have the potentials of giving higher tonnage of yield per unit area than other field crops. Fruit production plays an important role in the prosperity of nation's economy.

Among the numerous challenges facing agricultural system is the problem of orchids; this faced as a result of adulteration of several agricultural soils due to the contamination caused by chemical fertilizers. This halts the productivity and quality of many fruit trees. In order to achieve and enhance food safety and security, many research have been focusing toward finding an alternative for the frequently used synthetic fertilizer to organic or biofertilizers with minimal or zero pollution potential. Nanochitosan is a pragmatic tool for orchid farming and other agricultural developments, especially in the aspect of fertilizer development programs. Nanochitosan can be designed with nutrients of interest and applied in the farm as redefined biofertilizers, which can help in boosting the bioavailability and soil fertility more effectively than standard fertilizers [18]. The nano-sized particles of chitosan can aid in increasing sufficient nutrient uptake, distribution, utilization, and regulation of residual influence on soil biodiversity [19]. Improving the fertility of orchard using the nanochitosan as fertilizer or nutrient delivery agent will enhance the quality, quantity, and cost effectivity of the farming activities. Furthermore, it also helps to maintained fruits sustainability so as to feed the ever-growing world population and also increases income to the farmers.

Another most promising problems faced by the horticulturist in in fruit production is the post-harvest disease; this causes a significant economic lost to the farmers by affecting the viability of the fruits. Almost all efforts applied in controlling this menace is the use of synthetic chemical fungicides and other pesticides. Most of the chemically synthesized products used in curtailing the malignant post-harvest are hazardous to human health due to their bioaccumulation potentials. Forthwith, a lot of efforts are geared toward the rediscovery and redefining of substances that are safe, nontoxic, biodegradable, and biocompatible coating agent in especially from organic origin in preventing post-harvest havocs affecting numerous fruits. Various criteria were put in place in the screening and accepting any agent to be used as a coating supplement in enhancing the shelf life of fruits and other farm products; these include water resistance, non-stickiness, thermostability during cold storage, preventing oxygen and carbon dioxide depletion, minimum permeability to water vapor, enhancing the gloss appearance of the fruits, stable drying performance, maintaining the original taste and the nutritional content of the fruit, low viscosity, and easily affordable. Chitosan is a polysaccharide obtained from the exoskeleton of crustaceans, such as shrimps, crabs, and some fungal species that have a promising future in ensuring food security, extend the shelf, and improve the quality of the fruits due to its potent antimicrobial properties. Nanochitosans are among such compounds that can easily be manipulated and be used as coating agents due to their biodegradability, environmentally friendly, and biocompatible nature. Nanochitosans have the potential to improve the appearance of foods and retard or inhibit the growth of pathogenic and spoiled microorganisms in fruits, thereby controlling decay and extends their shelf life; this occurs as a result of their natural elicitation activities in the plant tissues [20–23].

In enhancing the shelf life of fruits, several studies indicate the use of nanochitosan as shelf-life enhancers among numerous fruits, and these includes the studies conducted by Gad *et al.* [24], in which the ability of nanochitosan improves the shelf life of *Prunus persica* (peach fruits). According to the research, the fruits were covered with a certain concentration of the nanochitosan formulation and stored for 28 days at 0°C and 90%–95% relative humidity. The data obtained showed a lowest possible percentage of the fruit deterioration, thereby providing the finest quality fruits. The shelf life of guava fruit was showed to have increased by 20 days via appropriate coating of nanochitosan with nano ZnO and alginate, which proves to have an antibacterial activity, thereby extending the shelf life of the Guava fruit [25]. In Thailand, studies by Plainsirichai *et al.* [26] show a significant of chitosan treatment on the quality of against *Syzygium agueum* in terms of disease control. The *Syzygium agueum* fruit coated with 2% chitosan showed a higher disease resistivity against the control fruits. Coating Banan with 1% solution of chitosan can prolong its shelf life for about 27 days when stored at 13°C [27]. In carrots' processing, nanochitosan powder was proven efficient when applied as alternative to spray and dip-coating methods for quality maintenance of shredded carrots. The shredded carrot samples treated with 0.2% chitosan powder showed a significant low weight loss, a higher moisture retention, and a better vitamin C retention throughout the 10-day storage period. Tomato fruit is one of the most notable perishable fruits with minimal shelf life; studies of Liu *et al.* [28] showed the effectivity of 2% chitosan in the control of gray and blue rots in tomatoes stored at varying temperatures. The study demonstrated the inhibitory activity of the chitosan particles against germ tube elongation and spore germination of the spoilage fungi due to the ability of the chitosan to cause damage to the plasma membrane of conidia.

6.4 Guidelines in Effective Application of Nanochitosan

1. Optimal Timing: Apply nanochitosan at the appropriate stage of fruit development to maximize its benefits. This can vary depending on the specific crop and fruit variety. Generally, it is recommended to apply nanochitosan during the early stages of fruit development or at the onset of flowering.
2. Proper Dilution and Concentration: Follow the manufacturer's instructions for diluting and preparing the nanochitosan solution. Use the recommended concentration to ensure optimal efficacy without any negative effects on the plants or fruits. It is important not to exceed the recommended dosage, as excessive nanochitosan application may lead to undesirable effects.
3. Uniform Coverage: There is need to ensure uniform coverage of the fruit surface with the nanochitosan solution. This can be achieved by using appropriate spray equipment, such as handheld sprayers or sprayers attached to irrigation systems, to ensure even distribution across all fruits, paying attention to both sides of the fruit, ensuring that the nanochitosan solution reaches all surfaces.
4. Avoid Excessive Runoff: Minimize excess runoff of the nanochitosan solution by adjusting the application rate and ensuring proper absorption. Excessive runoff can lead

to ineffective utilization of nanochitosan and unnecessary waste. Implementing techniques like fertigation (incorporating nanochitosan into the irrigation system) can help reduce runoff and increase absorption efficiency.

5. Application Frequency: Determine the appropriate frequency of nanochitosan application based on the specific crop and fruit development cycle. In some cases, a single application during the early stages of fruit formation may be sufficient. However, for longer fruit development cycles, multiple applications at regular intervals may be necessary to maintain consistent protection and growth benefits.
6. Post-Harvest Applications: Consider applying nanochitosan post-harvest to extend the shelf life of harvested fruits. Post-harvest treatments, such as dipping or coating the fruits with nanochitosan solutions, can help prevent decay, maintain freshness, and reduce post-harvest losses. It is important to follow recommended guidelines for post-harvest applications to maximize the benefits of nanochitosan.
7. Monitor and Evaluate: Regularly monitor the progress and health of the fruit plants throughout the growing season. Assess the impact of nanochitosan application on fruit growth, yield, and quality by comparing treated plants with control plants. Keep records and evaluate the results to determine the effectiveness of nanochitosan and make any necessary adjustments for future applications. However, while nanochitosan has been shown to enhance fruit yield, it is just one component of a comprehensive fruit production strategy. Proper irrigation, fertilization, pest management, and overall crop management practices should also be considered for optimal fruit production [1].

The study of Abd-Elrahman *et al.* [1] revealed that the application of nanochitosan-K at 1,000 mg L^{-1} in the absence of K_2SO_4 or adding 50% of the optional dosage gave the highest percentage of constant potassium in the sandy soil.

In this regard, nano-potassium sprayed on wheat plants used the greatest amount of exchangeable potassium from sandy loam soil, resulting in the least amount of potassium loss through leaching. Nanochitosan has cation biopolymer, making it one of the major sensitive biopolymers mainly due to the presence of amino and hydroxyl functional groups on its backbone structure that embrace chemical connecting with nutrients like potassium [29].

The new nanomaterial works in the soil as a slow-release fertilizing agent, preserving potassium from leakage and ensuring its bioavailability. They concluded that the combination of nanochitosan and nano-potassium produced a better result on soil and harvested plants [9].

Chitosan nanoparticles were found to be easily taken up by plants via translocation and boosting plant growth and crops yield [30]. Rameshaiah *et al.* [31] reported that nanochitosans had increased absorption volume and are discrete and are optimally taken up by plants, and plants treated with these nanomaterials were found to contain all the nutrients that they need for proper growth [32]. The size and chemical properties of nanomaterials made it to be more efficient [1, 33, 34]. Due to some features attributed to nanochitosan, Ishkeh *et al.* [35] reported it having the potential to replace agrochemicals by reducing of abiotic stress [36].

6.5 Application of Nanochitosan for Different Fruits

Nanochitosan can be used in a wide variety of fruit crops. It has been studied and shown to be effective in enhancing fruit production in various fruits including apples, citrus fruits, strawberries, grapes, bananas, and tomatoes. The specific benefits and effects of nanochitosan can vary depending on the fruit crop and the application method used [37, 38]. Several studies have investigated the use of nanochitosan in fruit production. These studies have demonstrated positive effects on plant growth, yield, and quality. Here are a few examples:

1. Apple Trees: Research conducted on apple trees treated with nanochitosan showed improved nutrient uptake, increased photosynthetic activity, enhanced plant growth, and higher fruit yield compared to control trees [39].
2. Citrus Fruits: Studies on citrus fruits treated with nanochitosan revealed improved fruit quality, increased antioxidant activity, and enhanced resistance against diseases and pests.
3. Strawberries: Nanochitosan applications on strawberry plants resulted in increased fruit weight, improved fruit firmness, prolonged shelf life, and reduced susceptibility to diseases.
4. Grapes: Research on grapevines treated with nanochitosan reported increased yield, improved berry quality attributes (such as weight, diameter, and color), and enhanced resistance against fungal diseases.

5. Tomatoes: Nanochitosan applications on tomato plants showed increased fruit yield, improved size and weight, enhanced resistance to fungal diseases, and prolonged shelf life. These studies provide scientific evidence supporting the use of nanochitosan in enhancing fruit production. However, it is important to note that the specific effects of nanochitosan can vary depending on factors such as crop variety, growing conditions, application method, and timing [1, 40–42].

6.6 Conclusion

Global accentuation in the discovery of environmentally friendly technology has stimulated interest in biopolymers, which are more versatile and far more biodegradable than their synthetic counterparts. Chitosan and its nanoparticles had been studied as a successful edible coating material for whole fruits. They have made an enormous contribution to the agricultural orchard system through formation and formulation of nanocarriers, nanofertilizers, and nanopesticides, leading to improve yield, enhance disease resistivity, and increase the shelf life of fruits.

References

1. Abd-Elrahman, S.H., El-Gabry, Y.A., Hashem, F.A., Ibrahim, M.F.M., El-Hallous, E., II, Abbas, Z.K., Doaa Darwish, B.E., Al-Harbi, N.A., Al-Qahtani, S.M., Noura Mohamed Taha, N.M., Influence of Nano-Chitosan Loaded with Potassium on Potassium Fractionation in Sandy Soil and Strawberry Productivity and Quality. *Agronomy*, 13, 4, 1126, 2023. https://doi.org/10.3390/agronomy13041126.
2. Abbasi, H., Jamil, M., Haq, A., Ali, S., Ahmad, R., Malik, Z., Parveen, Salt stress manifestation on plants, mechanism of salt tolerance and potassium role in alleviating it: A review. *Zemdirb. Agric.*, 103, 229–238, 2016.
3. Martins, M., Veroneze–Junior, V., Carvalho, M., Carvalho, D.T., Barbosa, S., Doriguetto, A.C., Magalhaes, P.C., Ribeiro, C., Santos, M.H., Souza, T.C., Physicochemical characterization of chitosan and its effects on early growth, cell cycle and root anatomy of transgenic and non–transgenic maize hybrids. *Aust. J. Crop Sci.*, 12, 56–66, 2018.
4. Saqib, S., Zaman, W., Ayaz, A., Habib, S., Bahadur, S., Hussain, S., Muhammad, S., Ullah, F., Postharvest disease inhibition in fruit by synthesis

and characterization of chitosan iron oxide nanoparticles. *Biocatal. Agric. Biotechnol.*, 28, 101729, 2020.
5. Dos Reis, C.O., Magalhães, P.C., Avila, R.G., Almeida, L.G., Rabelo, V.M., Carvalho, D.T., Cabral, D.F., Karam, D., De Souza, T.C., Action of N-Succinyl and N, O-Dicarboxymethyl chitosan derivatives on chlorophyll photosynthesis and fluorescence in drought-sensitive maize. *J. Plant Growth Regul.*, 38, 619–630, 2019.
6. Madzokere, T.C., Murombo, L.T., Chiririwa, H., Nano-based slow-releasing fertilizers for enhanced agricultural productivity. *Mater. Today Proc.*, 45, 3709–3715, 2021.
7. Zulfiqar, F., Navarro, M., Ashraf, M., Akram, N.A., Munne-Bosch, S., , Nano fertilizer use for sustainable agriculture: Advantages and limitations. *Plant Sci.*, 289, 110270, 2019.
8. Elgala, A.M., Abd-Elrahman, S.H., Saudy, H.S., Nossier, M.I., Exploiting Eichhornia crassipes shoots extract as a natural source of nutrients for producing healthy tomato plants. *Gesunde Pflanzen*, 74, 457–465, 2022.
9. Abdel-Aziz, H.M.M., Hasaneen, M.N.A., Omer, A.M., Nano chitosan-NPK fertilizer enhances the growth and productivity of wheat plants grown in sandy soil. *J. Agric. Res.*, 14, e0902, 2016.
10. Yan, W. and Rajcan, I., Biplot analysis of test sites and trait relations of soybean in Ontario. *Crop Sci.*, 42, 11–20, 2002. [Google Scholar] [CrossRef].
11. Guan, Y.J., Hu, J., Wang, X., Shao, C., Seed priming with chitosan improves maize germination and seedling growth in relation to physiological changes under low temperature stress. *J. Zhejiang Univ. Sci.*, 10, 427–433, 2009.
12. Aune, D., Giovannucci, E., Boffetta, P., Fadnes, L.T., Keum, N., Norat, T., Greenwood, D.C., Riboli, E., Vatten, L.J., Tonstad, S., Fruit and vegetable intake and the risk of cardiovascular disease. *Int. J. Epidemiol.*, 46, 3, 1029–1056, 2017.
13. Río-Celestino, M. and Font, R., The Health Benefits of Fruits and Vegetables. *Foods*, 9, 369, 1–4, 2020.
14. Talga, G. and Ameji, B.N., An Analysis of government spending on agricultural sector and its contribution to economic growth in Nigeria. *Int. J. Bus. Social Sci.*, 2, 20, 244–250, 2020.
15. Okorie, G.C. and Nwachukwu, N., Impact of Fruits and Vegetables Export on Economic Growth in Nigeria (1981-2019). *Afr. J. Econ. Sustain. Dev.*, 5, 1, 47–58, 2022.
16. Birthare, S., Scope, Status and Importance of Fruits and Vegetables Industries in INDIA & Role of Fruits and Vegetables in Indian Economy. *Researchgate*, 2018, Doi: 10.13140/RG.2.2.10140.39044, publication/324758156.
17. Agricultural resource and economic growth in Nigeria. *Eur. Sci. J.*, 8, 22, 111–119, 2017.
18. Sidorowicz, A., Maqbool, Q., Nazar, M., Future of Nanofertilizer, in: *Nanotechnology for Agriculture: Crop Production & Protection*, pp. 143–152, Springer, Singapore, 2019.

19. Dimkpa, C.O. and Bindraban, P.S., Nano-fertilizers: new products for the industry. *J. Agric. Food. Chem.*, 66, 6462–6473, 2017.
20. Miranda-Castro, S.P., Application of Chitosan in Fresh and Minimally Processed Fruits and Vegetable, in: *Chitosan in the Preservation of Agricultural Commodities*, Eds. S. Bautista Baños, G. Romanazzi & A. Jimenez-Aparicio, pp. 67–113, Academic Press Springer, 2016, https://doi.org/10.1016/B978-0-12-802735-6.00003-3.
21. Pereira, A.E.S., Silva, P.M., Oliveira, H.C., Oliveira, J.L., Fraceto, L.F., Chitosan nanoparticles as carrier systems for the plant growth hormone gibberellic acid. *Colloids Surf. B Biointerfaces*, 150, 141–152, 2017.
22. Kheiri, A., Jorf, S.M., Malihipour, A., Saremi, H., Nikkhah, M., Synthesis and characterization of chitosan nanoparticles and their effect on Fusarium head blight and oxidative activity in wheat. *Int. J. Biol. Macromol.*, 102, 526–538, 2017.
23. Ingle, P.U., Shende, S.S., Shingote, P.R., Mishra, S.S., Sarda, V., Wasule, D.L., Rajput, V.D., Minkina, T., Rai, M., Sushkova, S., Mandzhieva, S., Gade, A., Chitosan nanoparticles (ChNPs): A versatile growth promoter in modern agricultural production. *Heliyon*, 8, e11893, 2022.
24. Gad, M.M., Zagzog, O.A., Hemeda, O.M., Development of nano-chitosan edible coating for peach fruits cv. *Desert Red. Int. J. Environ.*, 5, 4, 43–55, 2016.
25. Arroyo, B.J., Bezerra, A.C., Oliveira, L.L., Arroyo, S.J., Melo, E.A., Santos, A.M.P., Antimicrobial active edible coating of alginate and chitosan add ZnO nano particles applied in guavas (Psidium guajava L.). *Food Chem.*, 12, 55–66, 2019.
26. Plainsirichai, M., Leelaphatthanapanich, S., Wongsachai, N., Effect of chitosan on the quality of rose Apples (Syzygium Agueum Alston) cv. Tabtim Chan stored at an ambient temperature. *APCBEE Procedia*, 8, 317–322, 2014.
27. Wattanakorn, C., Effect of konjac gum coatings on storage time of banana cv, Kluai hom thong, M.Sc Thesis, King Mongkut's University of Technology, Thonburi, Bangkok, Thailand, 2003, (abstract).
28. Liu, J., Tian, S., Meng, X., Xu, Y., Effects of chitosan on control of postharvest diseases and physiological responses of tomato fruit. *Postharvest Biol. Technol.*, 44, 300–306, 2007.
29. Mubarak, M.U., Zahir, M., Ahmad, S., Wakeel, A., Sugar beet yield and industrial sugar contents improved by potassium fertilization under scarce and adequate moisture conditions. *J. Integr. Agric.*, 15, 2620–2626, 2016.
30. Malerba, M. and Cerana, R., Chitosan effects on plant systems. *Int. J. Mol. Sci.*, 23, 17, 7, 996, 2016. Doi: 10.3390/ijms17070996
31. Rameshaiah, G.N., Pallavi, J., Shabnam, S., Nano fertilizers and nano sensors an attempt for developing smart agriculture. *Int. J. Eng. Res. Gen. Sci.*, 3, 314–320, 2015.

32. Raliya, R. and Tarafdar, J.C., ZnO nanoparticle biosynthesis and its effect on phosphorous-mobilizing enzyme secretion and gum contents in cluster bean (Cyamopsis tetragonoloba L.). *Agric. Res.*, 2, 48–57, 2013.
33. Torabian, S., Zahedi, M., Khoshgoftar, A.H., Effects of foliar spray of nano-particles of FeSO4 on the growth and ion content of sunflower under saline condition. *J. Plant Nutr.*, 40, 615–623, 2017.
34. Abd-Elrahman, S.H. and Mostafa, M.A.M., Applications of nanotechnology in agriculture: An overview. *Egypt. J. Soil Sci.*, 55, 197–214, 2015.
35. Ishkeh, S.R., Shirzad, H., Asghari, M., Alirezalu, A., Pateiro, M., Lorenzo, J.M., Effect of chitosan nanoemulsion on enhancing the phytochemical contents, health-promoting components, and shelf life of raspberry (Rubus sanctus Schreber). *Appl. Sci.*, 11, 2224, 2021.
36. Hasanuzzaman, M., Bhuyan, M.H.M.B., Nahar, K., Hossain, M.S., Mahmud, J.A., Hossen, M.S., Masud, A.A.C., Moumita, F.M., Potassium: A vital regulator of plant responses and tolerance to abiotic stresses. *Agronomy*, 8, 31, 2018.
37. Cvelbar Weber, N., Koron, D., Jakopič, J., Veberič, R., Hudina, M., Baša Česnik, H., Influence of nitrogen, calcium and nano-fertilizer on strawberry (Fragaria × ananassa Duch.) fruit inner and outer quality. *Agronomy*, 11, 997, 2021.
38. Rahman, M.D.H., Hasan, M.D.N., Khan, M.D.Z.H., Study on different nano fertilizers influencing the growth, proximate composition and antioxidant properties of strawberry fruits. *J. Agric. Food Res.*, 6, 100246, 2021.
39. Yang, F., Hu, J., Li, J., Wu, X., Qian, Y., Chitosan enhances leaf membrane stability and antioxidant enzyme activities in apple seedlings under drought stress. *Plant Growth Regul.*, 58, 131–136, 2009.
40. Wang, X.G., Zhao, H.Z.X., Jiang, J.C., Li, H.C., Cong, S., Wu, D., Chen, Y.Q., Yu, H.Q., Wang, C.Y., Effects of potassium deficiency on photosynthesis and photoprotection mechanisms in soybean (Glycine max (L.) Merr.). *J. Integr. Agric.*, 14, 856–863, 2015.
41. Sun, J., Li, A., Jing, Q., Huang, Y., Han, J., Lin, L., Effects of chitosan on soluble sugar content in Prunus davidiana seedlings, in: *E3S Web of Conferences; EDP Sciences*, vol. 136, p. 06001, 2019.
42. Parveen, Anwar-Ul-Haq, M., Aziz, T., Aziz, O., Maqsood, L., Potassium induces carbohydrates accumulation by enhancing morpho-physiological and biochemical attributes in soybean under salinity. *Arch. Agron. Soil Sci.*, 67, 946–959, 2021.

7

Application of Nanochitosan in the Detection of Mycotoxins

Abdulmajid Musa Maku[1]*, Suleiman Imam Malik[1], Tsado Priscilla Yetu[1], Oyegbade Samuel Adeniyi[2], Oluwafemi Adebayo Oyewole[1], Charles Oluwaseun Adetunji[3], Eniola K. I. T.[4] and Mohammed Bello Yerima[5]

[1]*Department of Microbiology, Federal University of Technology, Minna, Nigeria*
[2]*Department of Biological Sciences Covenant University, Ota, Nigeria*
[3]*Department of Microbiology, Edo State University, Uzairue, Nigeria*
[4]*Department of Biological Sciences, Joseph Ayo Babalola University, Ikeji Arakeji, Osun State, Nigeria*
[5]*Department of Microbiology, Sokoto State University, Sokoto, Sokoto State, Nigeria*

Abstract

Mycotoxins, toxic secondary metabolites produced by molds, are a serious hazard to both human health and food safety. In order to assure the safety of food and feed products, effective detection techniques are essential. Nanochitosans are naturally occurring polysaccharides generated by the alkalescent deacylation of chitin and comprise a series of 2-deoxy-2-acetyamino glucose connected by β-1,4-gycosidic linkages. These are naturally formed by deacylation of shellfish shell and exoskeleton of aquatic arthropods and crustaceans as well as cell wall of fungi. Nanochitosan possesses unique qualities like biocompatibility, biodegradability, and a large surface area. It is suitable for the creation of sensitive and specific mycotoxin detection platforms because of these characteristics. Several methods like ionic gelation, reverse micelle, microemulsion, and precipitation are employed in the formation of nanochitosan. The application of nanochitosan-based detection methods for mycotoxin analysis, including electrochemical sensors and optical sensors like surface plasmon resonance and colorimetric assay, is explored in this chapter. This chapter will also explore the potential advantages and challenges associated with using nanochitosan-based approaches in mycotoxin detection.

**Corresponding author*: abdulmajeedmoosa3@gmail.com

Charles Oluwaseun Adetunji, Maulin P. Shah, Yerima Mohammed Bello, Daniel Hefft, Jay Singh, Shyam S. Pandey and Ravindra Pratap Singh (eds.) Nanochitosan Applications for Enhanced Crop Production and Food Security, (137–154) © 2025 Scrivener Publishing LLC

Additionally, it may assess the current state of research in this field, identify gaps, and suggest future directions for research and applications.

Keywords: Nanochitosan, molds, mycotoxin, detection, biosensor

7.1 Introduction

Mycotoxins are secondary metabolites produced by a wide range of filamentous fungi, including species from the genera *Aspergillus*, *Fusarium*, *Penicillium*, *Alternaria*, and *Claviceps* that contaminate major food crops and have a detrimental impact on the health of both animals and humans [1]. Fungi that create mycotoxins are known as mycotoxigenic fungi. According to Donat *et al.* [2], few of them have the capacity to form multiple mycotoxins, whereas certain mycotoxins are produced by various fungi. Mycotoxins can have a harmful influence on both human and animal health because they can lead to a number of illnesses such cancer development, mutagenicity, and estrogenic, gastrointestinal, and kidney issues. Some mycotoxins can impair resistance to infectious illnesses and suppress the immune system [3]. Molds produce mycotoxins when certain conditions are met, including high humidity, subpar agricultural practices, damaged or tainted crops, and deplorable environments. The possibility of mycotoxin synthesis remains even though the presence of mould on grains does not always imply the presence of mycotoxins. Additionally, the absence of mold on long-term-stored food and animal feed does not imply that the grain is free of mycotoxins. Because of this, the problem of mycotoxin danger is complicated and needs to be taken into account by both the scientific community generally and agrotechnology particularly [4]. Mycotoxin has severe economic repercussions and crops with high mycotoxin levels frequently need to be destroyed [2]. Foods including peanuts, grains, cereals, corn, wheat, and dairy products as well as animal products like milk, eggs, and meat can all contain mycotoxins, either singly or in combination. These have significant influence on public health and the economy due the fact that these mycotoxins can be found in various foods [5]. The common mycotoxins produced in food and feedstuffs are aflatoxin, ochratoxin, fumonisin, deoxynivalenol, and zearalenone [6]. Aflatoxin, ochratoxin A, and fumonisins (FB 1 and FB 2) are the deadliest mycotoxins that pose the biggest health risk across tropical Africa, Asia, and the rest of the world, according to research that has identified more than 400 different forms of them [7]. The fact that contaminated food is intended for consumption by farm animals, which results in animal products contaminated with these compounds' metabolites and increases human exposure to mycotoxins, is a crucial factor to take into account in the research of mycotoxin poisoning. The capacity of mycotoxins to cause the emergence of various pathologies makes monitoring mycotoxins in foods vital for the development of new technologies. Chronic exposure to small concentrations of mycotoxins in food is a significant global health hazard. Detection and removal of these compounds concentrate on lowering exposure and, consequently, the diseases that they cause [5]. For the accurate determination of mycotoxins, numerous analytical techniques have been developed. The methods that are being utilized in practice the most frequently are enzyme-linked immunosorbent assay (ELISA), mass detection, and high-pressure liquid chromatography. Current research has concentrated on enhancing the detection limit, time consumption, sample consumption, and usability, because early detection is essential to safeguard health [4]. Mycotoxin contamination has received international attention [4]. Therefore, the creation of precise and sensitive detection methods as well as highly effective management strategies for mycotoxins is essential to safeguard public health and the food industry. Numerous innovative nanomaterials have evolved as

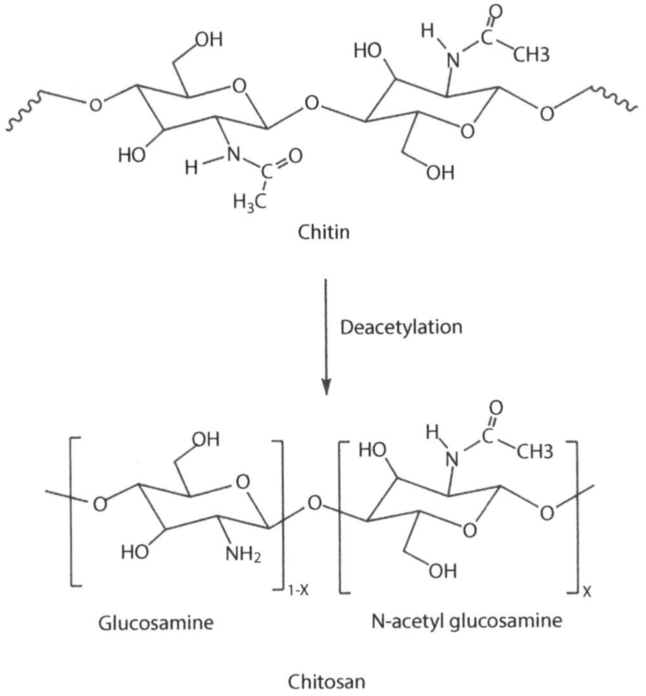

Figure 7.1 Schematic representation of chitosan [9].

a result of the swift growth of nanotechnology, and, because of their special features, they provide fantastic prospects for considerably increasing the detection and control performance of [8]. Figure 7.1 shows the schematic representation of chitosan.

7.2 Nanochitosan

Nanochitosans are naturally occurring polysaccharides generated by the alkalescent deacylation of chitin and comprise a series of 2-deoxy-2-acetyamino glucose connected by β-1,4-gycosidic linkages. These are naturally formed by deacylation of shellfish shell and exoskeleton of aquatic arthropods and crustaceans as well as cell wall of fungi [10]. These qualities were shown to be impacted by the degree of deacetylation (DDA), as measured by the molar fraction of deacetylated units or percentage of deacetylation [11]. Chitosan and its derivatives have been the subject of extensive research due to their wide range of uses in engineering, biomedicine, and pollution remediation as well as their non-toxic, biocompatible, palatable, and adsorptive characteristics in pharmacology for controlled drug delivery and wound healing [12]. It is now used for disinfection due to its antibacterial properties [11]. But, in recent years, chitosan's varied qualities have caught the attention of scientists, sparking a large amount of attention in the discipline of biomedical engineering [10]. Furthermore, due to their unique properties, chitosan nanoparticles (CSNPs) have refuted the contested influence that nanoscale materials have on natural and human systems [11]. In fact, interest in the use of CSNPs has increased due to their well-known polymeric and cationic characteristics [12]. One of the biopolymers used in the manufacture of metallic nanoparticles as a reducing agent and a protective polymer is chitosan [13]. Chitosan nanoparticles (CSNPs) are a particularly intriguing potential for use as a biomaterial in food-related applications. Because of their tiny size and high surface-to-weight ratio, CSNPs have been the subject of substantial investigation in the biomedical sector [14]. Nanochitosan has been produced using a variety of techniques, including ionotropic gelation, microemulsion, emulsification solvent diffusion, polyelectrolyte complex, and reverse micellar approach [9].

7.2.1 Preparation Methods of Nanochitosan

Ohya *et al.* [15] first described nanochitosan for the orderly transport of the chemotherapy medication 5-fluorouracil. Scientists have since conducted considerable research on nanochitosan and developed a wide range of procedures while taking into account a number of criteria, including as size, stability, drug loading potential, and retention period [16]. According to chitosan micro-particle technology, a technique for creating CSNP has been developed during the past 30 years [17]. Ionotropic gelation, reversed micelles (microemulsion), emulsification solvent diffusion, and polyelectrolyte complex are some of the ways that can be used to prepare nanochitosan [18]. Ionotropic gelation and self-assemble polyelectrolytes are the two approaches that have received the most development [19]. These techniques have a lot of benefits, including a simple, moderate preparation process without the need for an organic solvent or a lot of shear power [20].

7.2.1.1 Ionotropic Gelation Method

The ionotropic gelation process was used to create nanochitosan, which were first described by Calvo *et al.* [21] and have since undergone extensive research and development. The electrostatic contact between the amine group of chitosan and a negatively charged group of a polyanion, such as tripolyphosphate, is the basis for the creation of CSNPs [22]. In the aquatic environment, this procedure offers a straightforward and gentle preparation process [22]. Prior to or following the addition of polyanion, stabilizing agents, such as poloxamer, can be added to the chitosan solution and still allow the chitosan to dissolve in acetic acid. Nanoparticles spontaneously formed at room temperature while being mechanically swirled after adding the polyanion or anionic polymers [22]. One can alter the size and surface charge of particles by adjusting the surface tension [20]. Chitosan gelation and acrylic or methacrylic acid polymerization are both produced when ionic gelation and radical polymerization are combined [23]. Potassium persulfate is used to initiate the polymerization process, which requires 6 h of stirring at 60°C–70°C [16].

7.2.1.2 Emulsification and Crosslinking Method

The original method for creating CSNPs used the amino group of the chitosan and the aldehyde group of a crosslinking agent and was called emulsification and crosslinking [24]. Using span 80 as a stabilizing agent, toluene as the solvent, and glutaraldehyde as the crosslinker, an emulsion made up of an aqueous chitosan solution and an oil phase is created [24]. Droplets are the building blocks from which nanoparticles generated after crosslinking are formed once the phases are vigorously combined. Centrifugation, many washing processes, and vacuum- or freeze-drying are routinely used to separate nanoparticles from the emulsion [24]. It is feasible to obtain nanoparticles with a restricted distribution that are of small size [16]. Because of the challenges and overt toxicity of glutaraldehyde, this approach is no longer employed [16].

7.2.1.3 Reverse Micellar Method

This technique can be used to produce ultrafine polymeric nanoparticles in a certain size range [25]. To make reverse micelles, the surfactant is dissolved in an organic solvent. A chitosan aqueous solution is added while being continuously stirred in order to prevent any turbidity [26]. The reversed micelles (micro-emulsion) approach is based on covalent crosslinking, much like emulsification and crosslinking, but it now uses water-in-oil reverse micelle structures to aid in the production of CSNPs [26]. Chitosan and glutaraldehyde from the aqueous phase are combined with a lipophilic surfactant and an organic solvent [25]. Typically, n-hexane is employed as the preferred organic solvent and either sodium bis (2-ethylhexyl) sulfosuccinate or cetyltrimethylammonium bromide are used as surfactants [26]. Chitosan-containing micelles undergo crosslinking to produce CSNPs [16].

7.2.1.4 Precipitation-Based Methods

Precipitation-based techniques can also be used to create CSNPs [16]. Emulsification and precipitation are the foundations of the phase inversion precipitation method [16]. A stabilizer (poloxamer), an organic phase (dichlorome+thane and acetone), and an aqueous solution of chitosan are used to create the oil-in-water emulsion [27]. High-pressure homogenization is used to produce well-dispersed emulsion droplets that are nanoscale in size [28]. Then, acetone diffuses out of the droplets, and nanoparticles precipitate at the same time that methylene chloride is removed from the emulsion by evaporation at low pressure and room temperature [29]. The emulsion-droplet coalescence method, also known as desolvation, is an alternative strategy that relies on the coalescence of two water-in-oil emulsions and precipitates nanoparticles because one of the emulsions contains NaOH, which functions as a precipitation agent [16]. The continuous phase is first mixed with liquid to form the two emulsions—one with chitosan and one with NaOH. A high-speed homogenization process is used to create emulsions containing chitosan. The solubility of the chitosan is decreased and nanoparticles precipitate when two emulsions are combined because of the diffusion of NaOH into the ultrafine droplets [16]. The three steps utilized to make nanoparticles are centrifugation, washing with various solvents toluene, ethanol, and water, respectively, and freeze-drying. Nanoparticles produced by nanochitosan precipitation techniques typically have a size greater than 600–800 nm. Although both methods are not particularly preferred due to the use of organic solvents and high energy required homogenization applications and there are only a few studies in the literature, CSNPs obtained by the phase inversion precipitation method have high encapsulation efficiency for hydrophobic drugs such as cyclosporin A [16].

7.2.2 Nanochitosan-Based Sensors for Mycotoxin Detection

The use of optical biosensors such as surface plasmon resonance (SPR), photography, and colorimetry has grown in popularity, and nanotechnology-based biosensor–based methods have caught the attention of researchers worldwide. Transducers and bio-receptors (bio-capturing molecules and bio-recognition components) are necessary for the creation of accurate nanotechnology-based biosensor–based methods [30]. These bio-receptor surfaces were created with the intention of identifying and starting binding to defined analytes. It has been claimed that a variety of bio-receptors are used as functionalizing agents on nanomaterials. For target analytes to be detected in a complex environment with active bio-compatibility, high selectivity, and sensitivity, the choice of appropriate, pertinent, and corresponding bio-receptors on nanomaterials is crucial [31].

7.2.2.1 Surface Plasmon Resonance Technique

The determination of mycotoxins in wine, cereals, and milk has been done using a chitosan-based SPR technique, with limits of detection between 0.155 ng/mL and 0.55 ng/mL and gold film serving as the sensing material in the sensor chip due to its ability to excite the SPR response [32]. The binding mechanism of SPR is quite similar to that of immunological methods like the immunochromatographic technique and ELISA. The sensitivity of SPR, which causes fluctuations in resonance angle, is dependent on the nature of biological molecules on the sensor chip [32]. The weak sensor response (insignificant change in refractive index) makes it difficult to detect tiny compounds like ochratoxin A (Mw = 403.8 Da) [32]. Because of that, signal amplification techniques include the use of molecularly imprinted polymers, self-assembled monolayers, and selective polymers containing gold (Au) nanoparticles [32]. With regard to immobilizing small biological molecules on the surface of an SPR chip, chitosan is one of the most promising biopolymers due to its superior film-forming skills, chemical stability, affordability, biocompatibility, and environmental friendliness. In SPR biosensors for the determination of mycotoxins, chitosan has been employed as a matrix. Chitosan's strong affinity for binding to biomolecules is caused by the existence of protonated amino acids on its surface [32].

7.2.2.2 Colorimetric Assay

Colorimetric detection of mycotoxins using nanochitosan is a promising technique. Nanochitosan, a derivative of chitosan with nanoscale dimensions, can be functionalized to selectively bind to mycotoxins [33]. The surface of nanochitosan can be functionalized with specific receptors or aptamers that have high affinity for the target mycotoxin. This step enhances the selectivity of the detection [32]. Functionalizing nanochitosan with aptamers for mycotoxin detection is a promising approach that combines the unique properties of nanochitosan with the specificity of aptamers for sensitive and selective detection [34]. When the functionalized nanochitosan comes into contact with the sample containing toxins, it selectively binds to the mycotoxin of interest [33]. This binding event often results in a change in color or absorbance properties of the nanochitosan. The color change can be observed visually or measured using a spectrophotometer. The degree of the color change is directly related to the amount of the mycotoxin in the sample. Calibration curves are used to relate the colorimetric response to the mycotoxin concentration, allowing for quantitative analysis [34].

7.2.2.3 Chitosan-Based Electrochemical Sensors

A potential method in food safety and environmental monitoring is the electrochemical detection of mycotoxins using nanochitosan [16]. A chitosan derivative called nanochitosan is well suited for sensing applications due to its biocompatibility and simplicity of functionalization [35]. To improve its ability to bind mycotoxin, nanochitosan has been tweaked. This may entail bonding particular recognition components, such as antibodies or aptamers, to the nanochitosan surface [36]. The sensing layer based on nanochitosan is added to electrodes, which are frequently comprised of materials like gold or carbon. The platform for mycotoxin detection is provided by this layer. The sample solution is placed over the modified electrode [36]. Mycotoxins in the sample will bind to the nanochitosan surface's recognition elements. The electrode's properties are altered by this binding event [30]. Multiwalled carbon nanotubes (MWCNTs) have been functionalized to take use of their electrical capabilities as well as their high surface area-to-volume ratios and the existence of functional groups that can bind molecules in electrochemical immunosensors. They could be mixed with chitosan, a biopolymer that works particularly well as a matrix in biosensors due to its unique nature. The reactive functional groups in chitosan, such as the primary amine (–NH2) and the hydroxyl groups (–OH), can also bind antibodies. Therefore, chitosan and carbon nanotubes may be combined to provide a matrix with numerous functional sites for immobilizing molecules and increased electrochemical characteristics [37]. The working electrode was then coated with chitosan/MWCNTs and coated with an antibody layer that was covalently immobilized. The immunosensor's shape, spectral characteristics, and electrochemical characteristics were examined. The amount of Aflatoxin B 1 was measured using electrochemical impedance spectroscopy [38].

7.3 Advantages of Nanochitosan

Nanochitosan, a derivative of chitosan, has gained significant attention in various fields due to its unique properties and numerous advantages [39]. Nanochitosan can be used in biomedical applications such medication delivery, tissue engineering, and wound healing because it is biocompatible. Because of its low toxicity, there is less chance of negative effects when it comes into touch with biological systems. Nanochitosan exhibits inherent antimicrobial properties, which make it effective against a wide range of pathogens. In order to prevent infections, this characteristic is useful in wound dressings, medical devices, and food preservation. Chitosan, and, consequently, nanochitosan, is environmentally friendly and biodegradable. It can be discarded safely without harming ecosystems, allaying worries about plastic pollution [11]. In comparison to its macro-scale predecessor, chitosan's solubility in both aqueous and organic solvents has improved in the nanoscale form. This greater solubility increases its adaptability to a wider range of uses. Nanochitosan is an excellent candidate for drug delivery systems due to its ability to encapsulate and release pharmaceutical compounds gradually. This controlled release can enhance drug efficacy and reduce side effects [40]. Applications for nanochitosan in wound healing have showed promise. It can facilitate wound healing by encouraging tissue regeneration, lowering inflammation, and maintaining a moist environment. Because of its large surface area and chelating qualities, nanochitosan is efficient at removing heavy metals from water and industrial effluents, aiding in attempts to restore the environment. In agriculture, nanochitosan can be used to encapsulate and release fertilizers and insecticides. This improves plants' ability to absorb nutrients and lessens environmental pollution. Nanochitosan-based coatings can extend the shelf life of perishable foods by acting as a barrier against oxygen, moisture, and microbial contamination. Nanochitosan scaffolds in tissue engineering offer a favorable environment for cell adhesion, proliferation, and differentiation. This encourages the repair of damaged or missing tissues [29]. Low Cost and Widely Available: Chitosan, the predecessor to nanochitosan, may be made from chitin, which is widely distributed in fungi and crustacean cell walls. Because of that, nanochitosan is an economical substance. Nanochitosan is incredibly adaptable; it is simple to modify and combine with other materials, such nanoparticles and polymers, to modify its properties for particular uses. Nanochitosan is appropriate for use in medical and cosmetic applications because it is less likely to trigger allergic reactions in people than some synthetic materials [14].

7.3.1 Disadvantages of Nanochitosan

The production of nanochitosan can be expensive, primarily due to the need for chitosan extraction and the energy-intensive processes involved in nanoscale particle production. Because crab shells are the main source of chitosan, their availability may be constrained and influenced by changes in seafood production. Achieving consistent quality might be difficult because the characteristics of nanochitosan can differ depending on elements, such as the source of chitosan and production techniques. Despite being generally regarded as safe, there may be questions about the biocompatibility of nanochitosan when utilized in biomedical applications, necessitating additional research and testing [16].

The extraction of chitosan from crustacean shells can result in waste, and the disposal of nanochitosan products that are no longer required could have an impact on the environment. The commercialization and use of nanomaterials can be hampered by regulatory obstacles related to their safety and labeling [14]. Because of its restricted solubility in some solvents and conditions, nanochitosan cannot be used under all circumstances. According to certain studies, there may be toxicity issues with nanoparticles, such as nanochitosan, in some situations, particularly when inhaled or consumed. Nanochitosan's performance in a variety of applications, such as aggregation or deterioration, may be affected by stability difficulties [9].

7.3.2 Factors That Affect Nanochitosan Formation

The DDA, molecular weight (MW), concentration, kind, and quantity of the crosslinking agent, as well as environmental factors like pH and temperature, all have an impact on the creation of nanochitosan. These characteristics determine the effectiveness and range of applications for nanochitosan [41].

7.3.2.1 Molecular Weight and Degree of Deacetylation

The DDA and MW of chitosan affect the properties and usefulness of nanochitosan. The number of free amino groups in chitosan and its level of deacetylation were correlated. The free amino group engages the anion groups of the crosslinker to produce nanochitosan. In contrast to chitosan with a low DDA, which tends to disintegrate more quickly and has a faster drug release profile, high-MW chitosan has been found to generate nanochitosan with a reduced drug release rate [42]. The distribution of acetamide in the molecule varies, which is one of the factors that affects chitosan breakdown. Chitosan has a MW ranging from 300 to more than 1,000 kDa and a linear, nonbranched structure. The most common and simple technique is the viscosimetric method; however, it may also be detected by light scattering and high-performance liquid chromatography [43]. Higher-MW chitosan exhibits pseudoplastic behavior in acid solutions and has a higher viscosity. Chitosan with a higher MW had a stronger capacity for encapsulation, according to Xu and Du, because the molecule had longer chains that could entrap more biomolecules. High-MW materials were used to prepare CSNPs. Chitosan also has a lower binding ability. Depending on the number of monomeric units present in the biopolymer, chitosan has different MW. The control, evaluation, and change of viscosity and solubility, two qualities that are affected by MW, are essential. Zeta potential, encapsulation effectiveness, and delivery rate of CSNPs were affected by the DDA of the substance. The DDA, which depends on the ratio of acetylated amino groups in the chitosan molecules, indicates the percentage of deacetylated amino groups in the molecules. Zeta potential is influenced by the density of the positive charge on the surface of nanochitosan, which is correlated with the DDA. When the positive charge increased, chitosan's capacity to form crosslinks with tripolyphosphate increased, eventually resulting in smaller CSNPs [42].

7.3.2.2 pH

The electrostatic interaction between the polyanions and chitosan during the formation of nanochitosan depend on the pH of the mixing solution. Only at certain pH ranges that match the anions' natural characteristics do the electrostatic interactions take place, such as 1.0 to 7.5 for CSNP-sulfate, 4.5 to 7.5 for CSNP-citrate, and 1.9 to7.5 for CSNP-tripolyphosphate. Chitosan nanoparticles zeta potential and particle size are both influenced by pH. From pH 1 to pH 3.5, the particle size increased quickly, and, from pH 3.5 to pH 5.5, it progressively dropped. The relationship between pH and zeta potential was comparable, with pH 4 showing a slightly stronger transition. Zeta potential raised from pH 1 to pH 4 and then decreased gradually from pH 4 to pH 5.5. Because chitosan's amine groups were protonated under acidic conditions, this resulted in substantial charge repulsion and molecule extension [43]. Tripolyphosphate molecules protonated in low pH conditions and exhibited low charge densities. Therefore, in an acidic environment, the chitosan cannot properly bond with tripolyphosphate to generate stable nanochitosan. It was decided that pH 3.5 was the ideal level for the stable CSNP to develop. At higher pH levels, the tripolyphosphate molecule's charging degrees are improved and more effectively neutralize the chitosan charge, which causes the size of the nanochitosan to shrink [44]. Between pH levels of 4.5 and 6.0, nanochitosans' zeta potential and particle size tended to decrease, resulting in the creation of somewhat variable particles. It was hypothesized that the positive surface charge of chitosan acts as a shield due to the rearrangement of the molecule's structure or/and the adsorption of additional negatively charged ions at low pH [42].

7.3.2.3 Temperature

Temperature has a significant impact on the formation of nanochitosan; when chitosan-sodium hexametaphosphate was produced by agitation at 600 rpm at a concentration of 5 mg/mL, fewer CSNPs were produced than at higher temperatures (40°C, 50°C, and 60°C) [45]. Nanochitosan produced at higher temperatures is simpler to agglomerate because of its larger size and higher production. Smaller particles were produced at lower temperatures [42]. For production at 4°C with chitosan concentrations of 0.2 mg/mL and 0.4 mg/mL, CSNPs were smaller; however, when the concentration was 0.8 mg/mL, there was no appreciable difference in particle size when compared to nanochitosan formed at 27°C and

35°C [45]. Compared to CSNPs generated at 35°C, which had a larger size (130–190 nm), had a larger size spectrum, and were also more easily collected, those formed at 4°C were insignificant (60–80 nm) and had a limited size spectrum [45].

7.3.2.4 Crosslinker

Chitosan is an amino polysaccharide consisting of 2-deoxy-d-glucosamine and 2-deoxy-N-acetyld-glucosamine units that is glycosidic (1,4)-linked [41]. To create Schiff base, amide, or ester bonding linkages that result in the production of chitosan hydrogels, it has $-NH_2$ and –OH groups as the binding sites [42]. The amine groups of chitosan molecules cause it to behave in a pH-sensitive manner. Therefore, chitosan dissolves and depolymerizes in acidic solution. During regulated oral administration, depolymerization is a significant drawback that can be overcome by using crosslinking agents [43]. The dimensional and chemical stabilities, mechanical characteristics, swelling, water permeability, and solubility of polymers are affected by crosslinking. Polyanions are crosslinkers that are employed in the ionic gelation process to produce CSNPs [44]. The amine groups of chitosan and the crosslinker interact electrostatically, weakening the surface charge of chitosan and decreasing its solubility, resulting in the production of self-generated nanochitosan [43]. Three categories of polyanions can be distinguished: (a) low-MW ions, hydrophobic ions, and high-MW ions. Polyanions have been shown to affect the nanoparticle production time, particle size, polydispersity, drug-carrying capacity, stability, and muco-adhesiveness [43]. According to a study, the polyanions' variations in MW and binding sites have an impact on the CSNPs' sizes, polydispersity indices, and loading capabilities [42].

7.4 Conclusion

The application of nanochitosan in the detection of mycotoxins represents a promising and innovative approach to addressing the significant issue of mycotoxin contamination in food and feed. High surface area, biocompatibility, and the capacity to functionalize it for particular mycotoxin detection are just a few benefits of this nanomaterial. Its promise as a quick and sensitive detection technique has been shown by research and studies done so far.

7.5 Recommendations

- To assess the accuracy and applicability of nanochitosan-based tests, more research should concentrate on evaluating them under real-world circumstances with a range of food and feed samples.
- Collaborative efforts should be made to establish standardized protocols for using nanochitosan in mycotoxin detection. This will ensure reliability and comparability of results across different laboratories.
- It is imperative to optimize the production processes of nanochitosan in order to lower costs and increase the technology's accessibility for a larger group of stakeholders.

References

1. Sandlin, N., Kish, D.R., Kim, J., Zaccaria, M., Momeni, B., Current and Emerging Tools of Computational Biology To Improve the Detoxification of Mycotoxins. *Appl. Environ. Microbiol.*, 21, 3–5, 2022.
2. Donat, P.V., Marín, S., Sanchis, V., Ramos, A.J., A review of the mycotoxin adsorbing agents, with an emphasis on their multi-binding capacity, for animal feed decontamination. *Food Chem. Toxicity*, 114, 246–259, 2018.
3. Manzanares, N.A., Pérez, J.H., Campaña, A.G., Gracia, L.G., Mycotoxin Analysis: New Proposals for Sample Treatment. *Adv. Chem.*, 547506, 12, 2014.
4. Horky, P., Skalickova, S., Baholet, D., Skladanka, J., Nanoparticles as a Solution for Eliminating the Risk of Mycotoxins. *Nanomaterials*, 8, 727, 2018.
5. Da Silva, J.A., De Oliveira, C.A., Ramalho, L.Z., An overview of mycotoxins, their pathogenic effects, foods where they are found and their diagnostic biomarkers. *J. Food Sci. Technol.*, 42, 48520, 2022.
6. Twarużek, M., Skrzydlewski, P., Kosicki, R., Grajewski, J., Mycotoxins survey in feed materials and feeding stuffs in years 2015–2020. *Toxicon*, *202*, 27–39, 2021.
7. Assefa, T. and Geremew, T., Major mycotoxins occurrence, prevention and control approaches. *Biotechnol. Mol. Biol.*, 12, 1, 1–1, 2018.
8. Lin, X., Yu, W., Tong, Z., Duan, N., Wang, Z., Shijia, W., Application of Nanomaterials for Coping with Mycotoxin Contamination in Food Safety: From Detection to Control. *Crit. Rev. Anal. Chem.*, 7, 1, 876, 2022.
9. Chandrasekaran, M., Kim, K.D., Chun, S.C., Antibacterial Activity of Chitosan Nanoparticles: A Review. *Processes*, 8, 9, 1173, 2020.

10. Ahuekwe, E.F., Adelodun, C.A., Isibor, P.O., Eziegbe, O., Adekeye, B.T., Salami, A.O., Akinyosoye, A.D., Akinhanmi, F., Oyewale, J.O., Taiwo, O.S., Akinwunmi, R.A., Ajiboye, I., Akinpelu, S.A., Kuye, A.D., Bello, A.O., Poopola, T.S., Oranusi, S.U., Margaret, I.O., Nwinyi, O.C., Obembe, O.O., Application in Animal Husbandry, agriculture and food conservation. *Next Gener. Nanochitosan*, 12, 1, 147–168, 2023.
11. Ingle, P.U., Shandel, S.S., Shingote, P.R., Mishra, S.S., Sarda, V., Wasule, D., II, Rajput, B.D., Minkina, T., Rai, M., Sushkovad, S., Mandzhievad, S., Grade, A., Chitosan nanoparticles: A versatile growth promoter in modern agricultural production. *Heliyon*, 8, 11893, 2022.
12. Gal, M.R., Rahmaninia, M., Hubbe, M.A., A comprehensive review of chitosan applications in paper science and technologies. *Carbohydr. Polym.*, 3, 120665, 2023.
13. Verma, D.K., Malik, R., Meena, J., Rameshwari, R., Synthesis, characterization and applications of chitosan based metallic nanoparticles: A review. *J. Appl. Nat. Sci.*, *13*, 2, 544–551, 2021.
14. Nwalieji, A.V., Bankole, A.B., Daberechukwu, A., II, Francisca, M.O., Characterization of Nanochitosan/Nanosilver-Particles from Commercial Low Molecular Weight Chitosan and Extracted Chitosan from Snail Shell: A Comparative Study. *Res. Square*, 2, 1, 1–12, 2022.
15. Ohya, Y., Shiratani, M., Kobayashi, H., Ouchi, T., Release behavior of 5-fluorouracil from chitosan-gel nanospheres immobilizing 5-fluorouracil coated with polysaccharides and their cell specific cytotoxicity. *J. Macromol. Science—Pure Appl. Chem.*, *31*, 5, 629–642, 1994.
16. Yanat, M. and Schroën, K., Preparation methods and applications of chitosan nanoparticles; with an outlook toward reinforcement of biodegradable packaging. *React. Funct. Polym.*, 161, 104849, 2021.
17. Atia, G.A.N., Shalaby, H.K., Zehravi, M., Ghobashy, M.M., Attia, H.A.N., Ahmad, Z., Najda, A., Drug-Loaded Chitosan Scaffolds for Periodontal Tissue Regeneration. *Polymers*, *14*, 15, 3192, 2022.
18. Jha, R. and Mayanovic, R.A., A Review of the Preparation, Characterization, and Applications of Chitosan Nanoparticles in Nanomedicine. *Nanomaterials*, *13*, 8, 1302, 2023.
19. Luo, Y., Wang, Q., Zhang, Y., Biopolymer-based nanotechnology approaches to deliver bioactive compounds for food applications: a perspective on the past, present, and future. *J. Agric. Food. Chem.*, *68*, 46, 12993–13000, 2020.
20. Tiyaboonchai, W., Chitosan Nanoparticles: A Promising System for Drug Delivery. *Naresuan Univ. J.*, 11, 3, 51–66, 2003.
21. Calvo, P., Remunan-Lopez, C., Vila-Jato, J.L., Alonso, M.J., Novel hydrophilic chitosan-polyethylene oxide nanoparticles as protein carriers. *J. Appl. Polym. Sci.*, *63*, 1, 125–132, 1997.
22. Abdelgawad, A.M. and Hudson, S.M., Chitosan nanoparticles: Polyphosphates cross-linking and protein delivery properties. *Int. J. Biol. Macromol.*, *136*, 133–142, 2019.

23. Ross, M., Hicks, E.A., Rambarran, T., Sheardown, H., Thermo-sensitivity and erosion of chitosan crosslinked poly [N-isopropylacrylamide-co-(acrylic acid)-co-(methyl methacrylate)] hydrogels for application to the inferior fornix. *Acta Biomater.*, *141*, 151–163, 2022.
24. El-Naggar, M., Medhat, F., Taha, A., Applications of chitosan and chitosan nanoparticles in fish aquaculture. *Egypt. J. Aquat. Biol. Fisheries*, 26, 1, 23–43, 2022.
25. Dhapake, P.R., Baheti, J.R., Suruse, P., Nanoparticulate Drug Delivery: A Promising Drug Delivery for Protein Drug. *Asian J. Res. Pharm. Sci.*, 8, 38–42, 2023.
26. Zhao, L.M., Shi, L.M., Zhang, Z.L., Chen, J.M., Shi, D.D., Yang, J., Tang, Z.X., Preparation and application of chitosan nanoparticles and nanofibers. *Braz. J. Chem. Eng.*, 28, 3, 353–362, 2011.
27. Trabik, Y.A., Al-Afify, N.K., El-Kosasy, A.M., Magdy, N., Application of Precipitation-based and Nanoparticle-based Techniques for Fabrication of Potentiometric Sensors for Nano Molar Determination of Chitosan and Polyvinyl Pyrrolidone in Pharmaceutical Formulations and Biological Fluids. *Electroanalysis*, *33*, 5, 1233–1243, 2021.
28. Napagoda, M. and Witharana, S., Mycosynthesis of Chitosan Nanoparticles. *Mycosynthesis Nanomaterials Perspect. Challenges*, 99, 168–170, 2023.
29. Singh, A., Mittal, A., Benjakul, S., Chitosan nanoparticles: Preparation, food applications and health benefits. *Sci. Asia*, 47, 1–10, 2021.
30. Daramola, O.B., Omole, R.K., Akinwale, I.V., Otuyelu, F.O., Akinsanola, B.A., Fadare, T.O., George, R.C., Torimiro, N., Bio-Receptors Functionalized Nanoparticles: A Resourceful Sensing and Colorimetric Detection Tool for Pathogenic Bacteria and Microbial Biomolecules. *Front. Nanotechnol.*, 4, 1, 556–562, 2022.
31. Nguyen, Q.H. and Kim, M., II, Using Nanomaterials in Colorimetric Toxin Detection. *Biochip J.*, 15, 123–134, 2021.
32. Rehmat, Z., Mohammad, W.S., Sadiq, M.B., Somarapalli, M., Anal, A.K., Ochratoxin A detection in coffee by competitive inhibition assay using chitosan-based surface plasmon resonance compact system. *J. Colloid Surf. B.*, 147, 569–574, 2019.
33. Jiang, D., Huang, C., Shao, L., Wang, X., Jiao, Y., Li, W., Chen, J., Xu, X., Magneto inductively coupled plasma mass spectrometry with multiple metal nanoparticles as element controlled aptasensors for simultaneous detection of ochratoxin A and B fumonisin B1 using labels. *Anal. Chem. Acta*, 1127, 182–189, 2020.
34. Zheng, Y.T., Zhao, B.S., Zhang, H.B., Jia, H., Wu, M., Colorometric aptasensors for fumonisin B1 detection by regulating the amount of bubbles in closed bipolar platform. *J. Elecroanal. Chem.*, 877, 114584, 2020.
35. Ding, F., Fu, J., Tao, C., Yu, Y., He, X., Gao, Y., Zhang, Y., Recent advances of chitosan and its derivatives in biomedical applications. *Curr. Med. Chem.*, *27*, 18, 3023–3045, 2020.

36. Karrat, A. and Amine, A., Recent advances in chitosan-based electrochemical sensors and biosensors. *Arabian J. Chem. Environ. Res.*, 7, 2, 66–93, 2020.
37. Gupta, S., Murthy, C.N., Prabha, C.R., Recent advances in carbon nanotube based electrochemical biosensors. *Int. J. Biol. Macromol.*, 8, 108, 38–42, 687703.
38. Migliorini, F.L., Santos, D.M., Soares, A.C., Mattoso, L.C., Oliveira, O.N., Correa, D.S., Design of A Low-Cost and Disposable Paper-Based Immunosensor for the Rapid and Sensitive Detection of Aflatoxin B1. *Chemosensory*, 8, 87, 2020.
39. Mukhtar, M., Fényes, E., Bartos, C., Zeeshan, M., Ambrus, R., Chitosan biopolymer, its derivatives and potential applications in nano-therapeutics: A comprehensive review. *Eur. Polym. J.*, *160*, 110767, 2021.
40. Rizeq, B.R., Younses, N.N., Rasool, K., Nasrallah, G.K., Synthesis, Bioapplications, and Toxicity Evaluation of Chitosan-Based Nanoparticles. *Int. J. Mol. Sci.*, 20, 5776, 2019.
41. Román-Doval, R., Torres-Arellanes, S.P., Tenorio-Baraja, A.Y., Sánchez, A.G., Valencia-LazcanO, A.A., Chitosan: Properties and Its Application in Agriculture in Context of Molecular Weight. *Polymers*, 13, 15, 2867, 2023.
42. Warsito, M.F. and Agustiani, F., A review on factors affecting chitosan nanoparticles formation. *Mater. Sci. Eng. J.*, 1011, 012027, 2021.
43. Liu, H. and Gao, C., Preparation and properties of ionically cross-linked chitosan nanoparticles. *Polym. Adv. Technol.*, 20, 613–619, 2009.
44. Croisier, F. and Jérôme, C., Chitosan-based Biomaterials for Tissue Engineering. *Eur. Polym. J.*, 49, 4, 780–792, 2013.
45. Handani, W.R., Sediawan, W.B., Tawfiequrrahman, A., Wiratni, Kusumastuti, Y., The Effect of Temperature and Chitosan Concentration During Storage on the Growth of Chitosan Nanoparticle Produced by ionic Gelation Method. *AIP Conf. Proc.*, 1840, 1–7, 2016.

8

Application of Nanochitosan in Food Packaging Sectors

Ummulkhair Salamah Ilyasu[1]*, Faridah Laraba Tauheed[1], Sherifat Enejiyon Ozavize[1], Oluwafemi Adebayo Oyewole[1], Charles Oluwaseun Adetunji[2], Eniola K. I. T.[3] and Mohammed Bello Yerima[4]

[1]*Department of Microbiology, Federal University of Technology, Minnan, Nigeria*
[2]*Department of Microbiology, Edo State University, Uzairue, Nigeria*
[3]*Department of Biological Sciences, Joseph Ayo Babalola University, Ikeji Arakeji, Osun State, Nigeria*
[4]*Department of Microbiology, Sokoto State University, Sokoto, Sokoto State, Nigeria*

Abstract

Synthetic polymers generated from petroleum have long dominated the food packaging industry, and, by the year 2050, it is predicted that 500 million tons of plastic would have been used worldwide. Food packaging is designed to protect food products from the environment without compromising their intrinsic properties. In addition to plastics, food packaging has also traditionally included the use of glassware, metals, and paper. However, these traditional materials also have issues with degradability, disposal, and recycling. Due to their total biodegradability in nature, biodegradable materials have the potential to replace conventional polymers. Chitosan works wonders for food preservation. Due to its antibacterial qualities, chitosan is frequently utilized in antimicrobial films, edible protective coatings, dipping, and spraying for food products. Nanoparticles made of chitosan are possible. A variety of foods have been packaged using chitosan films to preserve their quality. Chitosan offers a lot of potential for a variety of applications due to its biodegradability, biocompatibility, antibacterial activity, non-toxicity, and varied chemical and physical properties. The use of nanochitosan in the food packaging industry is described in the current review.

Keywords: Chitosan, food, nanoparticles, nanochitosan, packaging

**Corresponding author*: ummulkhair3536@gmail.com

Charles Oluwaseun Adetunji, Maulin P. Shah, Yerima Mohammed Bello, Daniel Hefft, Jay Singh, Shyam S. Pandey and Ravindra Pratap Singh (eds.) Nanochitosan Applications for Enhanced Crop Production and Food Security, (155–180) © 2025 Scrivener Publishing LLC

8.1 The Evolution of Food Packaging

Glass was first used to store food and drink in Egypt approximately 7000 BC, where food packaging first emerged. China created flexible packaging in the second century BC using mulberry bark treated paper to package food. Wooden barrels were used in ships to transport food and water from 500 to 1500 AD [1]. The British Empire was in full swing with paper production in 1310. The increase in product demand throughout the industrial revolution (1760 to 1840) accelerated the need for high-quality packaging. One of the most often used packaging materials today was invented by Peter Durand in 1810 for iron cans that were covered with tin rather than glass to maintain the quality of food [2].

Sir Malcolm Thornhill created the cardboard box in 1817, although cardboard was first created by the Chinese [3]. Bristol, England, produced the first paper bags used for commerce in 1844 [4]. A paper bag manufacturing machine was then created in 1852. The first carton was inadvertently created by Robert Gair, who was the owner of a paper bag business. Branding, labels, and trademarks were all created for packaging. Smith Brothers formally registered trademarks for their cough medicines in 1866. NABISCO, an American producer of cookies and snacks, created the first branded packaging in 1890 [5]. To deliver their cereals, the Kellogg brothers created the cereal box in 1906 [2]. In 1908, Swiss chemist Jacques E. Brandenberger created cellophane, a material that repels liquids [6]. Aluminum foil containers first appeared on the market in 1910, followed by the aluminum can in 1959 [5].

The first commercially available modern plastics injection molding machine was patented by Eckert and Ziegler in 1926. Polyvinylidene chloride, which was used to package food after World War II, was discovered by Ralph Wiley in 1933. Styrene was first extracted from a balsam tree in 1831, and, after Germany improved the method in 1933 because the results were brittle, Styrofoam became widely available in 1950. The concept of frozen food packing was introduced between 1940 and 1949. Plastic manufacturing sectors started to expand significantly around 1970. Chemical engineer Nathaniel Wyeth received the patent for polyethylene terephthalate bottles in 1973. In 1980, hot foods were packaged in plastic. The Food Standards Act, which had an impact on the food packaging business, was introduced to the House of Commons in 1999 [5, 7, 8].

8.2 Standard Food Packaging

Food packaging is a vital phase in sustaining the safety of food. The primary objectives of food packaging are to avert contamination, minimize spoiling, and improve sensitivity by facilitating the activity of enzymes and curtailing weightiness. In order to facilitate easy food handling and maintain food quality for a longer period of time, conventional packaging for food is used to offer automated support and shield them from external impacts such as germs, light, oxygen, and off-odors [9]. Food spoiling typically happens as a result of microbial contamination, insect stings, endogenous enzyme degradation, or, less frequently, physical and chemical changes.

The majority of the times, microorganisms are what cause the deterioration, and there are numerous ways to stop, delay, or lessen microbiological decomposition [10]. Drying, freezing, fermenting, salting, low-temperature storage, canning, irradiation, changed environment, and the addition of antimicrobial agents are all common methods of food preservation. These methods are used to stop microbial growth [11]. The disadvantages of adopting these conventional materials are their high cost, susceptibility to breakage, and association with other hazardous materials. Each type of packaging requires a variety of materials, including wood, fibers, chemicals, minerals, energy, and water. Heavy metals, particulates, greenhouse gases, and wastewater or sludge containing harmful pollutants are also commonly released during its manufacture [9].

8.3 Types of Food Packaging

When discussing different package types, they are typically broken down into three groups: primary packaging, secondary packaging, and tertiary packaging.

8.3.1 Primary Packaging

This is often referred to as a consumer unit and is the kind of packaging that directly touches the product, for this reason. The major goal of primary packaging is to prevent the final product from contamination while also keeping it confined, preserved, and protected [12].

8.3.2 Secondary Packaging

When referring to a packaging system, secondary packaging refers to the exterior portion of the package that helps to protect the main packaging, and the main product is also known as reusable packaging [13].

8.3.3 Tertiary Packaging

This is also known as bulk or transit packaging. This type of packaging is used to bundle more stock keeping units for transportation from point A to point B. During this phase, products are handled as distribution units. With this kind of packing, large and/or heavy material might also be transported safely and securely. This kind of packing aids in protecting the goods from damage when handling and transporting it [14]. Packaging specifically for food has several different types such as the following:

Aseptic packaging: A paper and aluminum mixture with a polyethylene coating makes up aseptic packaging. It is used to preserve the sterility of products like milk and eggs and beverages with milk as well as liquid eggs.
Trays: The vast majority of this is obvious. Trays can be used to transport beverages, meats, and seeds.
Bags: Much like trays, bags are a common method for packing food. Fruit and bagged snacks are the most well-known options. The food is kept apart from its surroundings, mainly the air, by "bagging."
Boxes: Boxes are used to convey food items because they are the most practical. The most widely used box materials are metal, corrugated fiber board, and wood. Cans, cartons, flexible packaging, pallets, and wrappers are also used for packaging food [5].

8.4 Environmental Impacts of Food Packaging

Problems usually arise throughout the design process of packaging food. Resources like water, petroleum, energy, minerals, wood, textiles, and chemicals are needed to supply all the products and packaging [15]. The designing phase of the production process frequently results in the release of harmful pollutants such wastewater and sludge as well as some airborne particles and greenhouse gases. The role of packaging in preventing and reducing product waste is frequently disregarded, but, in recent years, its significance in waste reduction has started to gain attention [16].

Pollution issues can occasionally be caused by improper package disposal. Even society may perceive garbage to be a problem. More people than ever before are conscious. Others do their living, working, and playing in garbage-filled places. "Source reduction" is the process of creating packaging using less material. Examples of source reduction include lightweight plastic beverage bottles and thin goods like aluminum cans. Environmental factors are crucial to the production of new commodities and should remain so [17].

In the creation of biomimetic materials for nanotechnology, natural polymers are essential. In the field of biomaterials, natural polymers constitute a vital supply. It is anticipated that more cost-effective packaging designs will be recoverable. Throughout its life cycle, sustainable packaging is practical, safe, and economical and satisfies market and health expectations for consumers as well as society [17]. Renewable energy is used in their creation, supply, production, transportation, and recovery. It is made using the best practices and most eco-friendly production processes, and it is meant to have the greatest amount of recyclable or renewable components. It is made of all healthy elements and is physically constructed in accordance with all end of shelf-life scenarios in order to use the most suitable resources and energy efficiently.

Consumer awareness of recycling and "green" packaging is anticipated to increase over the next 10 years, according to the packaging industry. The best or least expensive designs in terms of social, economic, and environmental performance and costs are those that are sustainable. In order to address future human demands, sustainable design must be strategically employed while minimizing environmental risks and design flaws [5].

8.5 Significance of Food Packaging

The primary goal of packaging is to protect and preserve food products from outside elements that could spoil them or otherwise degrade their quality. Depending on preferences and the items being preserved, several materials might be utilized to package food. Most food packaging is made of glass, paper, steel, and plastic [18]. To protect food from contamination, food packaging is essential. The preservation of the product's shelf life is greatly influenced by food packaging. The trademark of the business or the product is often incorporated into food packaging, allowing consumers to recognize and distinguish it from competing goods. Labels on food packaging allow consumers to read information about the ingredients, expiry date, and the nutritional value [19].

8.6 Current Challenges in the Field of Food Packaging and Sustainability

Marketing changes particularly in the extremely vibrant universal food packaging business emphasize geniality, aesthetics, and usability to appeal to customers. Without a thorough and impartial evaluation of their overall environmental benefit, some business practices assert that they are sustainable either in the case of their resources (bio-based) or their shelf life (biodegradable) [20]. The vast majority of these eco-friendly technologies are less eco-friendly than expected, for instance, despite frequent assertions to the contrary, materials vary widely in the amount of renewable resources used during their manufacture and may or may not be readily biodegradable. None of these technologies made any claims to being sustainable for the utilization advantage of reducing food loss [21]. A thriving research and development sector is still necessary to close the critical societal gap of sustainable food consumption. This sector should provide a substantial pool of cutting-edge packaging innovations that will improve the sustainability of packaged foods. A lot of scientific consideration has been paid to the creation of bio-packaging solutions, or packaging made from renewable resources or biodegradable materials. However, overcoming specific technical issues with these bio-packaging materials that are currently preventing their wider market adoption presents a challenge for those working in packaging. These technological issues, particularly unnecessary raw material unpredictability and a too narrow processing window, prevent the scaling up and distribution of these products among package manufacturers when compared to normal oil-based alternatives. Additionally, stakeholders are unable to fully take advantage of the financial, societal, and environmental advantages of these innovations because there are not enough resources to assist users in customizing packaging according to food needs and comprehend the true sustainable development of bio packaging solutions and packaging in general. This is especially true in terms of curtailing food losses [22].

8.7 Current Scenario of Nanotechnology Application in Food Packaging

Numerous industries, including electronics, health, textiles, defense, the food and beverage industry, agriculture, and cosmetics, are investigating the use of nanoparticles. Food safety, processing, packaging, fortification, encapsulation, bioavailability, disease discovery, and other fields of the food sciences are all served by nanotechnology [23]. By improving a number of qualities that effectively convey active materials into biological systems at low prices with reduced environmental impact, such as temperature resistance, improved stability, barrier, flame resistance, recycling, and optical properties, nanotechnology-based food packaging offers many benefits over conventional food packaging resources [24].

8.8 Different Nanoparticles in Food Packaging Applications

8.8.1 Inorganic Nanoparticles in Food Packaging

Nanotechnology can be applied to food packaging to enhance food quality and act as antibacterial agents to prolong the shelf life of foods during storage and distribution. Silver, magnesia, selenium, zinc oxide, gold, copper, and titanium nanoparticles, among others, are glazed on the surfaces. The packaging and processing of food, the manufacture of medical devices, the production of synthetic materials, and water purification are only a few industrial uses for these nano-scale components [25]. Silver nanoparticles are widely employed in the food and healthcare industries, home products, biomedical and environmental applications, as well as other fields due to their specific physiochemical characteristics like thermal, biological, optical, high antimicrobial and electrical capabilities. According to [26], silver has numerous uses in the food system, including anti-inflammatory, antibacterial, anti-cancer, antifungal, anti-angiogenic, and antiviral properties. One of the most popular types of metal oxide nanoparticles used in packaging are zinc oxide nanoparticles due to their, photocatalytic activity, excellent stability and antibacterial action [27]. Compared to other metal oxides, ZnO has a stronger inhibitory effect on bacterial growth such *Listeria monocytogenes E. coli, S. aureus, Salmonella* spp., and *Campylobacter jejuni* [9].

8.8.2 Organic Nanoparticles in Food Packaging

Products made from renewable sources are valued more by consumers in the 21st century. The use of biodegradable materials in packaging and their application to food products to improve security by regulating the exchange of gas and/or moisture transmission is known as "bio-based packaging" [28]. Food is shielded from temperature changes, microbial growth, spoilage, and gas conditions by biodegradable packaging, which creates a hurdle between the materials and the food products. In recent commercial applications, bio-based polymers, which are naturally present in polysaccharides (such as collagen, starch, and chitosan), nucleic acids, and proteins, have become more prevalent [29].

8.8.2.1 Chitosan

Chitosan is a naturally occurring linear biopolysaccharide that is produced through alkaline deacetylation of chitin. The most common sources of chitin are the skeletons of crustaceans like lobster, crab, and shrimp. It was also discovered in the skeletons of coral and jellyfish, two types of marine zooplankton [30]. Yeast, mushrooms, and several other fungi also have significant amounts of chitin in their cell walls. Typically, as compared to other fungal classes, zygomycetes have the highest levels of chitin in their cell walls. Due to chitosan's biocompatibility, biodegradability, and safety, research into it has risen over the past few years. It is distinguished by its chelation, adsorption, film-forming, and antibacterial capabilities [31].

8.8.2.2 Nanochitosan as a Food Packing Material

Chitosan is a widely used, inexpensive, and a readily available polysaccharide and has disadvantages such as poor mechanical and thermal properties when compared to frequently employed non-biodegradable polymers [32]. Chitosan matrix's functional characteristics will be improved by adding nanoparticles or polymers. Chitosan films are a popular option for packaging materials due to their excellent antibacterial activity against a variety of infections and their capacity to maintain the integrity of food products [33]. Yanjun Tang along with his team, while comparing paper layered with chitosan to blank paper, demonstrated that covering cellulose paper with chitosan/titanium dioxide nanocomposites significantly improved the mechanical attributes and antibacterial activity of the paper. This finding supported the hypothesis that titanium dioxide and chitosan nanoparticles (CSNPs) would improve the coated paper's general qualities, including rip strength, brightness, tensile strength, antibacterial activity, air permeability, and opacity [34].

8.9 Preparation of Chitosan Nanoparticles

8.9.1 Ionic Gelation Method

The interaction of macromolecules with opposing charges can produce CSNPs. Tripolyphosphate (TPP) has frequently been utilized to make CSNPs because it is multivalent, nontoxic, and capable of making use of ionic interactions to form gels. The charge density of TPP and chitosan, which is influenced by the solution's pH, can be used to regulate the interaction [35].

8.9.2 Reverse Micellar Method

This method allows the production of ultrafine polymeric nanoparticles with specific range of sizes. The reverse micelles are produced by dissolving the surfactant in a natural solvent. Chitosan aqueous solution is added while being constantly stirred in order to prevent any turbidity. The aqueous phase is controlled to maintain an optically transparent microemulsion phase throughout the whole combination. Add extra water if larger-sized nanoparticles are to be created [36].

8.9.3 Nano-Based Food Packaging Methods

8.9.3.1 Active Packaging

Active packaging is a creative way to increase or maintain the shelf life of food goods. Furthermore, this will ensure their reliability, excellence, and security [37]. Contrary to normal food packaging, active food packaging acts as a passive barrier and promotes the release of antimicrobials and antioxidants. These will interact with food in a direct way and aid in the removal of particular pollutants like oxygen or water vapor [38]. Systems for packaging food contain antimicrobial ingredients such as bacteriocins, natural extracts, fungicides, enzymes, organic acids, polymers, and nanoparticles. Active packaging made of nanomaterials can interact directly with food or its surroundings to improve protection of the product and serve as an antibacterial agent [39]. Metal and metal oxide (nanocopper oxide, carbon nanotubes, nanotitanium dioxide, nanosilver, and nanomagnesium oxide) are utilized as packaging resources owing to their superior antibacterial properties [40].

8.9.3.2 *Smart Packaging*

Smart packaging is a new nanotechnology designed specifically to track microbiological, chemical, or biochemical changes in food or the environment [41]. To ensure the food's quality and safety, it also reacts to changes in the food's internal and external factors as they occur inside the food or in its surroundings. A variety of nanomaterials are utilized to enhance the functionality of packaging for various sensing events, such as freshness indicators, time and integrity indicators, radiofrequency identification, and temperature indicators [42].

Nanosensors can identify, quantify, and detect decaying materials, proteins that cause allergies, and pathogenic microbes; they can also be used to detect flavors or colors [9]. Nano-biosensors have proven effective for real-time monitoring of food quality, plummeting the need to resolve shelf life, and also retorting to some chemical markers, infections, and toxins present in the food [43].

8.9.3.3 *Intelligent Packaging*

The focus of intelligent packaging is to track the entire food supply chain, whereby a user can learn more about the product by scanning the code embedded inside the goods, QR code, or barcode, with a smartphone or any other type of scanner. It can also check the quality of food by printing thermochromic ink on packaging, which changes color as the temperature changes [9].

8.9.4 Food Application of Chitosan

8.9.4.1 *Edible Coating or Film*

The use of chitosan (CS) as well as its byproducts has extended the shelf life of a variety of foods. Therefore, the antimicrobial/antioxidant capabilities of food packaging materials can be improved by the capability of CSNPs to control the release of diverse bioactive constituents. Electrospray was used to create the gelatin film that included tea polyphenol-loaded CSNPs [18]. In combination with modified atmospheric packaging, fillets of beluga fish (aka Huso huso) coated with edible coatings containing CSNPs laden with fennel essential oils displayed decreased levels of peroxide, total volatile nitrogen, and thio barbituric acid. Additionally, CSNP-coated fillets showed reduced psychotropic and mesophilic *Pseudomonas* spp. and lactic acid bacteria populations than CSNP-uncoated fillets [44].

In a different study, *Carangoides coeruleopinnatus* fillets coated with chitosan and CSNPs at 4°C for 12 days were compared for qualitative attributes. Compared to fillets coated with Chitosan, fillets coated with nanoparticles more effectively slowed down lipid oxidation and the growth of microbes. Recently, it was discovered that soy-bean oil packaged in gelatin film with 30% (w/v) tea polyphenol-loaded CSNPs had the lowest peroxide value on day 14 [18]. In addition to extending the shelf life of different foods, CSNPs also enhanced the physicochemical and mechanical characteristics of films. Bulk CS and CSNPs were incorporated to create tara gum films. In comparison to CS added film, films supplemented with CSNPs displayed higher water solubility, mechanical strength, antibacterial activity, and lower hydrophilicity. Another study found a 42% increase in the solubility of the composite coating of CS-ZnO-NPs.

8.9.4.2 Bread

Bread often has a short shelf life due to microbial development and staling [45]. Staling is a term used to describe the gradual degradation of bread's flavor and texture over time. Multiple factors contribute to the complex phenomena of bread staling [46]. There are documented uses of chitosan to prolong the shelf life of bread by suspending starch retrogradation and/or by preventing microbiological progression. The impact of chitosan (493 kDa) coating on the shelf life of baguette was examined by [47]. After molding, the surface of the dough was painted with 0.5%, 1.0%, or 1.5% chitosan in 1.0% acetic acid using a brush. During storage at 25°C for 36 h, baguette coated with chitosan, particularly with 1% chitosan, showed less hardness, weight loss, and retrogradation compared to the control. Chitosan's ability to act as a moisture barrier is probably to blame for this [48]. The chitosan coating may provide a hindrance to prevent moisture from transferring through the surface of the bread, hence averting weight loss, delaying hardness, and retrogradation.

8.9.4.3 Egg

Egg storage problems include interior quality decline, weight loss, and microbiological contamination, to name just a few [49]. According to Akarca *et al.* [50], egg quality fluctuations in the yolk and albumen, as well as weight loss, are controlled by the moisture and carbon dioxide that flow from the albumen through the shell. It has been established that chitosan is capable of creating films, which can be utilized as edible coatings [51]. The shelf life of eggs may, therefore, be improved by the protective barrier that chitosan film may provide against gas transfer and moisture from the albumen through the shell.

8.9.4.4 *Vegetables and Fruits*

The primary causes of fruit post-harvest losses include mechanical damage, careless handling, incorrect temperature and relative humidity control, and hygiene problems. Applying edible coatings to the surface of these perishable goods and then storing them in the cold are two potential methods for extending their storageability [52]. Edible coatings could be utilized as a protecting barrier to lower transpiration and respiration rates through the surface of fruits, delay microbiological development, change in color, and enhance the textural feature of fruits [53]. By altering the endogenous O_2, CO_2, and ethylene levels of fruits, semipermeable film has typically been proven to delay fruit ripening [54]. Due to the fact that chitosan films preferentially permit more O_2 than CO_2 to pass through than CO_2, The interior environment may change due to the chitosan covering without causing anaerobic respiration [55]. Therefore, the fungistatic and internal atmosphere-modifying properties of chitosan covering have the prospect to extend fruit storage time and prevent deterioration.

8.9.4.5 *Juice*

Clarifying agents, such as tannins, gelatin, silica sol, bentonite, and polyvinylpyrrolidone, are frequently used in the processing of clarified fruit juices. It has been demonstrated that the partially positive charge on chitosan gives it the ability to bind acids and is useful for assisting in the separation of colloidal and isolated constituent parts from food processing wastes. These characteristics make chitosan a desirable processing service in the preparation of fruit juice [56].

8.9.4.6 *Meat*

Meat and meat products are very prone to lipid oxidation, which causes a rancid or reheated flavor to emerge very quickly. Chitosan has antioxidant and antibacterial properties, which may prevent the growth of spoilage microorganisms and delay the oxidation of lipids in meat during storage. Multiple studies have noted the beneficial effect of chitosan addition on the storage stability of meat. Youn *et al.* [57] reported that, by lowering the overall bacterial cell counts and preventing lipid oxidation during storage at 4°C for 10 days, spiced beef treated with 1.0% of chitosan (120 kDa,

DD = 85%) and dissolved in 0.3% of lactic acid was noticeably improved [58].

Chitosan's ability to prevent microbial growth in chilled comminuted pork products was shown by [59] and the effect of chitosan depended on concentration. Chitosan glutamate additions of 0.3% and 0.6% decreased total viable counts, lactic acid bacteria, yeasts, and molds, by 3-log CFU/g at 4°C for 18 days in comparison to the control. According to [60] adding 3% chitosan glutamate (DD = 86%) to turkey or ground beef may lower the likelihood that *Clostridium perfringens* spores will germinate and spread during harsh cooling in 12, 15, or 18 h from 54.4 to 7.2°C [61].

8.9.4.7 Milk

There have been several attempts to assess the viability of utilizing chitosan to enhance the shelf life and quality of milk. The impact of water-soluble chitosan having three distinct molecular weights (0.2–3 kDa, 3–10 kDa, and 10–30 kDa) on the sensory and physicochemical characteristics of milk was investigated by [62]. Chitosan concentrations and molecular weights (0.5%, 1.0%, and 1.5%) enhanced the consistency of milk that had chitosan added to it. No protein coagulation occurred during the 15 s of sterilization at 73°C for milk containing 0.5% and 1.0% chitosan, regardless of the molecular weight. Milk's sensory quality of taste, color, and flavor were severely impacted by adding 0.5% water-soluble chitosan [63], which resulted in 0.2- to 3-kDa, 3- to 10-kDa, and 10- to 30-kDa chitosan browning in color and chemical off-flavor, as well as a pungent taste. However, there were no differences in the sensory quality of the chitosan-infused coffee milk compared to the control; this may be because the coffee had a masking effect.

8.9.4.8 Noodles

Roy *et al.* (2023) explored at how incorporating chitosan (Mw = 37 kDa) affected the quality and shelf life of wet noodles. Wheat flour was mixed with chitosan that had been dissolved in 1% acetic acid at concentrations of 0.0%, 0.17%, 0.35%, 0.52%, and 0.7%. Regardless of chitosan concentrations, wet noodles' moisture content reduced somewhat over the course of 6 days at 18°C regardless of how long they were stored for. When chitosan concentration was increased from 0% to 0.70%, the amount of viable cells decreased during the course of 6 days at room temperature. Wet noodles with chitosan concentrations of 0.17%, 0.35%, 0.52%, and 0.70% had shelf lives that were 1, 2, 3, and 3 days longer than those of the control, respectively. In comparison to other wet-noodle samples, the wet noodle with 0.35% chitosan had the best sensory qualities. In a different study, wet noodles with chitosan (Mw = 37 kDa, 0.1% or 0.5% dissolved in 1% lactic acid) could be preserved for up to 80 days as opposed to just 7 days for the control [64]. As a result of its antibacterial action, these experiments amply proved that chitosan may be employed as a successful preservative in wet noodles.

8.9.4.9 Rice Cake

White rice cake's shelf life was significantly increased after chitosan treatment [65]. Prior to vacuum sealing, these authors coated a white rice cake with alcohol (95%) lactic acid (1%), 1% and/or 2% chitosan (Mw = 37 kDa, dissolved in 1% lactic acid) and left it in for 10 s. The total microbial counts for the untreated control, alcohol-treated, and 1% lactic acid–treated white rice cakes all went over the initial degeneration threshold level of 1×10^6 CFU/g after 6, 27, and 20 days of storage at 4°C. On the other hand, even after 76 days of storage, the white rice cake treated with 1% and 2% chitosan had total microbial counts (3.3×10^5 and 1.4×10^5 CFU/g, respectively) that were lower than the required limit. However, it was discovered that the sensory suitability of the white rice cakes was adversely impacted by the dipping procedure, regardless of the type of solution used. Hadidi *et al.* [66] also showed how adding chitosan to rice cakes could extend their shelf life. Rice cake was infused with water-soluble chitosan at concentrations of 0%, 0.05%, 0.1%, 0.3%, and 0.50%. As chitosan level increased during storage for 4 weeks at 5°C, the total number of microbes decreased. After 4 weeks of storage, the overall microbial counts of the chitosan-containing rice cakes were 2-log cycles less than that of the control (8.2×10^4 CFU/g).

8.9.4.10 Sausage

In Korea, sodium nitrite is typically employed as a curing agent to develop flavor and color for preservation effects. However, nitrosoamine, a potent toxin harmful to humans, is produced when nitrite combines with amine in meat. Numerous studies have looked into the use of chitosan as a curing agent in sausage [67, 68], and it was found that the presence of chitosan could minimize or completely eliminate the usage of sodium nitrite without compromising the preservation action or color development.

8.9.4.11 Seafoods and Seafood Products

The fish muscle contains a lot of metal ions and hematin molecules, which accelerate the lipid oxidation of unsaturated fatty acids and is a major cause of quality degradation in seafood products [69]. In addition, autolysis, microbial contamination and development, and protein functional loss all have a noteworthy influence on the quality of seafood [70].

8.9.4.12 Soybean Curd (Tofu)

Due to its high moisture level and rich vitamin content, freshly prepared soybean curd has an exceptionally short shelf life, typically only lasting 1 to 2 days even under professional refrigeration [71, 72]. Water soluble chitosan was used as a coagulant or as an immersion solution for the purpose of extending the shelf life of tofu. The shelf life of the tofu was increased by over 7 days at 4°C when combined with the coagulant $CaCl_2$ and compared to that prepared with just $CaCl_2$ [73]. Similar to this, Li and Yu [74] found that using a 0.5% water soluble chitosan solution as an immersion solution increased its shelf life by over 7 days at 4°C. The antibacterial properties of chitosan are what gave tofu an extended shelf life.

8.9.4.13 Vinegar

Chitosan's ability to clarify persimmon vinegar was studied by Kang *et al.* [75]. Chitosan of 150 kDa and 37 kDa was dissolved in 1% acetic acid, and concentrations of 100 mg, 200 mg, 300 mg, 400 mg, and 500 mg were added to vinegar. The coagulated solids increased as the chitosan concentrations increased, whereas the contents of browning, turbidity, soluble solids, and tannins decreased. The decrease in turbidity and soluble solids varied a little depending on the molecular weight of chitosan. No matter the molecular weight of the chitosan used, 400 mg/L of it was necessary to produce the best results in clarifying persimmon vinegar. The overall acceptability was lower and this was the outcome of the taste's pronounced astringency when chitosan concentrations (37 kDa and 150 kDa) were increased to 500 mg/L. It was discovered that the quality of persimmon vinegar that had undergone the aforementioned chitosan treatment was more consistent after storage at room temperature for 6 months in contrast to the control (no chitosan treatment) [74].

8.9.5 Other Applications of Chitosan Nanoparticles

8.9.5.1 Medicine and Pharmaceuticals

Chitosan NPs have been thoroughly investigated by researchers for use in pharmaceutics and medicine. The substance is biocompatible and enables the chain grafting of the active components and the drug encapsulation. Their use is extremely advantageous in biological imaging, drug administration, diagnosis, and cancer treatment due to notable properties such limiting enzymatic degradation of medicines and minimizing injury to non-targeted tissue or cells [75]. Additionally, it has been claimed that the slow biodegradation of CSNPs ensures continuous and controlled drug release [76]. The stable NPs that convey drugs using different methods in the human body are made possible by the substantially positive surface charge. Two major methods—nanoencapsulation and chemical modification—are typically used to create drug-loaded CSNP variants. According to Rajitha *et al.* [77], nanoencapsulation is the process of creating a nanostructure with an absorbed medication on its surface or inside the nanoparticle.

8.9.5.2 Wastewater Treatment

The investigation of bio-based substitutes has been driven by the dearth of an affordable, environmentally friendly, and functional sorbent to replace activated carbon, which is widely used. These nanoparticles are remarkable for the elimination of a variety of contaminants such dyes, heavy metals, and herbicides, because chitosan contains functional amino and hydroxyl groups. Additionally, due to their larger surface area, nanoparticles may have a higher capability than commonly employed micro-sized sorbents [77]. Water treatment has often been advised for derivatives of CSNP with improved magnetic and electrostatic characteristics. Amination is done by grafting substances like hexanediamine, ethylenediamine, or diethylenetriamine, whose NH_2 groups are in charge of the pollutants' electrostatic interactions [78]. Chitosan NPs have been utilized in conjunction with magnetic characteristics to enhance the removal of heavy metals and dangerous dyes.

8.10 Advantages of Nanotechnology in Food Packaging

8.10.1 Nanoparticles Protect Food Quality Decay Caused by Chemicals

The many chemical constituents of food and the environment interact, degrading the quality of the food as a result. Several nanomaterials have been shown in numerous studies to have the capacity to regulate unfavorable responses in food mediums. Nanotechnology can improve the functionality of food by removing chemical contaminants or increasing the availability of nutrients at the nanoscale. Numerous benefits of nano-encapsulating chemicals include continuous administration of various active compounds, pH-triggered release, and shelf-life extension [9]. Edible nano-coatings are used as gas and moisture barriers, as well as to add color, enzymes, flavor, antibacterial protection, anti-browning properties, and antioxidants and to extend the shelf life of manufactured goods in vegetables, fruits, meat, bakery goods, cheese, and fast-food [9]. Few nanomaterials are utilized directly as antibrowning agents, although ZnO nanoparticles were used as coated active packaging to extend the shelf life of fresh-cut apples [53].

8.10.2 Nanoparticles for Enhancing Physical Properties

The physical characteristics of both food and packaging materials will improve with the use of nanoparticles in food packaging. Common applications for nanocomposites include coating and packaging. To make polymer matrices lighter, with lower gas permeability and improved thermal properties, SiO_2, carbon nanotubes, clay and silicate nanoplatelets, starch nanocrystals, graphene, cellulose-based nanofibers, CSNPs, and other inorganic nanoparticles are included [79].

8.10.3 Nanoparticles for the Detection of Food Borne Pathogens

Foodborne pathogens and chemicals that cause food products to degrade and become contaminated are being detected with nano-biosensors [80]. As a nano-biosensing technology, surface enhanced raman scattering can swiftly and precisely identify harmful bacteria. Surface enhanced raman scattering, which uses materials such as plasmonic gold, silver nanocolloids, graphene oxide, silver nanoparticles, carbon nanotubes, and magnetic beads to boost Raman signals, has been used to identify bacteria. *E. coli*, *Salmonella* spp., *S. aureus*, and *S. epidermidis* are only a few of the foodborne pathogens that can be detected using nanoparticles [9].

8.10.4 Nanoparticles for Inhibiting Biofilm Formation

A biofilm is a complexly organized, densely filled mass of bacteria that has the ability to adhere to a variety of surfaces and build a polymeric extracellular matrix that is often made up of polysaccharides, proteins, and DNA, rendering it impenetrable [81]. The circulating bacteria adhere to the surface and cause buildup, biofouling, and biocorrosion issues in the food industry. Concentrations of silver nanoparticles as low as 0.2 ppm can enhance bacterial catabolism and inhibit the formation of several types of bacterial biofilms. This nanoparticles approach has been applied in many industrial settings due to the fact that it reduces the overall density of microbial biofilm without compromising cell proliferation [9].

8.10.5 Eco-Friendly

Several supplies used in contemporary food storing are practically non-biodegradable and pose a huge threat to the environment worldwide. A key strategy for extending the quality and shelf life of food while decreasing packaging waste and even food loss is the use of edible and biodegradable films [20]. However, issues related to performance (such as fragile nature, weak gas or moisture-proofing, and the production process, incorporating low-temperature heat distortion), and cost limit the adoption of edible and degradable polymers. As a biodegradable decomposer, starch, for instance, has drawn a lot of attention.

8.11 Conclusion

Technology has made packaging an essential component of contemporary living. The modern lifestyles of people, their circumstances, their hectic work schedules, their access to and competency in the kitchen, as well as many other variables, have significantly contributed to the rising demand for pre-packaged foods, which has, in turn, raised the entire output of food packaging. The food packaging industry also experiences change as a result of how swiftly technology alters our lives, makes them more active, and impacts each sector in a unique way. The majority of synthetic plastic materials used in the present food packaging system are derived from petroleum. However, this not only places an excessive amount of pressure on the use of non-renewable fossil resources, but also the non-degradability of these polymers makes it extremely difficult to dispose of municipal waste.

Search for alternatives, especially biodegradable ones, to address each of these issues at once. Nanotechnology-based food packaging offers many benefits over conventional food packaging materials by improving a number of properties, including flame resistance, temperature, resistance, barrier, recycling and optical properties, increased durability, processability due to lower viscosity, and successfully delivering active materials into the biological systems, at minimal costs with lessen environmental issue.

Chitin and chitosan, which have emerged as value-added products from the waste generated during the production of seafood, have been the subject of extensive investigation in recent years. Chitosan stands out in comparison to other biopolymeric counterparts because of its many attributes, including antibacterial activity, chelation property, film formation, and reasonable mechanical strength, to show enormous potential as a food packaging material. Chitosan films have so far demonstrated their capacity to increase food shelf life while preserving food quality. Additionally, on a lab scale, the biopolymer has been employed to extend the shelf life of freshly produced fruit, poultry, vegetables, and dairy items without impairing their sensory qualities. Therefore, it is certain that chitosan will eventually displace non-biodegradable polymers and establish itself as a natural substitute.

References

1. Teixeira-Costa, B.E. and Andrade, C.T., Natural polymers used in edible food packaging—History, function and application trends as a sustainable alternative to synthetic plastic. *Polysaccharides*, 3, 1, 32–58, 2021.
2. Human, U., A brief history of food packaging: from animal skins to plastic. *Farmlink Afr.*, 2016, 11, 44–45, 2016.
3. Hawkins, A.R., Bistline, E.N., Blackwell, C.S., Ives, M. (Eds.), editors *Playing Games in Nineteenth-century Britain and America*, State University of New York Press, New York, 2021.
4. Pathak, G., Plastic pollution and plastics as pollution in Mumbai, India. *Ethnos*, 88, 1, 167–186, 2023.
5. BT, C. and K Sanikop, N., Food packaging in India: an overview. *Int. Res. J. Adv. Sci. Hub.*, 3, Special Issue 7S, 103–110, 2021.
6. Amara, A.A.A.F., Natural Polymer Types and Applications, in: *Biomolecules from Natural Sources: Advances and Applications*, pp. 31–81, 2022.
7. Smil, V., *Materials and Dematerialization: Making the Modern World*, John Wiley & Sons, Canada, 2023.
8. Tickner, J., Geiser, K., Baima, S., Transitioning the chemical industry: the case for addressing the climate, toxics, and plastics crises. *Environ. Sci. Policy Sustain. Dev.*, 63, 6, 4–15, 2021.
9. Thirumalai, A., Harini, K., Pallavi, P., Gowtham, P., Girigoswami, K., Girigoswami, A., Nanotechnology driven improvement of smart food packaging. *Mater. Res. Innov.*, 27, 4, 223–232, 2022.
10. Sung, S.Y., Sin, L.T., Tee, T.T., Bee, S.T., Rahmat, A.R., Rahman, W.A.W.A., Vikhraman, M., Antimicrobial agents for food packaging applications. *Trends Food Sci. Technol.*, 33, 2, 110–123, 2013.

11. Ibarra, V.G., Sendón, R., De Quirós, A.R.B., Antimicrobial food packaging based on biodegradable materials, in: *Antimicrobial food packaging*, pp. 363–384, Academic Press, Spain, 2016.
12. Escursell, S., Llorach-Massana, P., Roncero, M.B., Sustainability in e-commerce packaging: A review. *J. Clean Prod.*, 280, 124314, 2021.
13. Mahmoudi, M. and Parviziomran, I., Reusable packaging in supply chains: A review of environmental and economic impacts, logistics system designs, and operations management. *Int. J. Prod. Econ.*, 228, 107730, 2020.
14. Alamri, M.S., Qasem, A.A., Mohamed, A.A., Hussain, S., Ibraheem, M.A., Shamlan, G., *et al.*, Food packaging's materials: A food safety perspective. *Saudi J. Biol. Sci.*, 28, 8, 4490–4499, 2021.
15. Nilsen-Nygaard, J., Fernández, E.N., Radusin, T., Rotabakk, B.T., Sarfraz, J., Sharmin, N., Sivertsvik, M., Sone, I., Pettersen, M.K., Current status of biobased and biodegradable food packaging materials: impact on food quality and effect of innovative processing technologies. *Compr. Rev. Food Sci. Food Saf.*, 20, 2, 1333–1380, 2021. https://doi.org/10.1111/1541-4337.12715.
16. Williams, H., Lindström, A., Trischler, J., Wikström, F., Rowe, Z., Avoiding food becoming waste in households–The role of packaging in consumers' practices across different food categories. *J. Clean. Prod.*, 265, 121775, 2020.
17. Shafi, H. and Bajpai, M., A review on importance of biodegradable packaging for foods and pharmaceuticals. *Curr. Nutr. Food Sci.*, 19, 1, 9–21, 2023.
18. Wang, Y.H., Zhang, R., Qin, W., Dai, J.W., Zhang, Q., Lee, K.J., Liu, Y., Physicochemical properties of gelatin films containing tea polyphenol-loaded chitosan nanoparticles generated by electrospray. *Mater. Des.*, 185, 108277, 2020.
19. Minooei, O., Mokshapathy, S., Zare, M., Zarei, M., Importance of food packaging and its relation to the consumer's demographic profile. *Int. J. Bus. Manag. Invention*, 4, 1, 8–11, 2015.
20. Iversen, L.J.L., Rovina, K., Vonnie, J.M., Matanjun, P., Erna, K.H., 'Aqilah, N.M.M.N., Felicia, W.X.L., Funk, A.A., The emergence of edible and food-application coatings for food packaging: a review. *Mol. (Basel Switzerland)*, 27, 17, 5604, 2022.
21. Kumari, S., Tiwari, S., Faisal, M., Eco-friendly management of northern Corn leaf blight of maize (Zea mays L.). *J. Pharmacogn. Phytochem.*, 9, 2, 1660–1663, 2020.
22. Berketova, L.V. and Volodina, S.S., Fudshering–kak ekologichnyy sposob ispolzovaniya produktov pitaniya [Food sharing–as an eco-friendly way to use food]. *Bull Nauki Practiki [Bulletin Sci. Practice]*, 1, 253–259, 2020.
23. Dos Santos, C.A., Ingle, A.P., Rai, M., The emerging role of metallic nanoparticles in food. *Appl. Microbiol. Biotechnol.*, 104, 6, 2373–2383, 2020.
24. De Sousa, M.S., Schlogl, A.E., Estanislau, F.R., Souza, V.G.L., dos Reis Coimbra, J.S., Santos, I.J.B., Nanotechnology in Packaging for Food Industry: past, present, and future. *Coatings*, 13, 8, 1411, 2023.

25. Hoseinnejad, M., Jafari, S.M., Katouzian, I., Inorganic and metal nanoparticles and their antimicrobial activity in food packaging applications. *Crit. Rev. Microbiol.*, 44, 2, 161–181, 2018.
26. Huang, Y., Mei, L., Chen, X., Wang, Q., Recent developments in food packaging based on nanomaterials. *Nanomaterials*, 8, 10, 830, 2018.
27. Kim, I., Viswanathan, K., Kasi, G., Thanakkasaranee, S., Sadeghi, K., Seo, J., ZnO nanostructures in active antibacterial food packaging: Preparation methods, antimicrobial mechanisms, safety issues, future prospects, and challenges. *Food Rev. Int.*, 38, 4, 537–565, 2022.
28. Gurunathan, T., Mohanty, S., Nayak, S.K., A review of the recent developments in biocomposites based on natural fibres and their application perspectives. *Compos. Part A Appl. Sci. Manuf.*, 77, 1–25, 2015.
29. Babu, R.P., O'Connor, K., Seeram, R., Current progress on bio-based polymers and their future trends. *Prog. Biomater.*, 2, 1–16, 2013.
30. Liu, T., Li, J., Tang, Q., Qiu, P., Gou, D., Zhao, J., Chitosan-based materials: an overview of potential applications in food packaging. *Foods*, 11, 10, 1490, 2022.
31. Darwesh, O.M., Sultan, Y.Y., Seif, M.M., Marrez, D.A., Bio-evaluation of crustacean and fungal nano-chitosan for applying as food ingredient. *Toxicol. Rep.*, 5, 348–356, 2018.
32. Yanat, M. and Schroën, K., Preparation methods and applications of chitosan nanoparticles; with an outlook toward reinforcement of biodegradable packaging. *React. Funct. Polym.*, 161, 104849, 2021.
33. Shariatinia, Z., Pharmaceutical applications of chitosan. *Adv. Colloid Interface Sci.*, 263, 131–194, 2019.
34. Tang, Y., Hu, X., Zhang, X., Guo, D., Zhang, J., Kong, F., Chitosan/titanium dioxide nanocomposite coatings: rheological behavior and surface application to cellulosic paper. *Carbohydr. Polym.*, 151, 752–759, 2016.
35. Singh, A., Mittal, A., Benjakul, S., Chitosan nanoparticles: preparation, food applications and health benefits. *Sci. Asia*, 47, 2021, 1–10, 2021.
36. Zhao, L.M., Shi, L.E., Zhang, Z.L., Chen, J.M., Shi, D.D., Yang, J., Tang, Z.X., Preparation and application of chitosan nanoparticles and nanofibers. *Braz. J. Chem. Eng.*, 28, 353–362, 2011.
37. Yildirim, S., Röcker, B., Pettersen, M.K., Nilsen Nygaard, J., Ayhan, Z., Rutkaite, R., Coma, V., Active packaging applications for food. *Compr. Rev. Food Sci. Food Saf.*, 17, 1, 165–199, 2018.
38. Karanth, S., Feng, S., Patra, D., Pradhan, A.K., Linking microbial contamination to food spoilage and food waste: the role of smart packaging, spoilage risk assessments, and date labeling. *Front. Microbiol.*, 14, 1198124, 2023.
39. Ashfaq, A., Khursheed, N., Fatima, S., Anjum, Z., Younis, K., Application of nanotechnology in food packaging: pros and cons. *J. Agric. Food Res.*, 7, 100270, 2022.

40. Awuchi, C.G. and Dendegh, T.A., Active, smart, intelligent, and improved packaging, in: *Application of nanotechnology in food science, processing and packaging*, pp. 189–202, Springer, Switzerland AG, 2022.
41. Richard, N.S., Edible coatings and the need for biodegradable polymers with focus on dairy products. *Online J. Anim. Feed Res*, 12, 4, 173–178, 2022.
42. Kumar, J., Akhila, K., Gaikwad, K.K., Recent developments in intelligent packaging systems for the food processing industry: a review. *J. Food Process Technol.*, 12, 895, 2021.
43. Sharma, C., Dhiman, R., Rokana, N., Panwar, H., Nanotechnology: an untapped resource for food packaging. *Front. Microbiol.*, 8, 1735, 2017.
44. Maghami, M., Motalebi, A.A., Anvar, S.A.A., Influence of chitosan nanoparticles and fennel essential oils (Foeniculum vulgare) on the shelf life of Huso huso fish fillets during storage. *Food Sci. Nutr.*, 7, 9, 3030–3041, 2019.
45. Taglieri, I., Macaluso, M., Bianchi, A., Sanmartin, C., Quartacci, M.F., Zinnai, A., Venturi, F., Overcoming bread quality decay concerns: Main issues for bread shelf life as a function of biological leavening agents and different extra ingredients used in formulation, a review. *J. Sci. Food Agric.*, 101, 5, 1732–1743, 2021.
46. Agunbiade, S.O., Ojezele, O.J., Barinemene, M., Storage environment/packaging materials impacting bread spoilage under ambient conditions: A comparative analysis. *Chem. Int.*, 6, 3, 131–6, 2020.
47. Park, I.K., Lee, Y.K., Kim, M.J., Kim, S.D., Effect of surface treatment with chitosan on shelf-life of baguette. *J. Chitin Chitosan*, 7, 4, 208–13, 2002.
48. Arumugam, K.N. and Zain, N.M., The effects of edible chitosan coating on prolonging the shelf life of green chilies (Capsicum annuum L.). *J. Trop. Resour. Sustain. Sci. (JTRSS)*, 10, 2, 1–5, 2022.
49. Sheng, X., Shu, D., Li, Y., Zhan, Z., Yuan, X., Liu, S., Zang, Y., Combined approach consisting of slightly acidic electrolyzed water and chitosan coating to improve the internal quality of eggs during storage. *J. Sci. Food Agric.*, 101, 6, 2355–2361, 2021.
50. Akarca, G., Istek, O., Tomar, O., The effect of resin coating on the quality characteristics of chicken eggs during storage. *J. Food Sci.*, 86, 4, 1243–1257, 2021.
51. Nair, M.S., Tomar, M., Punia, S., Kukula-Koch, W., Kumar, M., Enhancing the functionality of chitosan- and alginate-based active edible coatings/films for the preservation of fruits and vegetables: a review. *Int. J. Biol. Macromol.*, 164, 304–320, 2020.
52. Khatodiya, N. and Malik, M., Effects of edible coating on fresh-cut fruits. *J. Pharmacogn. Phytochem.*, 11, 1, 192–199, 2022.
53. Jafarzadeh, S., Nafchi, A.M., Salehabadi, A., Oladzad-Abbasabadi, N., Jafari, S.M., Application of bio-nanocomposite films and edible coatings for extending the shelf life of fresh fruits and vegetables. *Adv. Colloid Interface Sci.*, 291, 102405, 2021.

54. Ma, J., Zhou, Z., Li, K., Liu, L., Zhang, W., Zhang, H., Novel edible coating based on shellac and tannic acid for prolonging postharvest shelf life and improving overall quality of mango. *Food Chem.*, 354, 129510, 2021.
55. Roy, S., Mondal, A., Yadav, V., Sarkar, A., Banerjee, R., Sanpui, P., Jaiswal, A., Mechanistic insight into the antibacterial activity of chitosan exfoliated MoS2 nanosheets: membrane damage, metabolic inactivation, and oxidative stress. *ACS Appl. Bio. Mater.*, 2, 7, 2738–2755, 2019.
56. Cavallaro, G., Micciulla, S., Chiappisi, L., Lazzara, G., Chitosan-based smart hybrid materials: A physico-chemical perspective. *J. Mater. Chem. B*, 9, 3, 594–611, 2021.
57. Youn, S.K., Her, J.H., Kim, Y.J., Choi, J.S., Park, S.M., Ahn, D.H., Studies on the improvement of shelf-life in spicy beef meat using chitosan. *J. Korean Soc. Food Sci. Nutr.*, 33, 1, 207–11, 2004.
58. Lee, B.H., Wu, S.C., Shen, T.L., Hsu, Y.Y., Chen, C.H., Hsu, W.H., The applications of Lactobacillus plantarum-derived extracellular vesicles as a novel natural antibacterial agent for improving quality and safety in tuna fish. *Food Chem.*, 340, 128104, 2021.
59. Sagoo, S., Board, R., Roller, S., Chitosan inhibits growth of spoilage microorganisms in chilled pork products. *Food Microbiol.*, 19, 175–82, 2002.
60. Juneja, V.K., Thippareddi, H., Bari, L., Inatsu, Y., Kawamoto, S., Friedman, M., Chitosan protects cooked ground beef and turkey against Clostridium perfringens spores during chilling. *J. Food Sci.*, 71, 6, M236–40, 2006.
61. Stopforth, J. and Kudron, T., Sorbic Acid and Sorbates, in: *Antimicrobials in Food*, pp. 89–132, CRC Press, Tailor & Francis Group, London, 2020.
62. Olaniyan, O.T., Adetunji, C.O., Dare, A., Adeniyi, M.J., Ajayi, O.O., Application of nanochitosan and polymeric chitosan as antibacterial, antivirus and antifungal activities when incorporated into aquatic and animal-based food materials, in: *Next Generation Nanochitosan*, pp. 401–420, Elsevier, 2023.
63. Yaashikaa, P.R., Kamalesh, R., Kumar, P.S., Saravanan, A., Vijayasri, K., Rangasamy, G., Recent advances in edible coatings and their application in food packaging. *Food Res. Int.*, 173, 113366, 2023.
64. Baldelli, A., Ren, M., Liang, D.Y., Lai, S., Hartono, B., Sum, K., *et al.*, Sprayed microcapsules of minerals for fortified food. *J. Funct. Foods*, 101, 105401, 2023.
65. Tantala, J., Meethongchai, S., Suethong, W., Ratanasumawong, S., Rachtanapun, C., Mold-free shelf-life extension of fresh rice noodles by synergistic effects of chitosan and common food preservatives. *Food Control*, 133, 108597, 2022.
66. Hadidi, M., Jafarzadeh, S., Forough, M., Garavand, F., Alizadeh, S., Salehabadi, A., Jafari, S.M., Plant protein-based food packaging films; recent advances in fabrication, characterization, and applications. *Trends Food Sci. Technol.*, 120, 154–173, 2022.

67. Tang, T., Zhang, M., Law, C.L., Mujumdar, A.S., Novel strategies for controlling nitrite content in prepared dishes: Current status, potential benefits, limitations and future challenges. *Food Res. Int.*, 170, 112984, 2023.
68. Castro-Muñoz, R., Kharazmi, M.S., Jafari, S.M., Chitosan-based electrospun nanofibers for encapsulating food bioactive ingredients: A review. *Int. J. Biol. Macromol.*, 245, 125424, 2023.
69. Plati, F. and Paraskevopoulou, A., Micro-and nano-encapsulation as tools for essential oils advantages' exploitation in food applications: The case of oregano essential oil. *Food Bioprocess Technol.*, 15, 5, 949–977, 2022.
70. Wu, H., Tatiyaborworntham, N., Hajimohammadi, M., Decker, E.A., Richards, M.P., Undeland, I., Model systems for studying lipid oxidation associated with muscle foods: Methods, challenges, and prospects. *Crit. Rev. Food Sci. Nutr.*, 64, 1, 153–171, 2024.
71. Nikoo, M., Benjakul, S., Ahmadi Gavlighi, H., Protein hydrolysates derived from aquaculture and marine byproducts through autolytic hydrolysis. *Compr. Rev. Food Sci. Food Saf.*, 21, 6, 4872–4899, 2022.
72. Ahmed, S., Noor, A., Tariq, M., Zaidi, A., Functional improvement of synbiotic yogurt enriched with Lacticaseibacillus rhamnosus and aloe vera gel using the response surface method. *Food Prod. Process. Nutr.*, 5, 1, 1–19, 2023.
73. Das, A., Ringu, T., Ghosh, S., Pramanik, N., A comprehensive review on recent advances in preparation, physicochemical characterization, and bioengineering applications of biopolymers. *Polym. Bull.*, 80, 7, 7247–7312, 2023.
74. Li, Z. and Yu, F., Recent advances in lycopene for food preservation and shelf-life extension. *Foods*, 12, 16, 3121, 2023.
75. Kang, M., Ha, J.H., Lee, Y., Physicochemical properties, antioxidant activities and sensory characteristics of commercial grape vinegars during long-term storage. *Food Sci. Technol.*, 40, 909–916, 2020.
76. Yin, Y., Dang, Q., Liu, C., Yan, J., Cha, D., Yu, Z., Fan, B., Itaconic acid grafted carboxymethyl chitosan and its nanoparticles: preparation, characterization and evaluation. *Int. J. Biol. Macromol.*, 102, 10–18, 2017.
77. Rajitha, P., Gopinath, D., Biswas, R., Sabitha, M., Jayakumar, R., Chitosan nanoparticles in drug therapy of infectious and inflammatory diseases. *Expert Opin. Drug Deliv.*, 13, 8, 1177–1194, 2016.
78. Wong, C.Y., Al-Salami, H., Dass, C.R., Formulation and characterisation of insulin-loaded chitosan nanoparticles capable of inducing glucose uptake in skeletal muscle cells *in vitro*. *J. Drug Deliv. Sci. Technol.*, 57, 101738, 2020.
79. Sarangi, M.K., Padhi, S., Patel, L.D., Rath, G., Nanda, S.S., Yi, D.K., Tailoring of polymer and metal nanobiocomposites corroborated with smart food packaging systems—a Review. *Food Bioprocess Technol.*, 17, 4, 850–886, 2024.

80. Thakur, M.S. and Ragavan, K.V., Biosensors in food processing. *J. Food Sci. Technol.*, 50, 625–641, 2013.
81. Maan, A.M., Hofman, A.H., de Vos, W.M., Kamperman, M., Recent developments and practical feasibility of polymer-based antifouling coatings. *Adv. Funct. Mater.*, 30, 32, 2000936, 2020.

9

Application of Nanochitosan as Food Additive and Preservatives

Samuel Adeniyi Oyegbade[1]*, Gbotemi Jerry Oni[2], Andrew Alalade Kolawole[1], Tsado Priscilla Yetu[3], Oluwafemi Adebayo Oyewole[3], Charles Oluwaseun Adetunji[4], Eniola K. I. T.[5] and Mohammed Bello Yerima[6]

[1]*Department of Biological Sciences, College of Science and Technology, Covenant University, Ota, Nigeria*
[2]*Department of Molecular Biology and Bioinformatics, Africa Center of Excellence for Mycotoxins and Food Safety, Federal University of Technology, Minna, Nigeria*
[3]*Department of Microbiology, Federal University of Technology, Minna, Nigeria*
[4]*Department of Microbiology, Edo State University, Uzairue, Nigeria*
[5]*Department of Biological Sciences, Joseph Ayo Babalola University, Ikeji Arakeji, Osun State, Nigeria*
[6]*Department of Microbiology, Sokoto State University, Sokoto, Sokoto State, Nigeria*

Abstract
The rise in human population globally has created a pressing need to meet up with the ever-increasing food demand. More importantly, food quality and safety are paramount to human health. Unfortunately, the adverse impacts of food deterioration have contributed immensely to the food crisis, food insecurity, and health issues experienced in different parts of the world. This chapter extensively x-rays the applications, properties, and advantages of nanochitosan as suitable food additive and preservative source, as well as the role and limitations of traditional preservatives and approaches in mitigating the effects of food spoilage agents, and how nanochitosan addresses these limitations.

Keywords: Nanochitosan, food additive, food preservatives, agriculture, nanotechnology

**Corresponding author*: oyegbadesam@gmail.com

Charles Oluwaseun Adetunji, Maulin P. Shah, Yerima Mohammed Bello, Daniel Hefft, Jay Singh, Shyam S. Pandey and Ravindra Pratap Singh (eds.) Nanochitosan Applications for Enhanced Crop Production and Food Security, (181–216) © 2025 Scrivener Publishing LLC

9.1 Introduction

Chitin and chitosan are the major components of the exoskeletons of terrestrial insects like honeybees and silkworms, as well as the cell walls of fungi like yeast, molds, and ray fungi like Streptomyces [1, 2]. They are also found in the shells of crustaceans like squid, shrimp, lobster, and crab. A cationic biopolymer called chitosan is formed when chitin is partially or completely deacetylated. According to El-Naggar *et al.* [3], it is a linear polysaccharide made up of N-acetyl-D-glucosamine that has been deacetylated to 2-amino-2-deoxy-D-glucopyranose. Because of the degree of polymerization, molecular mass, deacetylation, viscosity, and acid dissociation constant of the co-polymers that make up chitosan [4], the term chitosan does not precisely refer to a single chemical. Rather, it refers to a group of copolymers. The chitosan obtained from microbial sources is thought to have promise because the underlying processes are easily modified to produce a pure and uniform product with desired particular features. It is a very adaptable biopolymer with diverse applications in the fields of agriculture, the food industry, health care, wastewater treatment, and cosmetics. ChNPs have a distinctive polymeric cationic character, absorption-enhancing properties, and a mucoadhesive nature; are biocompatible; and are biodegradable. Chitosan, a modified linear polysaccharide that permits ionic cross-linking of multivalent anions, is a crucial biopolymer for the production of nanoparticles (NPs). Chitosan has varied quantities of free amino groups in its polymeric chain and cationic characteristic [5]. *In vivo* site-specific targeting and greater affinity for the negatively charged biological membrane are characteristics of positively charged [5]. Lysozyme enzymes, which are found in chitosan, can break down the cell walls of organisms, reducing their capacity to spread illness. Gram-positive and gram-negative bacteria isolated from fisheries products are both susceptible to chitosan's ability to limit their growth [6]. In order to maximize the use of fishery products, chitosan made from leftover crustacean shell waste can be employed as a natural preservative in other fisheries products. By using the ionic gelation process, chitosan may be converted into nanochitosan, which has stronger antibacterial properties than regular chitosan [7].

9.2 Importance of Food Additives and Preservatives in the Food Industry

Foods have been preserved and enhanced with additives for ages. Chemicals such as butyl gallate and butylated hydroxyl anisole (BHA) are created when meats are smoked to preserve them, which have antibacterial and antioxidant properties. Throughout history, salt has been used as a preservative. In addition to preventing bacterial growth, the water activity of meats and other foods is decreased by salt, which also helps to preserve food. Many times, a high quantity of water in food might promote the development of bacteria and fungi. Lowering the pH of food through the process of pickling, which involves the addition of acids like vinegar, germs can be prevented from growing. According to Awuchi *et al.* [8], several spices and herbs, such curry, chili pepper, and cinnamon, contain antioxidants and may have bactericidal properties. Food additives are substances that are introduced into food to maintain flavor or enhance other qualities of the food, such as appearance or taste. The preservation of foods through pickling with vinegar, salting or brining, as with bacon, canning sweets, or employing sulfur dioxide, as in wines, are some examples of food additives that have been used for ages. More additives of both natural and synthetic origins were introduced with the widespread use of processed foods in the second half of the 20th century. Additionally, food additives include compounds that could be unintentionally added to food during production, packing, transit, or storage (sometimes referred to as indirect additives [8]. Any substance whose intended usage results in, or might reasonably be expected to result in, ending up as a constituent of, or otherwise affecting, the qualities of, any food is considered a food additive by the US Food and Drug Administration (FDA). Food additives, as defined by the WHO, are substances that are introduced into food to improve or preserve its flavor, texture, look, safety, or freshness. Some food additives have been used for food storage for hundreds of years, such as salt in meats like bacon or dried fish, sulfur dioxide in wine, or sugar in marmalade. In order to provide meals with a specific functional characteristic, food additives may also be used [9]. Numerous food additives have been created over time to meet the demands of the food production and food sectors because producing food on a large scale differs greatly from producing food on a smaller scale at home or for a family. In order to keep processed foods safe, in good shape, and fresh during their journey from manufacturing facilities or industries to retail establishments and warehouses, and finally to consumers who regularly seek out fresh, secure, healthful, and nutrient-dense foods, additives are required.

Food additives should only be used when they are necessary for technology, do not deceive customers, and fulfil a specific technological purpose, such as improving food stability or preserving nutritional value. Food additives may be made synthetically, from minerals, plants, or animals. Perfluoroalkyl compounds (PFCs), bisphenols, and phthalates are examples of indirect additives used in manufacturing or packaging. According to Awuchi *et al.* [8], the American Academy of Pediatrics advocated for greater research on food coloring, nitrates, phthalates, PFCs, and bisphenols in July 2018 because they may damage children's development. The national food rules must be put in place before a food additive can actually be used once the JECFA determines that it is safe to use at levels up to those allowed by the Codex General Standard for Food Additives. In the US, the governing bodies in charge of consumer protection, agriculture, and health tightly regulate the use of food additives. This is also true in other developed countries like Canada and Europe. Chemicals cannot be added to food in those countries unless regulators are persuaded that they are toxicologically safe and have a high margin of safety. In order to inform consumers about the ingredients in the foods they choose to eat, packaging for commercially produced foods must also indicate any chemicals that have been added. Due to the inherent nature of low-level, toxicological risk, especially those associated with diseases that may take a long time to develop, researchers and scientists have never been able to prove that trace concentrations of any chemical or additive are fully risk-free [8]. There is always some risk, no matter how small.

9.3 Transition to the Application of Nanochitosan in Food Preservation

The public's concern for food safety has spread around the world. Food can become contaminated and deteriorated by harmful microbes during the production, processing, transportation, storage, and sales processes [10]. This involves a number of microbial metabolism, enzyme action, and chemical reactions, as well as sensory quality degradation and toxicity accumulation, which seriously depletes food resources and poses a serious risk to human health [11]. The most pressing issue facing the food business is how to properly reduce food deterioration. However, research into alternative materials and methods of food preservation is essential to combating the existing safety issues due to concerns surrounding the safety of conventional chemical preservatives [12]. Chitosan (CS), one of the biopolymers, is a sustainable biological resource with a variety of sources and outstanding antibacterial, antioxidant, biodegradable, edible, non-toxic, and film-producing properties [13]. As a result, it is commonly used in the food, pharmaceutical, environmental protection, chemical, cosmetic, and other industries [14]. Because of its unique film-forming and preservation properties, it may be applied directly to food surfaces or used to package food, offering promising prospects for both food preservation and environmentally friendly packaging [15].

Recent research on the preservation of shrimp [16] and pork slices [17] has reported significant results with CS-based biofilms. According to the food species and storage conditions, the shelf-life of items like mackerel fillets, turbot fillets, strawberries [18], longans, and eggs can range from a few days to a few months. There are still obstacles in the way of practical implementation despite substantial study on CS preservation. There is currently little knowledge about the preservation mechanisms, which depend on the intricate nature of food matrix degradation, and most research only pay attention to changes in a few food evident indicators. The use of a high concentration of CS-induced coating, which would influence the sensory quality of food, is one example of the technical limitations connected with CS-induced food preservation that also have an impact on its widespread adoption. According to Stallard *et al.* [19], the mechanical characteristics, structural strength, tensile strength, oxygen permeability, and vapor transmission performance of typical packaging film materials currently available fall short of the standards for food packaging. As a result, derivatization, the creation of nanocomposites, and additional augmentation with biological extracts are required to improve its functionality as a food packaging medium. More scientists are now looking at this problem from all sides recently. By solution casting, layer-by-layer assembly, and extrusion technology, the blending film of CS and polysaccharide or protein was created, and its physical chemistry, function, and antimicrobial properties were investigated in order to make it suitable for use as food packaging materials [20, 21]. A substantial amount of recent study has been devoted to CS composite materials, which have a variety of natural features. Because of this, using CS as a substrate for food preservation is a popular issue right now. It is envisaged that further advancements in food preservation would be made possible by the ongoing addition of techniques and technologies like nanotechnology, microcapsule technology, and self-assembly technology. The relevant biological activities and recent advances in food preservation using chemical synthesis (CS) are the main themes of this review, along with potential applications for CS-based derivatization, nanocomposites, and extract composites in food packaging and the preservation of fruit, vegetable, meat, and fish products. According to Xiaoli Liu, Liao, and Xia [22], this work will probably give researchers the direction that they need to improve the superior qualities of CS in food preservation and broaden the use of this technology in this field.

9.4 Physicochemical Properties of Chitosan

Numerous D-glucosamine residues are joined by a -(1-4) linkage to form the structural components of chitin and chitosan. N-acetyl-D-glucoseamine units make up 15%–50% of chitosan, depending on how much chitin has been deacetylated. According to Singh *et al.* [23], chitin and chitosan have different biological roles because they both include amino groups and have nitrogen levels that range from 5% to 8%. The process of converting the acetamide groups into primary amino groups results in the N-deacetylated derivative of chitin known as chitosan. The presence of primary and secondary hydroxyl groups on each repeat unit, as well as the amine group on each deacetylated unit, makes chitosan more chemically reactive and water-soluble than chitin [24]. Because of the reactive groups in chitosan's structure, which are highly amenable to chemical alteration, it can have its physical and mechanical properties altered, making it a suitable material for a variety of uses. Below are some of its properties:

9.4.1 Solubility

Insoluble in neutral or alkaline solvents, chitosan can be dissolved in an acidic solution. Despite the fact that chitin is usually insoluble in solvents, deacetylating chitin produces soluble chitosan, which possesses primary amino groups and a pKa value of 6.5 [25]. Chitosan is made soluble by the protonation and positive charge of the amine that results from the dissolution of chitosan in acidic solutions. According to Zargar *et al.* [25], they lose their charge and become insoluble when the pH is 6 or above. The degree of deacetylation at each molecular weight, temperature, and polymer crystallinity all have an impact on chitosan's solubility in addition to pH [26].

9.4.2 Molecular Weight

The biological and physical characteristics of chitosan are significantly influenced by molecular weight. Depending on the parent material and how it is processed and extracted, chitosan's molecular weight might change [27]. Based on its spectrum of molecular weights, chitosan can be categorized as high, medium, or low molecular weight. For a number of industrial applications, chitosan is not desired because it gets more viscous and less soluble as the molecular weight rises. Low–molecular weight chitosan is recommended for usage in biological and commercial applications due to its higher solubility and stability [28].

9.4.3 Degree of Deacetylation

Chitosan's physicochemical characteristics, activity, and applications are also influenced by the degree of deacetylation. According to Matica *et al.* [29], the degree of deacetylation is determined by how amino groups are distributed along the polymer chain. Chitosan's amino group in the polymeric chain is what causes it to be cationic in acidic solutions. Therefore, the level of deacetylation has a significant impact on both solubility and viscosity. The amount of deacetylation indicates how many N-acetylglucosamine units are present in the chain as a whole.

9.4.4 Viscosity

The degree of deacetylation and the molecular weight of the chitosan are two important elements that affect the viscosity, which, in turn, affects the industrial usability of the substance. Alemu, Getachew, and Mondal [26] found that viscosity rose as the level of deacetylation rose and molecular weight fell. Additionally, it may be based on the chitosan's particle size and period of storage. When compared to a standard chitosan solution at the same concentration, nanochitosan has a viscosity that is roughly 30% lower. According to Chattopadhyay and Inamdar [30], viscosity is a good indicator of the stability of the polymer in solution as a reduction is seen during polymer storage due to polymer degradation [26] and nanocolloids dropped by 17% for the same storage time. Storage time also affects about a 10% drop in the viscosity of normal chitosan for a storage time of 24 h.

9.5 Mechanisms of Food Spoilage and Preservation

The most prevalent agent of food spoilage is microbes. Foods with relatively lower shelf-life are quite prone to attacks by microbes, which are also primarily responsible for food-borne diseases. Nonetheless, microbial growth can be obstructed via thorough food packaging, use of food preservatives, and by regulating key parameters such as moisture activity, pH, and temperature at the point of storing foods [31]. Microbes, which are food spoilage agents, are mainly classified into bacteria, yeasts, and moulds [32]. The inherent attributes of foods such as oxygen, substrates, and enzymes often influence the kind and extent of food deterioration caused by microorganisms. These attributes can be adjusted in the course of formulating food product in order to regulate its safety and quality [32]. Extrinsic factors responsible for food deterioration are activities of other microorganisms, temperature, and relative humidity, as well as temperature, presence, and activities of other microbes, whereas intrinsic factors are nutrient level, moisture activity, oxidation–reduction potential, and pH [32]. Diverse biochemical reactions often take place in foods, resulting in changes and unpleasant sensory attributes in food products. Change and decline in quality of fresh foods occurs as a result of sporadic growth and metabolism of microbes, toxic substances, and rancidification, which lead to changes of pH, color, flavor, and taste [32]. Chemical spoilage such as oxidation phenomena is connected to activity of microorganisms and also influenced by changes in temperature [32].

Rancidification implies oxidation, whereby lipids react with oxygen leading to changes in color, flavor, and build-up of toxic compounds. The presence of light and metal oxides are capable of speeding up the reaction rate of rancidification [32]. Series of anaerobic reactions known as putrefaction often lead to breakdown of amino acids into a mix of different compounds and organic acids. Bacteria play a crucial role in the biochemical process. This process is prevalent in foods that are particularly rich in proteins at temperature higher than 15°C, which promotes the activities of microorganisms [32]. Non-enzymatic browning causes food deterioration via by a reaction (known as Maillard), which occurs mainly in proteinous foods. Maillard reaction often results in unpleasant change in color, flavor, and lowered availability of nutrients and amino acids. This takes place in the course of dry milk storage, etc. Hydrolytic rancidity results when lipolytic enzymes degrade and break down lipids, with the release of free fatty acids. This phenomenon can be detected easily in butter, and the released volatile fatty acids have a rancid flavor, stiff malodor, and taste [32].

Several biological and chemical reactions bring about food deterioration, and attributes such as color, flavor, and nutritional value of foods are likely to reduce drastically as food deteriorates [32]. In order to minimize food spoilage as a result of microbes and toxins, diverse conventional approaches have been explored. These include freezing, drying, smoking, and chilling. In recent times, modern techniques such as irradiation are being employed in food preservation, in order to elongate its shelf life and ensure that the texture, flavor, taste, and food nutritive value are kept intact [32]. The merits of traditional food preservatives include extension of shelf life, maintenance of texture and sensory properties, and antimicrobial effect. However, several traditional food preservatives pose severe public health concerns due to their side effects. Substances such as benzoic acid and sorbic acid have been found to enhance carcinogenic and mutagenic compounds [33]. Meat preservatives and additives such as nitrate and nitrite have been discovered to be linked to health conditions such as cancer and leukemia [34, 35].

Traditional preservatives are produced in various forms and extracts including liquid and solid (powder form) via the application of essential oils, water, and organic solvents. Plant-derived traditional preservatives are applied with proper consideration to the kind of formulations applied to the food [36]. Plant-derived traditional preservatives have been produced from plants such as tumeric, grape seed, rosemary, and cranberry, among others. However, properties such as rapid release, low bioavailability, low solubility, as well as environmental conditions greatly limit the versatility and use and of traditional food preservatives. Encapsulation has been found to enhance the stability and versatility of these preservatives [37].

Conventional food preservation approaches are being widely adopted globally. However, recent advancements in science and technology is front lining the dynamic transition of existing technologies into new, improved and cutting edge innovative technological solutions [6].

Technologies such as heating, drying, canning, and freezing increase the shelf life of food substances by making microbes inactive, thereby contributing to reducing the extent of spoilage and deterioration of food and also minimize the outbreak of diseases associated with the food. Various traditional preservatives and processing techniques play significant roles in reducing the extent of food deterioration in food substances or products [32].

Drying is one of the most primitive food preservation techniques [32]. It entails eliminating water from food forms via evaporation. The process of drying leads to the formation of solid, dry food product with low moisture content. Because the presence of moisture is a major requirement for the activation of enzymes and microbes to facilitate food deterioration, drying reduces moisture content to level or limit that hinder the activation of these microbes [32]. Most microbes including bacteria become inactive when moisture activity is below 0.88, whereas the upper limit at 0.95 and above supports microbial growth [32]. Drying is cost effective, gives varying flavors, significantly lowers food volume and weight, and also enhances food packaging and storage. The major drawback is the remarkable loss of functional compounds, aroma, and flavor lost in the process of drying [32].

The technique of preservation, which entails heating up food up to a certain degree of temperature in order to eliminate microbes and enzymes that cause food spoilage, is known as pasteurization. This technique usually increases the food shelf-life because most of the microbes such as yeasts, mold, and bacteria responsible for food spoilage would have been destroyed. The effectiveness of this technique is greatly influenced by the blend between temperature and time [32]. The primary limitation of pasteurization process is the destruction of functional minerals and beneficial microbes at pasteurization temperature. The reduction in vitamin C and phosphorus has been found to be 20% and 5%, respectively, during pasteurization [32].

The changes in the physicochemical properties of chitosan NPs as a result of introducing hydrophobic groups helps in overcoming the limitations of traditional preservatives and lowered solubility at physiological pH, which is a significant drawback of chitosan. The application of functional nanocapsules as coatings finds relevance in the food sector, and it is expected that the physicochemical modification of chitosan will be an attractive and interesting research area, which will facilitate the development of chitosan applications for food and also enhance bioavailability in functional foods [38]. Chitosan-based NPs are versatile and have wide applications in various fields and aspects especially in its use as film coating materials for the preservation of food quality as a result of its antimicrobial and biodegradable potential. However, there is need to explore the quality and safety aspects for large scale establishment of nanochitosan-based products [39].

The conversion and processing of raw materials of biomass origin into valuable substances finds expression in the synthesis and adoption of chitosan-based NPs as preservative and additive in various food materials [40]. Nanochitosan, which is considered as chitosan NPs, has a greater antibacterial ability than chitosan. Nanochitosan is synthesized from chitosan via ionic gelation method. Nanochitosan exists in the morph of a sphere alongside functional groups such as amine, hydroxyl, and carboxyl groups found in chitosan. Nanochitosan can be used as a natural preservative because it has antibacterial properties [6].

9.6 Application of Nanochitosan as Food Additives

Food additives have the sole aim of maintaining and enhancing the nutritional parameters of food, improve consumer acceptability, minimize wastage, and foster availability of food [32]. Encapsulation technology has been widely and successfully deployed to address the issues of instability of compounds such as enzymes, essential oils, polyphenols, and vitamins, with nanochitosan acting as one of the primary agents of encapsulation. Encapsulation of nanochitosan with clove essential oil enhanced *in vitro* antifungal activity and reduced degradation of volatile compound degradation [41].

Bioactive components from chitosan NPs and extracts of clove buds were composited to eradicate pathogens of bacteria in food substances while preserving the sensory properties of stored shrimps. Chitosan NPs and clove bud extracts could serve as composites to preserve shrimps in the course of cold storage. The actions of this composite could be further explored in diverse food items as well as their health implications [42]. Chitosan is adopted as matrix-forming polymer for the encapsulation of active ingredients and as a substance for coating surface as a result of its biodegradability, non-toxicity, and biocompatibility. Chitosan-coated nanocarriers (CCNs), which are polymeric NPs mainly liposome and alginate, offer ample merits which include improved physicochemical properties. Nano-based chitosan have been utilized as a coating on alginate beads and liposome. It has also been adopted as a delivery system for bioactive ingredients such as essential oils, flavors, vitamins, antioxidants. Chitosan-based encapsulation is regarded as a workable approach in the promotion of functional foods, putting into consideration, safety regulations, and proceedings. Significant efforts are being geared toward preparation of chitosan-based micro- and nano-structures preparation, into adapting the properties of the polymeric constructs in order to fulfill the requirement of regulations [43]. The enhanced physicochemical properties of chitosan have facilitated its wide acceptability and utilization in the dairy industry, to treat dairy waste water, to extend the shelf-life of dairy products, and to immobilize and encapsulate several bioactive ingredients and enzymes. The derivation of nanochitosan derived from chitosan offers immense benefits to dairy product formulation via nanoencapsulation as well as safety and security of milk nutrients. The application of nanochitosan further necessitates the adoption of nanotechnology for nanofood dairy production and processing [44].

9.6.1 Nanochitosan as a Preservative Agent

The superior antimicrobial properties of nanochitosan have been found to be potent against a wide range of pathogenic microbes. Nanochitosan expressed antimicrobial activity on biofilms of *C. albicans and S. mutans* by reducing the rate of survival of the cells of the microbes [45]. Chitosan synthesized by isolating chitin obtained from scales of marine fishes via deacetylation reaction, on further modification of the synthesized chitosan led to nanochitosan production, which was confirmed to have higher antimicrobial natural preservative properties than chitosan [46]. Nanochitosan was found to impede bacteria growth in the course of storing smoked Skipjack Tuna, in food security with respect to fishery and aquaculture sector [40]. Chitosan NPs exhibited protective effects against oxidation of lipids as well as greater antimicrobial activity against microbial contamination of *C. arel* fillets in the course of super-chilling storage. Nanochitosan has huge potential to minimize the microbial growth. Nanochitosan can be utilized as a safe fish preservative for fish [47]. Nanochitosan and chitosan have been found to be useful in preserving fish quality parameters. The result revealed that nanochitosan was more effective than chitosan in minimizing oxidation of lipid and microbial contamination. Hence, antimicrobial activity of nanochitosan is greater than chitosan as a result of its charge density and greater surface per unit volume, which enables better interaction with the membrane of the bacteria cell [48].

The increasing attention and interest in attaining environmental friendliness and safety and heightened the need to create natural and organic-based polymers, solutions, or alternatives to chemical and artificial polymers such as biopolymer in recent times. Biopolymers such as starch, cellulose, and chitosan are obtained from living substances or renewable sources and are widely known for its biodegradability and non-toxic properties and macromolecular structure. Much attention was been drawn toward the application of chitosan various industries and disciplines such as nanotechnology, microbiology, industrial biotechnology, and food industries [49–51]. Despite the wide range of applications of natural substances and compounds, its use is often impeded by the sensory properties of the products. The greater surface area exposed and the small size, facilitating greater accessibility in food, is raising attention and interest in applying NPs in fish and meat products. Encapsulating these compounds will mitigate their sensory properties while permitting continuous and controlled release. Aspects relating to designing smart and intelligent food packaging as well as physicochemical attributes need to be reached further and improved upon, in order to enhance fish and meat product packaging [52]. Chitosan derivatives, like nanochitosan, can be used in a variety of ways to preserve food, extending its shelf life and preventing spoilage by preventing the growth of microorganisms and producing films that are both biocompatible and selectively permeable [53–55]. A number of scientists have investigated the effects of biofilms and chitosan NPs in food packaging in recent years. The reaction between positively charged amino groups and polyanions (cross-linkers) during the ionic gelation process [56]. The mechanical and barrier qualities of hydroxypropyl methylcellulose were enhanced by the addition of a chitosan-based nanocomposite. Antimicrobial activities of chitosan/gold nanocomposite were also found to express antimicrobial potential against *E. coli*, *S. aureus*, *P. aeruginosa*, and *Aspergillus niger* [57]. Chitosan NPs expressed higher antimicrobial effect at reduced concentrations against *Pseudomonas* sp. and *Bacillus* sp. [58].

The effects of using nanochitosan alone and in combination with crude onion extracts as materials for coating chicken fillet meat during refrigeration storage were investigated. The results showed that treating chicken breast fillets with nanochitosan coating could significantly and effectively reduce and inhibit mold growth compared to extract of crude onion because of the enhanced antimicrobial potential, charge density, and greater surface area/volume of nanochitosan, which, in turn, offer anionic cell membrane interaction [59]. The qualitative attributes of chicken fillet meat samples can be preserved by using nanochitosan, which significantly reduced the fungus count of the fillets compared to using crude onion extract. When combined with crude onion extract, nanochitosan can be used as an efficient natural preservative and flavoring to reduce food deterioration [59]. Consumers use sensory attributes including texture, flavor, aroma, taste, and sight to assess the quality of food. Food products are frequently compared to a reference food item before being purchased [60]. For several months, the physicochemical and sensory characteristics of nanochitosan in powder form were investigated. The addition of chitosan did not significantly alter the levels of fat or water; however, as the concentration of chitosan added rose, the amount of ash increased while the amount of protein decreased. According to Kim *et al.* [61], the sensory analysis result showed a modest rise in butyric off-flavor and bitter taste. Additionally, food items are classed into three classifications based on their shelf lives: perishable, non-perishable, and semi-perishable. This is where nanochitosan plays a major part. Semi-perishable foods have a shelf life of roughly 6 months, whereas perishable goods have a short shelf life of a few weeks. On the other hand, non-perishable foods have an endless shelf life that can last for many years [32]. The use of chitosan derivatives has extended the shelf life of a number of food products. The unique capacity of chitosan NPs to control the movement of various bioactive substances enhances the antibacterial and antioxidant properties of materials used in food packaging. Food coatings made of nanochitosan-based films are now widely used on a variety of foods, including cheese and meat. Lecithin and sodium caseinate were used to create food-grade stable NPs, and they were then used as the foundational systems for the synthesis of nanocapsules [62]. The shelf-life of white leg shrimp was increased to roughly 10 days by coating them with chitosan NPs [63]. Lipid oxidation is reduced when orange peel extract and nanochitosan-coated silver carp are combined [64]. In addition to prolonging the food's shelf life, nanochitosan improved the physicochemical and mechanical properties of distinct food products.

It was investigated how the powdered form of chitosan and nanochitosan affected the microbiological quality and shelf life of beef when it was refrigerated for 9 days. Compared to chitosan, nanochitosan showed lower inhibitory concentrations, suggesting a higher antibacterial activity against S. aureus and E. coli. Lower mesophilic and E. coli levels were seen when minced beef samples were combined with chitosan NPs. The chitosan NPs exhibited higher antioxidant properties and pH values in comparison to the powdered chitosan-treated samples. Both materials have outstanding antibacterial and antioxidant properties, which make them viable options for preserving minced meat and extending its shelf life [65].

The creation of tara gum films involved the use of both bulk and nanochitosan. When chitosan NPs and films were mixed, the results showed improved mechanical strength, water solubility, and antibacterial activity [66]. Due to their inadequate preservation capabilities and low mechanical qualities, gelatin films have had very little progress in their development as food packaging materials. Fish gelatin films that were created by filling them with chitosan NPs had better mechanical qualities and a 50% higher water vapor barrier. Bio-composite films can be made more bioactive by incorporating nanochitosan into them. By lowering heme iron transformation, chitosan NPs prevented fish myoglobin oxidation [67]. According to Solval *et al.* [67], the incorporation of chitosan NPs in glazing materials has the potential to reduce lipid oxidation while maintaining the texture and sensory qualities of frozen shrimp throughout storage. Chitosan NPs were added to shrimp diet to protect against the white spot syndrome virus and to act as an encapsulating agent for DNA. In order to incorporate additional additives, such as those found in shark liver oil when mixed with calcium alginate beads and tuna oil droplets, chitosan NPs were employed for encapsulation [62]. Two unique characteristics that may open up new possibilities and uses for chitosan NPs are their quantum size and high surface-to-volume ratio. The production procedure of nanochitosan is straightforward, reasonably inexpensive, and capable of being scaled up [62]. Because of chitosan's reactivity, vast surface area, and small size, the hazards associated with using it on a broad scale in the real world are still not fully understood, despite its safe use in laboratories.

9.7 Safety and Regulatory Considerations for Nanochitosan

Nanomaterials (NM) have unique functional properties that are primarily impacted by size because they have a substantially higher surface-to-mass ratio than larger-sized bulk materials. Nanomaterials are used in a variety of sectors, involving but not limited to the nutrition and agro-allied industries, and a rising amount of goods containing NM are already available on the market [68]. The primary motivations for using nanosized components as well as additives in food and food-processed products are to enhance nutrient absorption and bioavailability; increase the preservation of fresh foods; improve the persistency, texture, and durability of foods; and improve food production processes [69]. Future developments in NPs may have a significant effect on the food industry, potentially benefiting both businesses and consumers, albeit any potential health risks must be taken into account when evaluating the precise qualities and characteristics of NMs [70]. Also, possible health implications of exposure to NPs remain poorly understood, despite the research that has been conducted. There are less understanding in critical parts, which are crucial for forecasting possible health risks, including channels of exposure, the process of NP absorption into the body, the factors that control NP allocation once they are inside the body (i.e., toxico-kinetics), and the interactions between NPs and the biological systems (i.e., toxico-dynamics) [71]. Although there is mounting evidence in the scientific literature that some NPs may have harmful effects, there are only a few numbers of NPs for which complete hazard data are available [71]. It is likely that new methods that do not rely on conventional toxicity testing will need to be developed in order to assess the potential dangers of NPs. This is because NPs come in a wide variety and it has been observed that diverse nanoforms with the same chemical makeup can have diverse toxicological properties [72]. In addition to the many benefits that nanotechnology has for the food sector, there are safety concerns with NMs that must be taken into consideration [73]. Numerous researchers have emphasized the potential for NPs to migrate from packaging materials into food and their effects on consumer health while discussing safety issues related to NMs [74].

Even though NMs are appealing because of their distinctive qualities and the ability to perform specific cosmetic tasks, but they may also pose a risk to consumer health. Insoluble NP health risks are, indeed, a hotly disputed topic in scientific literature, mostly because of inconsistent findings and a dearth of long-term toxicological investigations [75]. The exposure path is of utmost importance. In terms of cosmetics, skin is the primary route of exposure; however, it is yet unknown whether or not NPs might pass past the stratum corneum and enter the essential layers. Even while studies in the literature generally point to limited skin absorption, this may rise in the case of skin injury [75]. Additionally, as exposure by inhalation is a possibility, particular consideration should also be given to sprays or aerosols that contain NMs. On the other hand, ingesting cosmetics that are applied to the mouth region (such as lipsticks) or by an unintentional transfer from hand to mouth can happen [76]. Because the EU has the most stringent regulations, it is the primary source of evidence for the safety issues about NPs. Any insoluble or bio-persistent substance is referred to as NMs in this context. This phrase may also apply to substances that are both water-insoluble and biodegradable, particularly when referring to lipid-based substances (such as vesicles, - solid lipid nanoparticles (SLN), and nanostructured lipid carriers (NLC)) [77]. However, these systems are able to fuse with cell membranes and are consequently regarded as soluble materials because of their chemical makeup and similarity to physiological lipids found in the stratum corneum. These insoluble and biopersistent compounds, for which there are more toxicological risks, are specially regulated by the law [75].

Majority of countries lack clear rules on marketing and usage of industrial products developed from nanotechnology; European policy is one of many that have attempted to provide clear and robust regulations [78]. The same laws that were taken into consideration for conventional food products were also taken into consideration for a wide range of nano-foods. Carbon black, titanium dioxide, and silicon dioxide have been approved for use in food packaging with plastic bags under current EU regulations [79]. While ZnO, nanoclay, and aluminum have been granted GRAS status by the US FDA, this does not apply to all of the materials that have already been used in food packaging. Some compounds on the GRAS list, like ZnO, have undergone testing and received approval for particle sizes in the micron range or larger [80]. However, by shrinking to NP size, these chemicals may generate novel toxicological behaviors that need to be examined and addressed [81].

"Nanotechnologies in the Food and Agriculture Sectors: Potential Food Safety Implications" was organized in 2009 by the WHO, FAO, and UN [82]. All participating nations were followed up with by the FAO and the WHO in 2012 regarding the status of their action plans and the management of nano-safety in relation to risk assessment and food safety regulations [83]. Rules and regulations have been introduced, many of which have specific provisions targeting NPs. However, not all food categories have unique regulations; it is, therefore, challenging for the food business and the private sector to have clear information on the applicable legal framework [84]. Despite these limited rules and regulations, the discussion of nanoregulation in European institutions has over time grown, ultimately leading to the creation of a soft EU legislation regarding the framework for nanoregulation [85]. It should be made clear that EU activities like messages, resolutions, suggestions, opinions, and guidelines are known as "soft legislation" because, in contrast to directives, regulations, and decisions (hard law), the individuals to whom they are aimed do not have to follow them. Soft laws are commonly referred to as private laws because they are created using a stakeholder approach, which believes that the state should collaborate with members of the private sector and civil society in participatory decision-making processes [86]. Literature and past research on the regulation of nano-foods point out that new regulations must be made quickly because the current regulatory systems cannot keep up with the pace of the market [87].

In the field of nanosafety, the fundamental tenet of toxicology, which holds that the dose causes the toxin, is still not widely accepted [88]. To adequately compare *in vitro* results with previously reported *in vitro* data and *in vivo* biological responses, it is required (but currently missing) to choose realistic exposure concentrations and physiologically pertinent dosage evaluations [89]. Natural materials (NMs) require a careful adaptation of traditional dose-metrics because administered mass alone is frequently insufficient to represent their property, in contrast to conventional materials, whose toxic doses can only be defined by administered mass or concentration. Dose-response relationship is dependent on the toxicity of the substance [90]. Prior studies on *in vitro* nanotoxicity provided dosages in mass units (l g/mL), ignoring effects linked to surfaces or numbers [91]. The demand for NMs to move beyond mass-only metrics has led to the suggestion of a number of dose-metrics, each with certain drawbacks, including body load, particle volume, surface area, and number of particles. In the absence of widely accepted dose metrics that may precisely indicate NM exposure, reporting concentrations in a range of dosage metrics will allow for a number of exposure interpretations. To achieve precise dosimetry and deliver particle dosages to cells, extra care should be taken during the synthesis and characterization of NM dispersion [89].

Traditional toxicity assessment generally depends on expensive, cumbersome, and unethical animal testing. The swift production of new materials and the quick development of existing technologies (such as biotechnologies) have increased the demand for quicker and less expensive test methods for non-animal regulatory purposes [92]. "Non-animal testing" in the context of hazard assessment refers to the use of human cells and tissues (*in vitro*) and computer modeling as an alternative to *in vivo* animal testing (*in silico*) [93]. Around the world, *in vitro* methods are used to identify NMs' negative effects, such as cytotoxicity. Through the use of *in silico* approaches, it is feasible to extract fresh information from the hazard data that may be applied to "engineer in" safety for novel materials without compromising desired functionality [93]. It is obvious that the majority of non-testing methods are data-driven, reliant on experimental data for direction, and presently unable to fully replace animal testing in toxicology. These methods can, however, make the best use of experimental data, which is frequently expensive and scarce, the identification of potentially problematic testing materials, the filling in of data gaps, and the reduction of animal testing by removing irrelevant experimental methods. A recent study summarizes the present status of research on the application of *in silico* methodologies and the difficulties that still need to be resolved [94]. The increasing demand for NMs across numerous industrial sectors has also brought attention to worries regarding safety issues and detrimental impacts on the health of humans. A better understanding of these properties depends on knowing the mechanisms of action and nanotoxicity of NPs. To get beyond NP constraints, the primary mechanisms of nanotoxicity have recently undergone extensive study. Reactive oxygen species and oxidative stress generation are two of the key mechanisms of nanotoxicity [95]. These chemically active free radical species also cause programmed cell death, DNA damage, and altered cellular signaling, which all lead to an imbalance in downstream processes. Numerous studies have demonstrated that NMs can cause cellular death, are geno-toxic, and are inflammatory. NMs are highly studied for their ability to cause cancer, which affects cells and tissues and makes it more difficult for them to function properly. Safer NPs can be created, and the negative effects of the NPs can be anticipated, by understanding the mechanism of nanotoxicity. The next sections go into great detail on how we now understand NP-mediated toxicity [96].

9.8 Case Studies and Practical Applications of Nanochitosan

A popular conservation method to avoid spoilage of food is presently driving up demand for biodegradable packaging. An experiment was conducted Homayounpour, Jalali, Shariatifar, Amanlou, and Khanjari [97] to assess the effects of a novel environmentally favorably coating containing unconfined and nanoencapsulated Cuminum Cyminum L. aqueous extract (CCE) on sardine fish standard features. With the use of scanning electron microscopy (SEM), zeta potential, extract (E), nano-extract (NE), and nanoliposome addition, the microstructural alterations were assessed. Fourier-transform infrared spectroscopy and measurements of the mean size of nanoliposomes were also used. The peroxide value (PV), total volatile basic nitrogen (TVB-N), pH, and thiobarbituric acid (TBA) all fell significantly after cumin ingestion (p 0.05). Lactic acid bacteria and total viable counts (TVC) have been demonstrated to be inhibited by cumin extract coatings, particularly those in the form of nanoliposomes (LAB). Due to the regulated release of the chemical mixture on the sardine's surface made possible by the use of nanotechnology to encapsulate CCE, the coating's antioxidant action lasted longer than that of other groups (nanochitosan, nanochitosan-E, and control groups). Therefore, effective active packaging can be achieved using a combination of CCE encapsulated nanoliposomes and nanochitosan (NCh) An experiment was carried out to produce a novel degradable coating with free and nano-encapsulated Cumino cyminum L. essential oil (CCEO) with nanochitosan (Nch) base, which was then tested on samples of sardine fillets over a 16-day period at 4°C. To create a novel degradable coating, an experiment was conducted using free and nano-encapsulated CCEO on a basis of nanochitosan (Nch), which was then tested on samples of sardine fillets over a 16-day period at 4°C. As demonstrated by reduced pH, PV, and reactive thiobarbituric acid compounds, the results demonstrated that coating treatments can effectively inhibit microbiological development and chemical deterioration (P 0.05). The samples treated with nanocitosan/nano essential oil (NEO) also shown the lowest value among other treatments during the trial period, followed by Nch-EO, Nch, and control, according to the results of chemical and microbiological characteristics. Additionally, it was discovered that encapsulating CCEO slows down diffusion, enhancing antibacterial and antioxidant activities as well as sensory qualities. Based on the study's findings, sardine's shelf life can be increased by using coatings made of CCEO-encapsulated nanochitosan.

Another study on the results of coating nanochithosan (NC) with different amounts of Zataria multiflora Boiss and Polylophium involucratum essential oil (PIEO) was conducted by Mohammadkhan *et al.* [98]. Within 12 days of silver carp fillets being stored in the refrigerator, the microbiological, chemical, and sensory features of essential oil (ZMEO) were examined. The results showed that the aerobic plate count (APC) for the control and samples coated with pure NC, respectively, was performed after days 4 and 6, exceeded 7 log CFU/g. The samples coated with NC that included ZMEO and PIEO at concentrations of 0.6% and 0.6%, respectively, had the lowest microbial count. After 8 days, a control sample of NC containing 0.6% ZMEO and 0.6% PIEO had a total volatile base of nitrogen (TVB-N) of 33.15 mg/100 g, but for the coated samples of NC containing 0.6% ZMEO and 0.6% PIEO, this value remained below 25 mg/100 g. In general, integrating the ZMEO and PIEO had little to no impact on the sensory properties of coated materials when compared to those of controls. According to the study's findings, fish flesh products could be packaged in unique ways using NC coatings containing ZMEO and PIEO without having their organoleptic qualities suffer.

The use of chitosan in edible films and coatings to lengthen the shelf life of seafood has been well studied in the literature [99]. Essential oils and chitosan have frequently been combined in applications [100, 101]. For illustration, by mixing chitosan with oregano essential oil, red porgy held under ice was given a 8- to 9-day shelf-life extension [102]. The case studies mentioned above exemplify the practical applications of biodegradable coatings with various natural extracts and essential oils for extending the preservation of seafood, specifically sardine fish and silver carp fillets. These coatings are developed using nanochitosan and nanotechnology to encapsulate the active ingredients, providing enhanced antioxidant and antimicrobial attributes.

Chitosan has been largely used in the food and processing business, particularly in the manufacturing of dairy products. In the dairy processing industries, cheese production, particularly cottage cheese production, is rising daily [103]. It is possible to utilize chitosan to enhance the shelf life of mozzarella cheese, which is highly perishable. Mozzarella cheese has a typical shelf life of 4–5 days. The starter culture's use of chitosan/lactic acid solution prevents the growth of spoilage microorganisms for up to 10 days while it is stored in the refrigerator [104]. With less yeast and mold growth, the mycological quality and shelf life of Kareish cheese treated with various doses of chitosan can be enhanced. Thus, it can be said that adding chitosan to Kareish cheese has improved the mycological quality and extended preservation period of the cheese [105]. Because it contains a variety of healthy, live microorganisms that fight with pathogenic bacteria for resources, yogurt is a well-known fermented milk product around the world with a number of health benefits. The addition of diverse culinary ingredients has improved yogurt's nutritional value. Chitosan in yoghurt can be fortified to increase its nutritional value. According to Ausar, Passalacqua, Castagna, Bianco, and Beltramo [106], the growth of three representative milk fermentative bacteria is influenced by the molecular weight and concentration of chitosan (Propionibacterium freudenreichii, Streptococcus thermophilus, and Lactobacillus delbrueckii ssp. bulgaricus). However, the inhibitory effects of chitosan were reversed when it was incubated with milk before coming into contact with bacteria. Thus, the presence of milk fat or casein micelles had an impact on the fermentation process of milk. Due to its poor solubility, chitosan's application in dairy products is restricted. In this regard, using nanosized chitosan may aid in improving solubility and bioavailability. Nano-powdered chitosan (NPC), a different type of chitosan, has the possibility to extend the preservation of yogurt because it can be utilized to make yoghurt with less cholesterol and lower lactic acid bacteria counts when refrigerated at 4°C for 20 days. Additionally, NPC concentrations (0.3%–0.5%) can be used to manufacture yogurt with lower cholesterol without adversely impacting the sensory qualities [107]. Ice cream is another frozen dairy product where chitosan can be effectively used. Chitosan improves the viscosity and over run of ice cream, making it feasible to replace gelatin to the extent of 60% in ice cream [108].

9.9 Future Prospects and Challenges

In the past 10 years, nanotechnology has been used as a cutting-edge method for creating nanoscaled organic and inorganic molecules with special features owing to their small size [109]. NMs, sometimes known as fillers, are known to enhance the physical-mechanical properties of polymers used as film-forming matrices for food packaging. The use of antimicrobial agents has been the subject of in-depth study and review, particularly antimicrobial nanostructures, in food packaging [110]. The main health issues stem from the potential movement of active substances from food contact materials toward food matrix. It is essential to consider all edible packing materials as an integral part of the food product, provided that doing so would not put your health at risk [110]. To incorporate antimicrobial substances into food-grade polymer matrices to create antimicrobial edible films, both nanoemulsions containing essential oils and nanoencapsulated antimicrobial peptides have been employed. Nanocapsules containing epigallocatechin gallate have been introduced to chitosan-based films to provide them additional antioxidant properties. This is demonstrated by the fact that they exhibit higher DPPH radical scavenging activity than pure chitosan [13]. Nanoencapsulation is advantageous because adding this green tea polyphenol directly into food systems could interact with other food ingredients and cause oxidative deterioration [111]. This active ingredient's hydrophilic nature would also preclude its direct addition to fatty foods. The addition of epigallocatechin gallate-loaded nanocapsules to chitosan-based films has also given them antioxidant characteristics, as evidenced by their improved DPPH radical scavenging activity in comparison to pure chitosan film [99]. Because of the potential for interactions with other dietary components and oxidative degradation, direct absorption of this green tea polyphenol into food systems is undesirable [112].

Additionally, this active ingredient's hydrophilic nature would prevent its direct addition to fatty foods [112].

Although NPs of chitosan have been labeled as fillers, several authors have reported that they have antibacterial potential against pathogenic microbes and food deterioration [113]. Many researchers have demonstrated that nano-structuring chitosan is likely to result in a reduction in its antibacterial potency [114]. This dual-action method has also been used to create essential oil nano-emulsions, which exhibit antioxidant and antibacterial characteristics in addition to changing the properties of edible films (particularly mechanical and barrier to moisture) [115]. In some cases, applying a chitosan coating or film to the food's surface during the preservation process can help the product withstand microbial attack, postpone spoilage, and so lengthen its shelf life. For instance, Wei *et al.* [116] investigated how varying chitosan concentrations affected the color and microbial communities on the surface of frozen duck skin [117]. The findings demonstrated that chitosan, at a concentration of 3%, could compete with chemicals that present color for oxygen and could significantly lower the amount of surface duck skin colonies during the dyeing process. The development of chitosan biofilm preservation technology has been greatly aided by the growing demand for food preservation and the standardization of the production of chitosan-based preservation films. The broad use of traditional coating and film-making techniques for food preservation has produced effective results in terms of operation. Biofilms made from chitosan are frequently created by co-blending with natural plant extracts, which are gaining popularity in the food industry due to their high nutritional value and environmental safety [117]. The addition of active substances increased the films' final mechanical qualities as well as their application and inherent bioactivity. Frozen meat's shelf life can be extended by adding spice flavor to chitosan films. Chitosan might interact intermolecularly with the polyphenols in the spices, and it may also lead to covalent connections forming between water and the hydrogen molecules of the polyphenols, boosting the hydrophilicity of the composite films. The findings showed that chitosan could interact intermolecularly with the polyphenols in the spices and that it may also cause covalent bonds to form between water and the polyphenols' hydrogen molecules, increasing the composite films' hydrophilicity [118].

The precise legislation around NMs use is still in flux, despite the fact that their use in the food industry is growing. Kuswandi [119] asserts that, in order to ensure the safe application or the use of nanotechnology's best practices in the food sector, there should be at least three regulations, namely those governing food, environment, and health [120]. In 2014, the

US FDA published a set of guidelines on nanotechnology that mandated manufacturers conduct an in-depth toxicological analysis of the NMs used in food-related products. Therefore, the sale of items containing unvalidated NPs is prohibited. For instance, the US Environmental Protection Agency (EPA) pulled food containers containing nanosilver out of circulation in 2014 due to the lack of testing or EPA registration for the products. In compliance with the European Union Regulation No. 10/2011, NPs used in food contact materials must also be approved and examined for any potential issues. Therefore, it is not acceptable to sell unapproved NP-containing products. Because the goods had not undergone testing or been registered with the EPA, the US EPA, for example, pulled food containers containing nanosilver from circulation in 2014. Additionally, in compliance with EU Regulation No. 10/2011, NPs used in food contact materials must be authorized and examined for any potential issues [121]. Numerous concerns about the security of products utilizing nanotechnology still exist. For instance, reports on the potential health impacts of nano-Ag and nano-TiO_2 in food packaging [122]. However, it is still unclear how more nano-Ag use would affect the environment and human health. The general public's impression of food products made using nanotechnology is that these products contain nano-inorganic components that may someday enter the human body [123].

Three categories of food safety issues connected to nanotechnology goods are listed by Capon *et al.* [124]: Food products or food packaging that contains naturally occurring NPs or easily digestible nanostructures and do not bio-accumulate (take up space in the body or the environment) is considered to be low risk. Food products or food packaging containing food additives or active compounds covered with a nano delivery technology are regarded as moderate risk. The environment or consumers are very likely to be exposed to NPs in this high-risk group whose hazardous qualities are not yet well understood [123]. Therefore, when assessing the health effects of a novel food contact element, it is vital to keep in mind that toxicity data must be balanced against the ease with which the drug may be released from packing materials into a range of meals. Less food waste produced by food packaged with nanopackaging is an important feature of customer behavior [122].

Zhang *et al.* [96] analysis of the relationships between food waste and shelf life for packaged goods and nanopackaging. The responses from 628 people showed that nanopackaging extends food products' shelf lives by 1 to 3 years, which lowers the carbon footprint of food preservation. Consequently, more than just an environmental evaluation will be needed for a sustainable route to a productive shell bio-economy.

Industry-standard processing technologies that make use of frameworks for monetizing, low-carbon procurement tools, and decision support tools must be assessed for sustainability. A total of 174 streams were found to be undervalued via economic analysis.

The majority of chitosan research has been on chemical and physical alterations so as to get around its limitations in mechanical strength, increase its solubility, and broaden its applications. Chitosan should be handled carefully because some alterations made to it can lead to derivatives that are somewhat harmful to cells [125]. More research is required on the toxicity and therapeutic effects of chitosan and its derivatives in both animal models and people. Chitosan and its derivatives create difficulties when employed in food applications because of their molecular structure and interactions with the food matrix [126]. For instance, the amount of fat may affect how effective these biopolymers are. Despite all of chitosan's advantages, its films exhibit poor mechanical qualities and provide only a marginal barrier to gases and water vapor in food packaging applications. A further difficulty that restricts the use of chitosan-based food additives or preservation is the substance's typical astringent or bitter flavor [127]. Many methods have been employed to enhance these qualities, including chitosan mixtures with different natural biopolymers or other organic substances, such as essential oils [128]. Despite several successful studies on chitosan in the nanoscale, additional study is still needed, particularly in the development of robust chitosan nanocomposites and NPs that can compensate for chitosan samples' shortcomings in terms of their physical, mechanical, and thermal stability [129]. Chitosan biopolymers could be further enhanced to create medications and functional meals that are secure, eco-friendly, and successful at treating and preventing man's illnesses [130].

9.10 Conclusion

There is still great concern about health hazards and risks associated with the use of chemicals as preservatives or additives and how to improve food shelf life without altering the properties of the food products despite the progress and advancements made in researching food preservation. Nanochitosan is considered a suitable candidate and alternative to traditional food preservatives. Professionals and researchers in the food safety and food processing sector would find this article resourceful in exploring nanochitosan for an improved approach to food preservation.

References

1. Perera, U.M.S.P. and Rajapakse, N., Chitosan Nanoparticles: Preparation, Characterization, and Applications, in: *Seafood Processing By-Products*, pp. 371–387, Springer New York, New York, NY, 2014.
2. Sivashankari, P.R. and Prabaharan, M., Deacetylation modification techniques of chitin and chitosan, in: *Chitosan Based Biomaterials Volume 1*, J.A. Jennings and J.D. Bumgardner (Eds.), pp. 117–133, Woodhead Publishing, Elsevier, 2017.
3. El-Naggar, N.E.-A., Saber, W., II, Zweil, A.M., Bashir, S., II, An innovative green synthesis approach of chitosan nanoparticles and their inhibitory activity against phytopathogenic Botrytis cinerea on strawberry leaves. *Sci. Rep.*, *12*, 1, 1–20, 2022. doi:10.1038/s41598-022-07073-y.
4. Malerba, M. and Cerana, R., Chitosan effects on plant systems. *Int. J. Mol. Sci.*, *17*, 7, 996, 2016. doi:10.3390/ijms17070996.
5. Dutta, P., Kumari, A., Mahanta, M., Chitosan nanoparticle: Synthesis, characterization, and use as plant health materials, in: *Chitin-Chitosan - Isolation, Properties, and Applications [Working Title]*, D.B. Kumar (Ed.), IntechOpen, London, England, 2022.
6. Sektiaji, R.G.B., Rochima, E., Pratama, R., II, Utama, G.L., Study of Nanochitosan (Definition, Manufacture, Analysis of Characteristics and Utilization): Review. *Asian J. Fish. Aquat. Res.*, *20*, 4, 21–28, 2022. https://doi.org/10.9734/ajfar/2022/v20i4501.
7. Luthfiyana, N., Bija, S., Anwar, E., Laksmitawati, D.R., Rosalinda, G.L., Characteristics and activity of chitosan from mud crab shells on acne bacteria: Staphylococcus aureus, S. epidermidis and Propionibacterium acnes. *Biodiversitas J. Biol. Divers.*, 23, 12, 2023. doi:10.13057/biodiv/d231263.
8. Awuchi, C.G., Ondari, E.N., Ogbonna, C.U., Upadhyay, A.K., Baran, K., Okpala, C.O.R., Guiné, R.P.F., Mycotoxins affecting animals, foods, humans, and plants: Types, occurrence, toxicities, action mechanisms, prevention, and detoxification strategies-A revisit. *Foods (Basel Switzerland)*, *10*, 6, 1279, 2021. doi:10.3390/foods10061279.
9. Godswill, A.C., Proximate composition and functional properties of different grain flour composites for industrial applications. *Int. J. Food Sci.*, *2*, 1, 43–64, 2019. doi:10.47604/ijf.1010.
10. Zhang, X., Guo, M., Ismail, B.B., He, Q., Jin, T.Z., Liu, D., Informative and corrective responsive packaging: Advances in farm-to-fork monitoring and remediation of food quality and safety. *Compr. Rev. Food Sci. Food Saf.*, *20*, 5, 5258–5282, 2021. doi:10.1111/1541-4337.12807.
11. Kamarudin, S.H., Rayung, M., Abu, F., Ahmad, S., Fadil, F., Karim, A.A., Abdullah, L.C., A review on antimicrobial packaging from biodegradable polymer composites. *Polymers*, *14*, 1, 174, 2022. doi:10.3390/polym14010174.
12. Liu, X., Xue, F., Li, C., Adhikari, B., Physicochemical properties of films produced using nanoemulsions stabilized by carboxymethyl chitosan-peptide

conjugates and application in blueberry preservation. *Int. J. Biol. Macromol.*, *202*, 26–36, 2022. doi:10.1016/j.ijbiomac.2021.12.186.
13. Flórez, M., Guerra-Rodríguez, E., Cazón, P., Vázquez, M., Chitosan for food packaging: Recent advances in active and intelligent films. *Food Hydrocoll.*, *124*, 107328, 107328, 2022. doi:10.1016/j.foodhyd.2021.107328.
14. Yu, Z., Jiang, Q., Yu, D., Dong, J., Xu, Y., Xia, W., Physical, antioxidant, and preservation properties of chitosan film doped with proanthocyanidins-loaded nanoparticles. *Food Hydrocoll.*, *130*, 107686, 107686, 2022. doi:10.1016/j.foodhyd.2022.107686.
15. Shaikh, S., Yaqoob, M., Aggarwal, P., An overview of biodegradable packaging in food industry. *Curr. Res. Food Sci.*, *4*, 503–520, 2021. doi:10.1016/j.crfs.2021.07.005.
16. Liu, Y., Yuan, Y., Duan, S., Li, C., Hu, B., Liu, A., Wu, W., Preparation and characterization of chitosan films with three kinds of molecular weight for food packaging. *Int. J. Biol. Macromol.*, *155*, 249–259, 2020. doi:10.1016/j.ijbiomac.2020.03.217.
17. Zhang, H., Liang, Y., Li, X., Kang, H., Effect of chitosan-gelatin coating containing nano-encapsulated tarragon essential oil on the preservation of pork slices. *Meat Sci.*, *166*, 108137, 108137, 2020. doi:10.1016/j.meatsci.2020.108137.
18. Liu, Y., Wang, S., Lan, W., Qin, W., Fabrication and testing of PVA/chitosan bilayer films for strawberry packaging. *Coatings*, *7*, 8, 109, 2017. doi:10.3390/coatings7080109.
19. Stallard, J.C., Tan, W., Smail, F.R., Gspann, T.S., Boies, A.M., Fleck, N.A., The mechanical and electrical properties of direct-spun carbon nanotube mats. *Extreme Mech. Lett.*, *21*, 65–75, 2018. doi:10.1016/j.eml.2018.03.003.
20. Wang, P., Zhang, C., Zou, Y., Li, Y., Zhang, H., Immobilization of lysozyme on layer-by-layer self-assembled electrospun films: Characterization and antibacterial activity in milk. *Food Hydrocoll.*, *113*, 106468, 106468, 2021. doi:10.1016/j.foodhyd.2020.106468.
21. Hidaka, M., Kojima, M., Nakahata, M., Sakai, S., Visible light-curable chitosan ink for extrusion-based and vat polymerization-based 3D bioprintings. *Polymers*, *13*, 9, 1382, 2021. doi:10.3390/polym13091382.
22. Liu, X., Liao, W., Xia, W., Recent advances in chitosan based bioactive materials for food preservation. *Food Hydrocoll.*, 140, 108612, 108612, 2023. doi:10.1016/j.foodhyd.2023.108612.
23. Singh, R., Shitiz, K., Singh, A., Chitin and chitosan: biopolymers for wound management: Chitin and chitosan: biopolymers for wound management. *Int. Wound J.*, 14, 6, 1276–1289, 2017. doi:10.1111/iwj.12797.
24. Islam, S., Bhuiyan, M.A.R., Islam, M.N., Chitin and chitosan: Structure, properties and applications in biomedical engineering. *J. Polym. Environ.*, 25, 3, 854–866, 2017. doi:10.1007/s10924-016-0865-5.

25. Zargar, V., Asghari, M., Dashti, A., A review on chitin and chitosan polymers: Structure, chemistry, solubility, derivatives, and applications. *ChemBioEng Rev.*, 2, 3, 204–226, 2015. doi:10.1002/cben.201400025.
26. Alemu, D., Getachew, E., Mondal, A.K., Study on the physicochemical properties of chitosan and their applications in the biomedical sector. *Int. J. Polym. Sci.*, 2023, 1–13, 2023. doi:10.1155/2023/5025341.
27. Abo Elsoud, M.M. and El Kady, E.M., Current trends in fungal biosynthesis of chitin and chitosan. *Bull. Natl. Res. Cent.*, 43, 1, 2019. doi:10.1186/s42269-019-0105-y.
28. Hu, S., Wang, Y., Wen, X., Wang, L., Jiang, Z., Zheng, C., Effects of low-molecular-weight chitosan on the growth performance, intestinal morphology, barrier function, cytokine expression and antioxidant system of weaned piglets. *BMC Vet. Res.*, 14, 1, 1–7, 2018. doi:10.1186/s12917-018-1543-8.
29. Matica, Aachmann, Tøndervik, Sletta, Ostafe, Chitosan as a wound dressing starting material: Antimicrobial properties and mode of action. *Int. J. Mol. Sci.*, 20, 23, 5889, 2019. doi:10.3390/ijms20235889.
30. Chattopadhyay, D.P. and Inamdar, M.S., Aqueous behaviour of chitosan. *Int. J. Polym. Sci.*, 2010, 1–7, 2010. doi:10.1155/2010/939536.
31. Li, T., Sun, X., Chen, H., He, B., Mei, Y., Wang, D., Li, J., Effect of the combination of vanillin and chitosan coating on the microbial diversity and shelf-life of refrigerated turbot (Scophthalmus maximus) filets. *Front. Microbiol.*, *11*, 462, 2020. doi:10.3389/fmicb.2020.00462.
32. Amit, S.K., Uddin, M.M., Rahman, R., A review on mechanisms and commercial aspects of food preservation and processing. *Agric. Food Secur.*, 6, 51, 2017, 2017. https://doi.org/10.1186/s40066-017-0130-8.
33. Chaleshtori, F.S., Arian, A., Chaleshtori, R.S., Assessment of sodium benzoate and potassium sorbate preservatives in some products in Kashan, Iran with estimation of human health risk. *Food Chem. Toxicol.*, 120, 634–638, 2018. doi: 10.1016/j.fct.2018.08.010. [PubMed] [CrossRef] [Google Scholar].
34. Lee, N.K. and Paik, H.D., Status, Antimicrobial mechanism, and regulation of natural preservatives in livestock food systems. *Korean J. Food Sci. Anim. Resour.*, 36, 547–557, 2016. doi: 10.5851/kosfa.2016.36.4.547. [PMC free article] [PubMed] [CrossRef] [Google Scholar].
35. Crowe, W., Elliott, C.T., Green, B.D., A review of the *in vivo* evidence investigating the role of nitrite exposure from processed meat consumption in the development of colorectal cancer. *Nutrients*, 11, 2673, 2019. doi: 10.3390/nu11112673. [PMC free article] [PubMed] [CrossRef] [Google Scholar].
36. Olszewska, M.A., Gędas, A., Simões, M., Antimicrobial polyphenol-rich extracts: Applications and limitations in the food industry. *Food Res. Int.*, 134, 109214, 2020. doi: 10.1016/j.foodres.2020.109214. [PubMed].
37. Zhu, Y., Li, C., Cui, H., Lin, L., Encapsulation strategies to enhance the antibacterial properties of essential oils in food system. *Food Control*, 123, 107856, 2020. doi: 10.1016/j.foodcont.2020.107856.

38. Yu, J.Y., Ko, J.A., Park, H.J., Kim, H.W., Application of nanochitosan in food industry: a review. *Korean Society of Food Science and Technology Food Science and Industry*, vol. 53, pp. 56–68, 2020.
39. Radhakrishnan, Y., Ganesh, G., Chittur, C.L., Krishna, S.N., Chitosan Nanoparticles for Generating Novel Systems for Better Applications: A Review, 2015.
40. Inneke, F.M., Rumengan, P.S., Stenly, W., Aldian, H., Luntungan, F.B., Sandana, N.S., Potential Application of Chitosan Nanoparticles Derived from Marine Fish Scales as Preservatives for Fishery Product University of Sam Ratulangi Manado; Bogor Agriculture Institute, Fisheries and Marine Science, Indonesia. *IOP Conf. Ser. Earth Enviormental. Sci.*, 855, 1, 012013, 2017. doi:10.1088/1755-.
41. Hasheminejad, N., Khodaiyan, F., Safari, M., Improving the antifungal activity of clove essential oil encapsulated by chitosan nanoparticles. *Food Chem.*, 275, 113–122, 2019.
42. Ahmed, A.T., Aml, F.E., Shaaban, H.M., Mohammed, S.A., Amany, M.D., Biopreservation of Shrimps Using Composed Edible Coatings from Chitosan Nanoparticles and Cloves Extract. *Special Issue: Natural Products and Nanomaterials for the Improvement of Aquatic Products*, 2020, Article ID 8878452 | https://doi.org/10.1155/2020/8878452.
43. Maleki, G., Woltering, E.J., Mozafari, M.R., Applications of chitosan-based carrier as an encapsulating agent in food. *Trends Food Sci. Technol.*, 120, February 2022, 88–99, 2022. https://doi.org/10.1016/j.tifs.2022.01.001.
44. Adetunji, J.B., Adewale, O.O., Adetunji, C.O., Hefft, D., II, Ejidike, I.P., Chapter 14 - Application of nanochitosan for enhanced milk production, in: *Next Generation Nanochitosan Applications in Animal Husbandry, Aquaculture and Food Conservation*, pp. 203–212, 2023, https://doi.org/10.1016/B978-0-323-85593-8.00029-1.
45. Ikono, R., Vibriani, A., Wibowo, I., Radyum, I., Saputro, K.E., Wibias, M., Bachtiar, B.M., Etik, M., Bachtiar, E.W., Rochman, N.T., Hideaki, K., Li, X., Nagamura-Inoue, T., Arinobu, T., Nanochitosan antimicrobial activity against Streptococcus mutans and Candida albicans dual-species biofilms. *BMC Res. Notes*, 12, 383 (2019), 2019. https://doi.org/10.1186/s13104-019-4422-x.
46. Saputra, D., Ula, F.R., Nur Fadhila, A.B., Sijabat, Y.Y., Romadoni, A.A., Windarto, S., Nano-chitosan Spray as a Preservative and Food Security of Fishery Products in The Middle of the Covid-19 Pandemic. *Sci. J. Fish. & Mar./Jurnal Ilmiah Perikanan dan Kelautan*, 14, 1, 2022 Apr 1.
47. Ghorabi, R.S. and Khodanazary, A., Effects of chitosan and nano-chitosan as coating materials on the quality properties of large scale tongue sole Cynoglossus arel during super-chilling storage, 2017, Received: August 2017 Accepted: December 2017.
48. de Azeredo, H.M.C., Antimicrobial nanostructures in food packaging. *Trends Food Sci. Technol.*, 30, 1, 56–69, 2013. doi:10.1016/j.tifs.2012.11.006.

49. Sashiwa, H. and Harding, D., Advances in marine chitin and chitosan. *MDPI AD*, 484, 2015.
50. Dima, J.B., Sequeiros, C., Zaritzky, N., E.A., Chitosan from marine crustaceans: production, characterization and applications, in: *Biological activities and application of marine polysaccharides*, Shalaby (Ed.), pp. 39–56, Springer, Rijeka, 2017, https://doi.org/10.5772/65258 (Chapter 3).
51. Philibert, T., Lee, B.H., Fabien, N., Current status and new perspectives on chitin and chitosan as functional biopolymers. *Appl. Biochem. Biotechnol.*, 181, 1314–1337, 2017. https://doi.org/10.1007/s12010-0162286-2.
52. Montero, M.P., López-Caballero, M.E., Alemán, A., Gómez-Estaca, J., Martínez-Álvarez, Ó., Gómez-Guillén, M.C., Instituto de Ciencia y Tecnología de los Alimentos y Nutrición (ICTAN). Consejo Superior de Investigaciones Científicas (CSIC), 2018.
53. Elsabee, M.Z. and Abdou, E.S., Chitosan based edible films and coatings: a review. *Mat. Sci. Eng. C.*, 33, 1819–1841, 2013. https://doi.org/10.1016/j.msec.2013.01.010.
54. Muñoz-Bonilla, A., Cerrada, M.L., Fernández-García, M. (Eds.), Antimicrobial activity of chitosan in food, agriculture and biomedicine, in: *Polymeric materials with antimicrobial activity. From synthesis to applications*, vol. Chapter 2, pp. 22–53, The Royal Society of Chemistry. RSC Polymer Chemistry Series no. 10. RSC Publishing, Cambridge, 2014, https://doi.org/10.1039/9781782624998-00022.
55. Van den Broek, L.A.M., Knoop, R.J., II, Kappen, F.H.J., Boeriu, C.G., Chitosan films and blends for packaging material. *Carbohydr. Polym.*, 116, 237–242, 2015. https://doi.org/10.1016/j.carbpol.2014.07.039.
56. Ahmed, T. and Aljaeid, B., Preparation, characterization, and potential application of chitosan, chitosan derivatives, and chitosan metal nanoparticles in pharmaceutical drug delivery. *Drug Des. Dev. Ther.*, 10, 483, 2016. https://doi.org/10.2147/DDDT.S99651.
57. Youssef, A.M., Abdel-Aziz, M.S., El-Sayed, S.M., Chitosan nanocomposite films based on Ag-NP and Au-NP biosynthesis by Bacillus Subtilis as packaging materials. *Int. J. Biol. Macromol.*, 2014, 69, 185–91, 2014. https://doi.org/10.1016/j.ijbiomac.2014.05.047.
58. Alarfaj, A.A., Preparation, Characterization and Antibacterial Effect of Chitosan Nanoparticles against Food Spoilage Bacteria. *J. Pure Appl. Microbiol.*, 13, 2, 1273–1278, 2019. Article Number: 5462, https://dx.doi.org/10.22207/JPAM.13.2.70.
59. Shaltout, F.A., El-Diasty, E.M., Hassan, M.A., Effect of Nano-Chitosan and Onion Extract as Coating Materials on the Quality Properties of Chicken Fillet Meat During Refrigeration. *Global Vet.*, 21, 6, 368–372, 2019. ISSN 1992-6197© IDOSI Publications, 2019, DOI: 10.5829/idosi.gv.2019.368.372368.
60. Neelesh, K.N., Nayana, R., Sachithra, S., Swapnil, D.K., Use of Nanotechnology for the Improvement of Sensory Attributes of Foods, in: *Application of*

Nanotechnology in Food Science, Processing and Packaging, pp. 31–46, 2022, DOI:10.1007/978-3-030-98820-3_3.

61. Kim, H.Y., Jeong, Y.T., Bae, I.H., Kwak, H.S., Physicochemical and Sensory Properties of Nano powdered Chitosan-Added Maribo Cheese during Ripening. *Korean Soc. Food Sci. Anim. Resour.*, 34, 1, 57–64, 2014. https://doi.org/10.5851/kosfa.2014.34.1.57.
62. Gomes, L.P., Vânia, M.F., Del Aguila, E.M., Cidade Universitária, CEP 21949-909, Rio de Janeiro-RJ, Brazil, 2017, * 2016 Revista Virtual de Química ISSN 1984-6835 Janeiro-Fevereiro 2017 388 Rev. Virtual Quim. |Vol 9| |No. 1| |387-409| DOI: 10.21577/1984-6835.20170022 Volume 9, NÃºmero.
63. Wang, Y.H., Zhang, R., Qin, W., Dai, J.W., Zhang, Q., Lee, K.J., Liu, Y., Physicochemical properties of gelatin films containing tea polyphenol-loaded chitosan nanoparticles generated by electrospray. *Mater. Des.*, 185, 108277, 2020. ID 108277.
64. Zarei, M., Ramezani, Z., Ein-Tavasoly, S., Chadorbaf, M., Coating effects of orange and pomegranate peel extracts combined with chitosan nanoparticles on the quality of refrigerated silver carp fillet. *J. Food Process Pres.*, 39, 2180–2187, 2015.
65. Abdelhady, E.M., Mostafa, N.Y., El-Magd, M.A., Kirrella, G.A., Effect of Chitosan Powder and Nanoparticles as Natural Preservatives for Beef Minced Meat Pakistan. *J. Zool.*, 1-5, 2022, 2022. DOI: https://dx.doi.org/10.17582/journal.pjz/20220315170358.
66. Antoniou, J., Liu, F., Majeed, H., Zhong, F., Characterization of tara gum edible films incorporated with bulk chitosan and chitosan nanoparticles: A comparative study. *Food Hydrocoll.*, 44, 309–319, 2015.
67. Solval, K.M., Rodezno, L.A.E., Moncada, M., Bankston, J.D., Sathivel, S., Evaluation of chitosan nanoparticles as a glazing material for cryogenically frozen shrimp. *LWT Food Sci. Technol.*, 57, 172–180, 2014.
68. Mekuye, B. and Abera, B., Nanomaterials: An overview of synthesis, classification, characterization, and applications. *Nano Select Open Access*, 4, 8, 486–501, 2023. https://doi.org/10.1002/nano.202300038.
69. Ayala-Fuentes, J.C. and Chavez-Santoscoy, R.A., Nanotechnology as a key to enhance the benefits and improve the bioavailability of flavonoids in the food industry. *Foods (Basel Switzerland)*, 10, 11, 2701, 2021. https://doi.org/10.3390/foods10112701.
70. Singh, R., Dutt, S., Sharma, P., Sundramoorthy, A.K., Dubey, A., Singh, A., Arya, S., Future of nanotechnology in food industry: Challenges in processing, packaging, and food safety. *Global Challenges (Hoboken NJ)*, 7, 4, 2200209, 2023. https://doi.org/10.1002/gch2.202200209.
71. Gupta, R. and Xie, H., Nanoparticles in daily life: Applications, toxicity and regulations. *J. Environ. Pathology Toxicol. Oncology Off. Organ Int. Soc. Environ. Toxicol. Cancer*, 37, 3, 209–230, 2018. https://doi.org/10.1615/jenvironpatholtoxicoloncol.2018026009.

72. Egbuna, C., Parmar, V.K., Jeevanandam, J., Ezzat, S.M., Patrick-Iwuanyanwu, K.C., Adetunji, C.O., Khan, J., Onyeike, E.N., Uche, C.Z., Akram, M., Ibrahim, M.S., El Mahdy, N.M., Awuchi, C.G., Saravanan, K., Tijjani, H., Odoh, U.E., Messaoudi, M., Ifemeje, J.C., Olisah, M.C., Ibeabuchi, C.G., Toxicity of nanoparticles in biomedical application: Nanotoxicology. *J. Toxicol.*, 2021, 1–21, 2021. https://doi.org/10.1155/2021/9954443.
73. Onyeaka, H., Passaretti, P., Miri, T., Al-Sharify, Z.T., The safety of nanomaterials in food production and packaging. *Curr. Res. Food Sci.*, 5, 763–774, 2022. https://doi.org/10.1016/j.crfs.2022.04.005.
74. Paidari, S., Tahergorabi, R., Anari, E.S., Nafchi, A.M., Zamindar, N., Goli, M., Migration of various nanoparticles into food samples: A review. *Foods (Basel Switzerland)*, 10, 9, 2114, 2021. https://doi.org/10.3390/foods10092114.
75. Ferraris, C., Rimicci, C., Garelli, S., Ugazio, E., Battaglia, L., Nanosystems in cosmetic products: A brief overview of functional, market, regulatory and safety concerns. *Pharmaceutics*, 13, 9, 1408, 2021. https://doi.org/10.3390/pharmaceutics13091408.
76. Nazarenko, Y., Zhen, H., Han, T., Lioy, P.J., Mainelis, G., Nanomaterial inhalation exposure from nanotechnology-based cosmetic powders: a quantitative assessment. *J. Nanopart. Research An Interdiscip. Forum Nanoscale Sci. Technol.*, 14, 11, 1–14, 2012. https://doi.org/10.1007/s11051-012-1229-2.
77. Naseri, N., Valizadeh, H., Zakeri-Milani, P., Solid lipid nanoparticles and nanostructured lipid carriers: Structure, preparation and application. *Adv. Pharm. Bull.*, 5, 3, 305–313, 2015. https://doi.org/10.15171/apb.2015.043.
78. Malik, S., Muhammad, K., Waheed, Y., Nanotechnology: A revolution in modern industry. *Mol. (Basel Switzerland)*, 28, 2, 661, 2023. https://doi.org/10.3390/molecules28020661.
79. Wagner, C., Nanomaterials. *Food Packag. Forum Food Packag. Forum Food Packag. Forum Foundation FPF*, 2013, June 25. https://www.foodpackagingforum.org/food-packaging-health/nanomaterials.
80. Espitia, P.J.P., Otoni, C.G., Soares, N.F.F., Zinc oxide nanoparticles for food packaging applications, in: *Antimicrobial Food Packaging*, pp. 425–431, Elsevier, 2016.
81. Tirumala, M.G., Anchi, P., Raja, S., Rachamalla, M., Godugu, C., Novel methods and approaches for safety evaluation of nanoparticle formulations: A focus towards *in vitro* models and adverse outcome pathways. *Front. Pharmacol.*, 12, 612659, 2021. https://doi.org/10.3389/fphar.2021.612659.
82. Beaudoin, S., Vandelac, L., Papilloud, C., Nanofoods: Environmental, Health, and Socioeconomic Risks or the Achilles' Heel of Nanotechnologies?. *Nanofoods, nanotechnol. hum. health*, 109–26, 2013 Aug 13
83. Schlundt, J., Institutions involved in food safety: World health organization (WHO), in: *Encyclopedia of Food Safety*, pp. 359–364, Elsevier, Rockville Pike Bethesda, MD, 2014.
84. Allan, J., Belz, S., Hoeveler, A., Hugas, M., Okuda, H., Patri, A., Rauscher, H., Silva, P., Slikker, W., Sokull-Kluettgen, B., Tong, W., Anklam, E., Regulatory

landscape of nanotechnology and nanoplastics from a global perspective. *Regul. Toxicol. Pharmacol. RTP*, 122, 104885, 104885, 2021. https://doi.org/10.1016/j.yrtph.2021.104885.

85. Flear, M.L. and Pickersgill, M.D., Regulatory or regulating publics? The European union's regulation of emerging health technologies and citizen participation, in: *Medical Law Review*, vol. 21, pp. 39–70, 2013, https://doi.org/10.1093/medlaw/fws039.
86. Terpan, F., Soft law in the European union-the changing nature of EU law: Soft law in the European union. *Eur. Law J.*, 21, 1, 68–96, 2015. https://doi.org/10.1111/eulj.12090.
87. Engelmann, W., Aldrovandi, A., Berger Filho, A.G., Prospects for the regulation of nanotechnology applied to food and biofuels, in: *Vigilância Sanitária Em Debate*, vol. 1, 2013, https://doi.org/10.3395/vd.v1i4.69en.
88. Lison, D., Vietti, G., van den Brule, S., Paracelsus in nanotoxicology. *Part. Fibre Toxicol.*, 11, 1, 35, 2014. https://doi.org/10.1186/s12989-014-0035-7.
89. Cohen, J.M., DeLoid, G.M., Demokritou, P., A critical review of *in vitro* dosimetry for engineered nanomaterials. *Nanomed. (London England)*, 10, 19, 3015–3032, 2015. https://doi.org/10.2217/nnm.15.129.
90. Delmaar, C.J.E., Peijnenburg, W.J.G.M., Oomen, A.G., Chen, J., de Jong, W.H., Sips, A.J.A.M., Wang, Z., Park, M.V.D.Z., A practical approach to determine dose metrics for nanomaterials: Dose metrics for nanomaterials. *Environ. Toxicol. Chem.*, 34, 5, 1015–1022, 2015. https://doi.org/10.1002/etc.2878.
91. Pal, A.K., Bello, D., Cohen, J., Demokritou, P., Implications ofin vitrodosimetry on toxicological ranking of low aspect ratio engineered nanomaterials. *Nanotoxicology*, 9, 7, 871–885, 2015. https://doi.org/10.3109/17435390.2014.986670.
92. Chinedu, E., Arome, D., Ameh, F., A new method for determining acute toxicity in animal models. *Toxicol. Int.*, 20, 3, 224, 2013. https://doi.org/10.4103/0971-6580.121674.
93. Silva, R.J. and Tamburic, S., A state-of-the-art review on the alternatives to animal testing for the safety assessment of cosmetics. *Cosmetics*, 9, 5, 90, 2022. https://doi.org/10.3390/cosmetics9050090.
94. Moradi, M., Golmohammadi, R., Najafi, A., Moosazadeh Moghaddam, M., Fasihi-Ramandi, M., Mirnejad, R., A contemporary review on the important role of *in silico* approaches for managing different aspects of COVID-19 crisis. *Inf. Med. Unlocked*, 28, 100862, 100862, 2022. https://doi.org/10.1016/j.imu.2022.100862.
95. Fu, P.P., Xia, Q., Hwang, H.-M., Ray, P.C., Yu, H., Mechanisms of nanotoxicity: Generation of reactive oxygen species. *J. Food Drug Anal.*, 22, 1, 64–75, 2014. https://doi.org/10.1016/j.jfda.2014.01.005.
96. Zhang, N., Xiong, G., Liu, Z., Toxicity of metal-based nanoparticles: Challenges in the nano era. *Front. Bioeng. Biotechnol.*, 10, 1001572, 2022. https://doi.org/10.3389/fbioe.2022.1001572.

97. Homayounpour, P., Jalali, H., Shariatifar, N., Amanlou, M., Khanjari, A., Protective effect of nanochitosan incorporated with free/nanoliposome Cumin (Cuminum cyminum L.) aqueous extract on sardine fish. *J. Aquat. Food Prod. Technol.*, *29*, 9, 949–961, 2020.
98. Mohammadkhan, F., Noori, N., Basti, A.A., Khanjari, A., Vanaki, E., Farzaneh, M., The Effects of Nanochitosan Coating Integrated to Zataria Multiflora Boiss and Polylophium Involucratum Essential Oils on the Shelf-Life Extension of Silver Carp Fillets. *J. Nutrition Fasting Health*, *10*, 3, 2022.
99. Muñoz-Tebar, N., Pérez-Álvarez, J. A., Fernández-López, J., & Viuda-Martos, M., Chitosan edible films and coatings with added bioactive compounds: Antibacterial and antioxidant properties and their application to food products: A review. *Polymers*, 15, 2, 396, 2023. doi:10.3390/polym15020396.
100. Hassoun, A. and Emir Çoban, Ö., Essential oils for antimicrobial and antioxidant applications in fish and other seafood products. *Trends Food Sci. Technol.*, 68, 26–36, 2017. doi:10.1016/j.tifs.2017.07.016.
101. Qiu, L., Zhang, M., Bhandari, B., Yang, C., Shelf life extension of aquatic products by applying nanotechnology: a review. *Crit. Rev. Food Sci. Nutr.*, 62, 6, 1521–1535, 2022. doi:10.1080/10408398.2020.1844139.
102. Vatavali, K., Karakosta, L., Nathanailides, C., Georgantelis, D., Kontominas, M.G., Combined effect of chitosan and oregano essential oil dip on the microbiological, chemical, and sensory attributes of red porgy (Pagrus pagrus) stored in ice. *Food Bioprocess Technol.*, 6, 12, 3510–3521, 2013. doi:10.1007/s11947-012-1034-z.
103. Difonzo, G., Antonino, C., Squeo, G., Caponio, F., Faccia, M., Application of agri-food by-products in cheesemaking. *Antioxid. (Basel Switzerland)*, 12, 3, 660, 2023. doi:10.3390/antiox12030660.
104. Gao, X., Zheng, Y., Zhong, Y., Zhou, R., Li, B., Ma, M., Preparation and characterization of novel chitosan coatings to reduce changes in quality attributes and physiochemical and water characteristics of Mongolian cheese during cold storage. *Foods (Basel Switzerland)*, 12, 14, 2731, 2023. doi:10.3390/foods12142731.
105. Sayed-Elahl, R.M.H., El-Shinawy, M., Nagy, K., A trial for improvement of kareish cheese quality by using chitosan nanoparticles. *Egypt. J. Vet. Sci.*, 50, 8, 69–80, 2019. doi:10.21608/ejvs.2020.19943.1135.
106. Ausar, S.F., Passalacqua, N., Castagna, L.F., Bianco, I.D., Beltramo, D.M., Growth of milk fermentative bacteria in the presence of chitosan for potential use in cheese making. *Int. Dairy J.*, *12*, 11, 899–906, 2002.
107. Seo, M., II, Chang, Y.-H., Lee, S., Kwak, H.-S., The physicochemical and sensory properties of milk supplemented with ascorbic acid-soluble nano-chitosan during storage. *Int. J. Dairy Technol.*, 64, 1, 57–63, 2011. doi:10.1111/j.1471-0307.2010.00630.x.
108. El-Sisi, A.S., Impact of replacement of gelatin with chitosan on the physicochemical properties of ice-milk. *Int. J. Dairy Sci.*, 10, 1, 36–43, 2014. doi:10.3923/ijds.2015.36.43.

109. Anjum, S., Ishaque, S., Fatima, H., Farooq, W., Hano, C., Abbasi, B.H., Anjum, I., Emerging applications of nanotechnology in healthcare systems: Grand challenges and perspectives. *Pharmaceuticals (Basel Switzerland)*, 14, 8, 707, 2021. doi:10.3390/ph14080707.
110. de Azeredo, H.M.C., Antimicrobial nanostructures, 2013, Food [DOR: 20.1 001.1.15622916.2020.19.5.11.1] [Downloaded from jifro.ir on 2023-08-26] 2255.
111. Hosseini, H. and Jafari, S.M., Introducing nano/microencapsulated bioactive ingredients for extending the shelf-life of food products. *Adv. Colloid Interface Sci.*, *282*, 102210, 2020.
112. Pateiro, M., Gómez, B., Munekata, P.E.S., Barba, F.J., Putnik, P., Kovačević, D.B., Lorenzo, J.M., Nanoencapsulation of promising bioactive compounds to improve their absorption, stability, functionality and the appearance of the final food products. *Mol. (Basel Switzerland)*, 26, 6, 1547, 2021. doi:10.3390/molecules26061547.
113. Alqahtani, F., Aleanizy, F., El Tahir, E., Alhabib, H., Alsaif, R., Shazly, G., Antibacterial activity of chitosan nanoparticles against pathogenic N gonorrhoea. *Int. J. Nanomed.*, 15, 7877–7887, 2020. doi:10.2147/ijn.s272736.
114. Khubiev, O.M., Egorov, A.R., Kirichuk, A.A., Khrustalev, V.N., Tskhovrebov, A.G., Kritchenkov, A.S., Chitosan-based antibacterial films for biomedical and food applications. *Int. J. Mol. Sci.*, 24, 13, 10738, 2023. doi:10.3390/ijms241310738.
115. Kong, I., Degraeve, P., Pui, L.P., Polysaccharide-based edible films incorporated with essential oil nanoemulsions: Physico-chemical, mechanical properties and its application in food preservation-A review. *Foods (Basel Switzerland)*, 11, 4, 555, 2022. doi:10.3390/foods11040555.
116. Wei, C., Liao, X., Huang, Y., Zhou, F., Liao, Y., Zhou, Z., Effects of different levels of chitosan and licorice antioxidants on skin color and surface microbial counts of frozen duck meat. *J. Food Sci. Technol.*, *45*, 280–285, 2020.
117. Liu, T., Li, J., Tang, Q., Qiu, P., Gou, D., Zhao, J., Chitosan-based materials: An overview of potential applications in food packaging. *Foods*, *11*, 10, 1490, 2022.
118. Liu, T., Liu, L., Gong, X., Chi, F., Ma, Z., Fabrication and comparison of active films from chitosan incorporating different spice extracts for shelf life extension of refrigerated pork. *Lebensmittel-Wissenschaft Und Technologie [Food Sci. Technology]*, 135, 110181, 110181, 2021. doi:10.1016/j.lwt.2020.110181.
119. Kuswandi, B., Environmental friendly food nano-packaging. *Environ. Chem. Lett.*, *15*, 2, 205–221, 2017.
120. Nile, S.H., Baskar, V., Selvaraj, D., Nile, A., Xiao, J., Kai, G., Nanotechnologies in food science: Applications, recent trends, and future perspectives. *Nano Micro Lett.*, 12, 1, 1–34, 2020. doi:10.1007/s40820-020-0383-9.
121. Franz, R., Bott, J., Störmer, A., Considerations for and guidance to testing and evaluating migration/release of nanoparticles from polymer

based nanocomposites. *Nanomater. (Basel Switzerland)*, 10, 6, 1113, 2020. doi:10.3390/nano10061113.

122. Siddiqui, S.A., Zannou, O., Bahmid, N.A., Fidan, H., Alamou, A.-F., Nagdalian, ..., Arsyad, M., Consumer behavior towards nanopackaging - A new trend in the food industry. *Future Foods A Dedicated J. Sustainability Food Sci.*, 6, 100191, 100191, 2022. doi:10.1016/j.fufo.2022.100191.
123. Amini, S.M., Gilaki, M., Karchani, M., Safety of nanotechnology in food industries. *Electron. Physician*, 6, 4, 962, 2014. doi:10.14661/2014.962-968.
124. Capon, A., Gillespie, J., Rolfe, M., Smith, W., Perceptions of risk from nanotechnologies and trust in stakeholders: a cross sectional study of public, academic, government and business attitudes. *BMC Public Health*, *15*, 1–13, 2015.
125. Zubareva, A., Shagdarova, B., Varlamov, V., Kashirina, E., Svirshchevskaya, E., Penetration and toxicity of chitosan and its derivatives. *Eur. Polym. J.*, 93, 743–749, 2017. doi:10.1016/j.eurpolymj.2017.04.021.
126. Ardean, C., Davidescu, C.M., NemeÅŸ, N.S., Negrea, A., Ciopec, M., Duteanu, N., Musta, V., Factors influencing the antibacterial activity of chitosan and chitosan modified by functionalization. *Int. J. Mol. Sci.*, 22, 14, 7449, 2021. doi:10.3390/ijms22147449.
127. Hu, Z. and GÃ¤nzle, M.G., Challenges and opportunities related to the use of chitosan as a food preservative. *J. Appl. Microbiol.*, 126, 5, 1318–1331, 2019. doi:10.1111/jam.14131.
128. Kumar, A., Yadav, S., Pramanik, J., Sivamaruthi, B.S., Jayeoye, T.J., Prajapati, B.G., Chaiyasut, C., Chitosan-based composites: Development and perspective in food preservation and biomedical applications. *Polymers*, 15, 15, 3150, 2023. doi:10.3390/polym15153150.
129. Mohammed, M., Syeda, J., Wasan, K., Wasan, E., An overview of chitosan nanoparticles and its application in non-parenteral drug delivery. *Pharmaceutics*, 9, 4, 53, 2017. doi:10.3390/pharmaceutics9040053.
130. Picos-Corrales, L.A., Morales-Burgos, A.M., Ruelas-Leyva, J.P., Crini, G., GarcÃa-Armenta, E., Jimenez-Lam, S.A., Inzunza-Camacho, L.N., Chitosan as an outstanding polysaccharide improving health-commodities of humans and environmental protection. *Polymers*, 15, 3, 526, 2023. doi:10.3390/polym15030526.

10

Applications of Chitosan Nanocomposites in Packaging of Food Products

Tsado Priscilla Yetu[1*], NengakYunana[1], Soda Gaius Baga[1], Innocent Ojeba Musa[2], Ibrahim Dawud[3], Oluwafemi Adebayo Oyewole[1], Charles Oluwaseun Adetunji[4], Eniola K. I. T.[5] and Mohammed Bello Yerima[6]

[1]*Department of Microbiology, Federal University of Technology, Minna, Nigeria*
[2]*Department of Microbiology, Skyline University Nigeria, Kano State, Nigeria*
[3]*Department of Integrated Science, Federal College of Education Zaria, Kaduna State, Nigeria*
[4]*Department of Microbiology, Edo State University, Uzairue, Nigeria*
[5]*Department of Biological Sciences, Joseph Ayo Babalola University, Ikeji Arakeji, Osun State, Nigeria*
[6]*Department of Microbiology, Sokoto State University, Sokoto, Sokoto State, Nigeria*

Abstract
The utilization of synthetic and non-degradable oil-derived substances in various nutrition sector for packaging poses significant ecological risks. Consequently, there has been a focus on creating eco-friendly and recyclable constituents obtained from nature for the packaging industry. Chitosan has proven to be one of the most widely acceptable biopolymer among biopolymers because of its ease of use, biodegradability, biocompatibility, and antibacterial properties. Chitosan possesses the potential to inhibit the proliferation of yeast, gram-positive and gram-negative bacteria, and various filamentous fungi commonly found in food. Consequently, the food packaging sector requires additional components beyond chitosan to meet its requirements. In this chapter, the summary of chitosan composites useful in food packing application that increase the longevity of food and improve its storage condition were highlighted. Active compounds with various nanoparticles, polysaccharides, and essential oils with chitosan are summarized. This work provides useful information on recent research developments on the choice of composites to be selected in food industries for packaging purposes

**Corresponding author*: tsado.priscilla@st.futminna.edu.ng

Charles Oluwaseun Adetunji, Maulin P. Shah, Yerima Mohammed Bello, Daniel Hefft, Jay Singh, Shyam S. Pandey and Ravindra Pratap Singh (eds.) Nanochitosan Applications for Enhanced Crop Production and Food Security, (217–248) © 2025 Scrivener Publishing LLC

when embedding chitosan. In recent times, attention has been drawn to nanochitosan composites, and so much progress has been made. This research provides a roadmap for the development of novel biodegradable that is friendly to the environment that will be useful in food packaging applications.

Keywords: Chitosan, nanocomposites, packaging, food products, biodegradable

10.1 Introduction

Every day, there is a growing trend in the popularity of non-renewable and petroleum-derived synthetic plastics that serve as the primary material for single-use shopping bags [1, 2]. Synthetic plastics are non-biodegradable and, hence, take several centuries to decay. Most of these non-biodegradable materials end up in oceans or landfills and become pollutants that can cause significant damage to the local ecosystem when they can be recycled. The used of synthetic plastics in food industry for food packaging is not ecosystem friendly. As a result, food industries are exploring packaging materials that are biodegradable. The attention is tending toward biodegradable materials of natural origins such as chitosan, pectin, whey, gelatin, soybean, proteins, pollulan, and alginate, which are all derivatives of cellulose [3].

Different celluloses made from plants and bacterial cellulose have been utilized for food packaging but the lack mechanical or functional properties. Gelatin exhibits the capacity to create films, yet its performance is constrained when it comes to enduring high temperatures. Alginates, on the other hand, have a good tensile strength with great flexibility and excellent film-forming ability but have a disadvantage of high rate of water vapor transmission. Chitosan and its derivatives stood out among all these biopolymers to possess good film-forming ability; antifungal, antibacterial, and metal chelating properties; biodegradability; and non-toxicity [4]. These excellent properties have increased the demand for chitin and chitosan [5, 6].

Chitosan is a polysaccharide manufactured through the deacetylation of chitin found in the shells of crustaceans. Chitosan has gain wide application in several industries such as medicine, food, and agriculture as a good alternative material. Because chitosans are obtained from natural sources and possess antimicrobial potential, biodegradability and non-toxicity, there have been an extensive study on their coatings and films [7, 8]. The general method for obtaining chitosan film is through casting. Using this method, chitosan is dissolved in a solvent that are suitable, and the plasticizer is simultaneously incorporated along with the active compounds and nanofiller of choice.

Chitosans are generally known to have positively charged amino group (CH_3^+) at the carbon-2 (C-2) in the glucose monomers. The making of chitosan from chitin through deacetylation is shown below.

As we advance in research, several techniques have been used recently to modify chitosan to reach certain packaging qualities in food industries. In doing this, alginate, starch, and pectin have been utilize to increase the mechanical properties, whereas the antioxidant and antibacterial potentials are increased by the incorporation of different additives such as plant extracts and essential oils aimed at improving its shelf life [2, 9].

Through research, a lot of improvement have been made on the physiochemical and mechanical properties through the incorporation of bioactive additives such as oils, plasticizers, and polyphenols [2, 10, 11].

A series of efforts have been put in place to produce chitosan-based composites with nanoparticles, citric acid, plants extracts, metal ions, tannic acid, phenols, etc., for the enhancement of the product's pharmacological properties. This present work focuses on chitosan composites for packaging applications. The article seeks to summarize the present developmental strides of chitosan composites for modern day food packaging applications.

10.2 The Chitosan Antimicrobial Potential

Chitosan has been shown to possess activity against wide range of microorganisms including fungi, bacteria, and algae. The antimicrobial potential of chitosan is embedded in the substrate composition, chitosan composition, the host, and the degree of polymerization. There are several hypotheses on the mechanism of chitosan activity on microbes as they are still not known. Research has shown chitosan to having greater bacteriostasis than fungistasis at a temperatures of −20°C [12]. One of the hypothesis on the mechanism of chitosan antimicrobial activity is that chitosan films form a cellphone like structure that build a layer of protection on the surface of the food, preventing external microorganism from gaining entrance [13]. The chitosan-based film that forms protective layer on the food may also prevent oxygen permeability to inhibit bacterial growth in the food [14].

Another mechanism of its antimicrobial activity includes the adsorption of the polymers of the chitosan on the microbes' surface to cause blockage.

A positively charged NH_3^+ group present in chitosan undergoes an interaction with the posphoryl group, which is negatively charged, which lead to intercellular electrolyte leakage, resulting to death of the bacteria [15]. Chitosan can penetrate the nuclei of the microbe inhibiting it from replicating and can bind to the DNA of the microorganism to inhibit RNA synthesis. Metals and essential nutrients can be chelated by chitosan making them unavailable for microbial used.

10.3 Chitosan Composites for Food Applications

Chitosan gotten from the deacetylation of chitin is a by-product of sea food industry. The preservative power of chitosan is in its film-forming ability, which forms edible coating of food products. The fungi and bacterial growth is prevented through the interaction between the negatively charged membranes of cationic chitosan film microorganisms. For years, pristine chitosan films have been researched on as a potential material for packaging of food. Nonetheless, current work on chitosan centers on different functional components of chitosan film to enhance its properties.

10.3.1 Chitosan Enhanced with Nano-Sized Metals

10.3.1.1 Nano-Sized Zinc Oxide Particles (ZNPs)

Oxide of zinc stands out as an economical choice, among metals that are nanoscale. It has a widespread utilization across various domains due to its exceptional antimicrobial properties, lack of toxicity, effective UV absorption, robust stability, and potent photo-catalytic characteristics [15]. Currently, the US Food and Drug Administration (FDA) regards it as a harmless substance for humans. It is used for various purposes such as a substance for preserving food, safeguarding against the sun, covering surfaces, adding color to paint, as a material for packaging, and in the purification of water. It has also further gained recognition in the realm of coatings and food packaging films where its purpose is basically to maintain color of food, inhibit the proliferation of harmful microorganisms and disease-causing bacteria, and boost the characteristics of these packaging materials; this includes their strength, thermal resistance, ability to block substances especially harmful ones, and overall durability [16]. In the typical procedure for creating chitosan-ZnO nanocomposites, researchers commonly employ a two-stage method. In the first stage, they introduce Zn particles into a chitosan solution with a low pH, resulting in the formation of chitosan-Zn ion combinations. Subsequently, these combinations are transformed into chitosan-ZnO nanocomposites through an *in situ* crystallization process involving a hot alkaline treatment that produces ε-$Zn(OH)_2$ [17].

Numerous studies have examined the use of ZnO and chitosan (CH) nanoparticles in the field of antibacterial food packing. For example, one study demonstrated that creating nanocomposite pouches using CH-ZnO with antioxidant and antibacterial effects would potentially extend the duration of storage of raw beef meat. Another finding highlighted that incorporating nano-ZnO into CH improves its antimicrobial effectiveness and reduces water vapor transmission rates when compared to pure CH films. Furthermore, ZnO-dispersed CH films exhibited remarkable resistance against *Staphylococcus aureus* (*S. aureus*) and *Escherichia coli* (*E. coli*). Additionally, the manufactured CH-ZnO nanocomposite film displayed suitable antimicrobial properties a variety of harmful microorganisms, including *Pseudomonas aeruginosa* (*P. aeruginosa*), *Escherichia coli*, *Bacillus subtilis* (*B. subtilis*), and *Staphylococcus aureus* in fruits of the blackgrape, among other research [16].

10.3.1.2 Titanium Dioxide Nanoparticles (TNPs)

Titanium dioxide (TiO_2), an oxide of a metallic element, stands as an innovative photocatalytic substance due to its extraordinary catalyzing ability, chemical durability, minimal harm to health, and affordability [18]. It has been documented that there is substantial effectiveness against bacteria as a result of the creation of highly reactive chemicals called reactive oxygen species (ROS). TiO_2 has been widely employed as agents to prevent bacterial growth. They encourage oxidative damage to the phospholipids found in the membranes of microbial cells, resulting in a breakdown of membrane stability [19].

Nanocomposites of chitosan/TNP have drawn attention of scientists because of its combined antimicrobial ability. ROS is emitted as a result of TNPs photo-reactive function, whereas, as a positively charged polymer, chitosan inflicts damage upon the membranes of microorganisms, thereby increasing the shelf life of food products, as depicted in Figure 10.1 [20].

In the research carried out by Lin *et al.* [20], they observed that CH-TiO_2 nanocomposites exhibited a remarkable reduction of *E. coli* population by 6 logarithmic units following a 24-h incubation period. Furthermore, these nanocomposites demonstrated a minimal inhibitory concentration of 0.38 mg/mL when stored in refrigerated conditions during fruit storage.

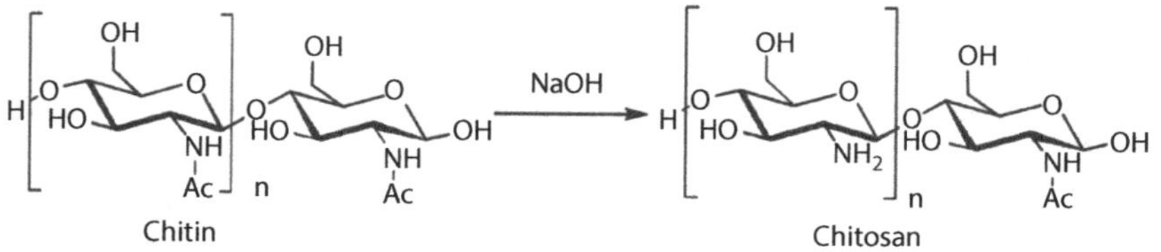

Figure 10.1 The preparation of chitosan from chitin through deacetylation [2].

Similarly, Zhang *et al.* [21] conducted an evaluation of chitosan-TNP nanocomposites to measure their effectiveness in inhibiting microbial growth, with the aim of exploring their suitability for application in food packaging, particularly when exposed to visible light. The chitosan-TNP coatings exhibited remarkable antimicrobial activity when tested against different strains, namely, *Candida albicans* (*C. albicans*), *E. coli*, *Aspergillus niger* (*A. niger*), and *S. aureus*, achieving complete inhibition within a period of 12 h. Chitosan-TNP films measuring 20 mm × 20 mm were fabricated in these antimicrobial experiments. Subsequently, about 0.1 mL of a microbial cell suspension with an approximate concentration of about10^6 CFU/mL was applied onto this film. These samples were then subjected to controlled conditions, including exposure to 20-W daylight lamp visible light, a humidity level of 50 ± 5%, and a temperature of 22 ± 2°C, After a period of 4 h, these microbial cells were then thoroughly rinsed off the film surfaces using around 2 mL of saline solution (0.85%). Finally, this remaining viable cells were then quantified using spread plate method.

10.3.1.3 Silver Nanoparticles (AgNPs)

Silver nanoparticles (AgNPs) are minute fragments of silver falling within the size spectrum of 1 nm to 100 nm, showing distinct attributes like electrical, optical, and magnetic characteristics, which find a broad scope of value [22]. Silver nanoparticles (AgNPs) exhibit high effectiveness against a broad range of microorganisms that are detrimental, encompassing viruses, yeast, mold, and bacteria [23]. Multiple approaches to counteract microorganisms through the utilization of AgNP activity have been recorded. These techniques involve the production and release of Ag+ ions, the development of ROS in both the inner and outer microbial membranes, disturbance of cellular membranes, destabilization of ribosomes, and damage caused to mitochondria and nucleic acids [24]. Qin *et al.* [25] in their research engineered smooth and receptive packaging films using chitosan, integrating around 2% AgNPs also extracts from purple corn.

They investigated the efficacy of these films in stopping the growth of *Listeria monocytogenes* (*L. monocytogenes*), *E. coli*, Salmonella, and *S. aureus*. According to their findings, chitosan nanocomposites fortified with AgNPs exhibited a commendable increase in antimicrobial potency against the four foodborne microorganisms when put side by side with unadulterated chitosan films. These chemical compounds have recently gained an approval from the US FDA as supplementary substances in food [16]. Pandey *et al.* [26] devised nanocomposite materials using chitosan-AgNP for the purpose of food packaging. They constructed this nanocomposite by blending AgNPs of dimensions measuring 80 ± 10 nm into chitosan within a mixture, which led to the formation of electrospun fibrous composite nano-layers. Subsequently, they assessed antibacterial properties of these fibrous layers using agar disc diffusion against *L. monocytogenes* and *E. coli*. Certainly, the polyvinyl alcohol (PVA) nano-layer exhibited no inhibitory effects. On the other hand, the PVA-chitosan and AgNP nano-layers demonstrated a significant inhibitory impact on the two tested organism. The most extensive inhibition zones, measuring 21 mm for *L. monocytogenes* and 20 mm for *E. coli*, and were observed in PVA (70%)–chitosan (30%)–AgNP nano-layers, potentially owing to the combined antimicrobial potency of chitosan and AgNPs. Moreover, chitosan (30%) and PVA (70%) nano-layers lacking AgNPs exhibited inhibition zones (15 mm for *L. monocytogenes* and 16 mm for *E. coli*) as a result of electrostatic interactions among the positively charged chitosan molecules and the negatively charged membranes of the bacterial cells.

10.3.1.4 Silicon Dioxide and Silica Nanoparticles

Another common component in the earth's crust is silicon [27]. Silicon dioxide (SiO_2) and silica nanoparticles are antibacterial, are stable, and are unable to be degraded by the digestive tract. Positively charged nanoparticles interact with the negatively charged cell surfaces of bacteria, resulting in a phenomenon suggestive of the early days of microscale particles with electrical charges and chemical reactions, leading to the formation of aggregates [28].

Chitosan, SiO_2, and β-acids were effectively combined to produce an active packaging film. The incorporation of β-acids (0.3%) with SiO_2 (0.3%) may possibly enhance the characteristics of this food packing film, including antibacterial property, water vapor permeability (WVP), mechanical, and antioxidant. The results revealed that the inclusion of β-acids and SiO_2 improves the anti-ultraviolet capacity of the film, which is critical for food to escape ultraviolet influence [29].

10.3.1.5 Copper Nanoparticles (CuNPs)

Cu NPs limit the growth of microorganisms in a way that is primarily dictated by the size of the particles and quantity of nanoparticles in the solution. Cu nanoparticles are efficient against bacteria, aside having good stability with antifungal properties, by penetrating through bacterial cell membranes and disrupting their key enzymes [30].

The incorporation of nano-CuO/ZnO to chitosan produced nanocomposite coatings with enhanced antibacterial activity, reduced moisture accessibility, and boosted the nutritional attributes of guava fruits, prolonging the duration of storage by 1 week. Furthermore, CuO added into chitosan killed over 99% of germs [31].

10.3.1.6 Magnesium Nanoparticles (MgNPs)

Magnesium oxide, a crystalline, pale mineral found in nature with an extreme melting point, exists commonly for use in a variety of industries. Although its heat conductivity is great, its electrical conductivity is poor. Recent research has focused on MgO's outstanding antibacterial properties and its greater stability to organic antimicrobial compounds. This combination of qualities makes it a flexible solid material. Numerous building components employed in manufacturing of these food packaging are gradually being replaced in United States and Europe with packaging made of MgO nanoparticles. Hydrothermal, laser ablation, microemulsion, microwave-assisted, sol-gel, ultrasound-assisted, and wet chemical reaction techniques can be utilized to produce it because it is thermally robust, reusable, adaptable, and not permeable to gas and has antibacterial activity [32].

10.3.1.7 Sulfur Nanoparticles (SNPs)

Sulfur nanoparticles (SNPs) offer promising potential in numerous industries due to their unique characteristics, which include their substantial surface area, catalytic ability, and compatibility with biological systems [33]. Sulfur nanoparticles (SNP) have been regarded to be effective antibacterial substances. According to studies, a chitosan/SNP composite with a higher elastic modulus and enhanced tensile strength has been created. This SNP concentration enhanced the tensile robustness/strength of the chitosan/SNP films. Additional mechanical testing revealed that SNP decreased the elongation at the point of break of the chitosan films. Chitosan/SNP composite films displayed improved thermal stability than chitosan films, according to a thermogravimetric examination. According to the findings, chitosan/SNP composite films totally stopped the development of *E. coli* and *L. monocytogenes* within 6 h and also within 12 h, respectively. This finding may have an impact on active food packaging materials. A method known as sequential stratum fabrication was employed to produce chitosan and PVA layers with ferulic acid (FA) serving as a cross-linking agent. As a result of the consistent distribution of nanoparticles, the antioxidant and antibacterial activities, tensile strength, and resistance to water enhanced. It was discovered through experiments that the resulting films could increase the duration of storage of cherry tomatoes [2]. Figure 10.2 shows the structural mechanism of chitosan supplemented with nano-metals.

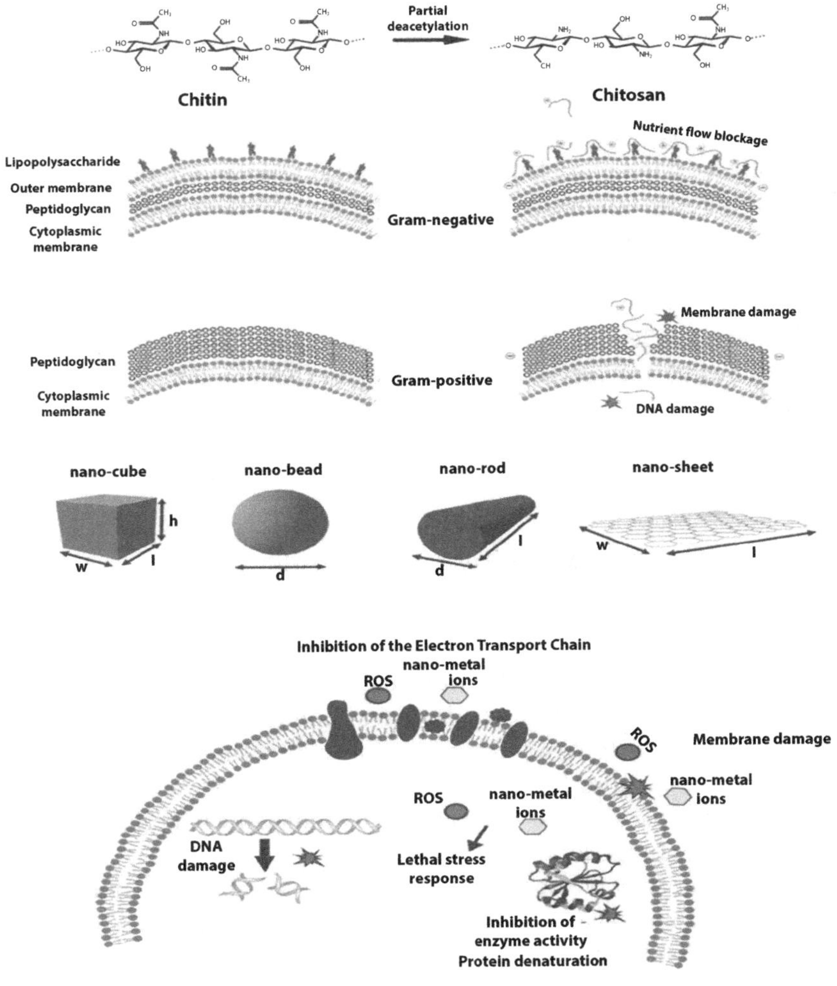

Figure 10.2 Mechanism of chitosan supplemented with nano-metals [17].

10.3.1.8 Chitosan Enhanced with Carbon

Graphite crystals, readily available and cost-effective, yield graphene oxide (GO), a single-layered material consisting of carbon, hydrogen, and oxygen molecules, when they undergo oxidation. It may be processed easily and dissolves in water. Most significantly, the GO can be (partially) converted to sheets that resemble graphene, when the oxygen-containing groups are removed and a conjugated structure is recovered. The reduced GO sheets, also known as rGO, are typically regarded as a type of chemically produced graphene. There are ways to modify rGO once it has been created so that it can be used in various applications. One can improve the characteristics of the compound to suit commercial uses by treating rGO with other chemicals or by mixing rGO with other materials to create novel substances. The functional modification of GO holds the key to numerous applications in addition to its crucial role that is to regulate the exfoliating behavior of GO also rGO [34]. Due to its significant antibacterial qualities, grapheme is currently used in the packaging of food in companies. Graphene oxide's (GO) ability to be uniformly spread in chitosan polymer nanocomposite films makes it unique. The mechanical performances and thermal stability were greatly improved by GO-chitosan nanocomposite films [2]. Chitosan and GO cross-linking may significantly alter the mechanical characteristics. The tensile strength and thermal characteristics of this chitosan film were enriched by adding GO. This inclusion of GO also contributed to the antibacterial activities against *B. Subtillis* and *E. coli*. Alcohol, ketones, and also aldehydes that are included in food flavors can be preserved well using carbon/chitosan packaging and laser-induced micropores [2].

10.3.2 Polysaccharide-Chitosan Composite

Chitosan-based functional films can be made by combining polysaccharides and chitosan, according to research. The use of starch has a number of benefits, including availability, biodegradability, renewability, and affordability. Chitosan and starch also have good film-forming properties that make it possible to create composite films [35]. The film's antioxidant activity and water vapor barrier properties were improved, and bacterial adherence to the packing was decreased. This composite material indicates its potential uses in the food packaging sector.

The characteristics of food-grade cellulose-based sheets manufactured from maize husk and sugarcane bagasse covered with chitosan and a beeswax-chitosan emulsion were studied by Bhardwaj *et al.* [36]. Depending on the temperature, the mold and yeast growth was slowed down by 8–12 days on the WC (beeswax-chitosan emulsion–coated) and CC (chitosan-coated) papers. In addition, enhanced barrier properties against air and moisture are attributed to the quality of the bread. The chitosan/cellulose nanocrystals (CNCs) displayed fungicidal, gram-positive, and gram-negative bacterial antibacterial activities. The growth of *Pseudomonas* and *Enterobacteriaceae* bacteria in meat was like wise inhibited by the chitosan films. The chitosan films that were produced had improved thermal stability and an improved oxygen barrier, but their water permeability remained unchanged. In addition, chitosan/CNC films showed improved Young's modulus and tensile strength [37]. In a different investigation, it was found that adding CNC to a chitosan solution enhanced the water resistance of the films created using the solvent casting technique. After being placed in a petri dish, the nanochitosan (NCH)–CNC suspension was dried for 2 days at 50% relative humidity. The final NCH-CNC films displayed a surface that was smooth and continuous. According to Pinem *et al.* [38], the incorporation of nanocellulose resulted in a smoother film morphology and a reduction in the film thickness.

Starch has demonstrated potential in the ongoing development of biodegradable materials to create active food packaging materials because it is readily available, inexpensive, useful, and, most significantly, biodegradable. Colorless, tasteless, odorless, non-toxic, and biodegradable are the typical characteristics of starch-based films. Starch films cannot be used in the food packaging sector because they do not have the same mechanical or barrier properties as synthetic polymers. As a result, researchers are looking into how to combine starch with other natural biopolymers while maintaining its properties. The thermal stability, tensile strength, water adsorption, hydrophobicity, and gas barrier capabilities of food packaging were all enhanced by the intermolecular interactions between the amino functional groups of chitosan [39]. The cast-method–produced corn starch and chitosan film had outstanding elongation at break and tensile strength. Composite films made of thermoplastic corn starch (TPS) and chitosan oligomer (CO) increased food shelf life. As an antibacterial agent, the COs diffused through the polymer and onto the food surface. The film was found to be able to preserve perishable goods for a week. The resulting TPS/CO composite sheet demonstrated strong antibacterial properties and offered a practical strategy for microbial growth inhibition [40].

The composite film made of sugar palm starch (SPS), chitosan, and extra virgin olive oil (EVOO) had the lowest second-stage degradation temperature (184.2°C). In contrast, chitosan/SPS-EVOO (2%) had the best improvements in elongation at break and tensile strength and displayed the greatest second-stage degradation temperature (223.1°C).

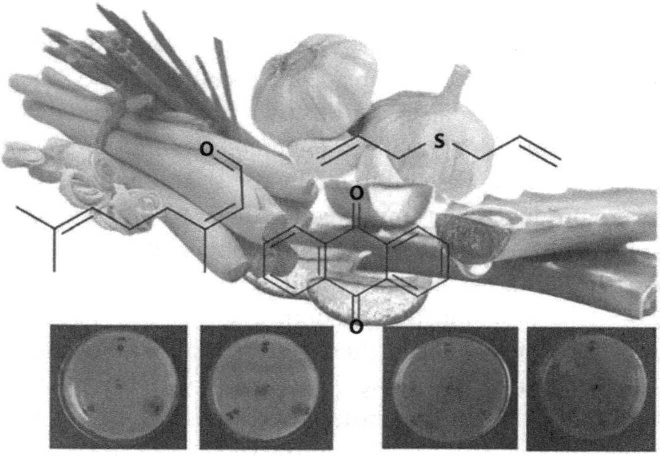

Figure 10.3 Pictorial representation of a composite film made of chitosan and starch that has antibacterial properties [42].

Response surface methodology was used to create a chitosan-Dioscoreaa-lata starch composite film that exhibited favorable food packaging characteristics such biodegradability, moisture content, elastic modulus, and tensile strength. Furthermore, the antibacterial effects against *Salmonella typhi*, *E. coli*, *S. aureus*, and *S. epidermidis* were strengthened by the addition of essential oils from garlic, lemon grass, and aloe vera.

Additionally, the starch-chitosan and montmorillonite filler nanoparticles in the composite film greatly improved the water vapor barrier characteristics and tensile strength [41]. Figure 10.3 shows the pictorial representation of a composite film made of chitosan and starch that has antibacterial properties.

Chitosan nanoparticles (CNPs) were added as a natural filler to starch films, which enhanced their mechanical characteristics and water sorption [43]. The starch/CNP nanocomposite films increased the shelf life of perishable goods while preventing the growth of *S. aureus*, *B. cereus*, *E. coli*, and *S. typhimurium* [44]. To create a composite film, Zhao *et al.* [45] used subcritical water technology to combine chitosan and gallic acid with cassava starch. The mechanical characteristics of each individual film were influenced by hydrogen bonding and crystallinity. The tensile strength and elongation (100.1%) of the films made of chitosan and starch were found to be the greatest. The shelf life of goods can be increased by using cassava starch, which has also been demonstrated to be an effective transporter of antioxidants and antimicrobials. The composite film made of garlic acid, chitosan, and cassava starch extended the shelf life of cooked ham for 25 days and may be used as a food packaging material [45]. Due to the intermolecular interactions between the dipolar ions, pectin, another polysaccharide, can be combined with chitosan to create a stable and homogenous composite film. The film's water vapor permeability and moisture content rose when pectin, tea polyphenols, and chitosan were added.

According to Gao *et al.* [46], adding TP increased antioxidant activity and decreased bacterial intermolecular growth. Similar to this, Schnell *et al.* created a film containing colloidal suspensions of xylan (Xyl) and chitosan's cationic polyelectrolyte complex (CatPEC). Lower oxygen permeability and water vapor transmission rates could be seen in the resulting film. Additionally, it prevented the growth of *S. aureus* and *E. coli.*

Amorphophalluskonjac is the source of the polysaccharide and water-soluble dietary fiber known as konjacglucomannan (KGM). KGM enhanced tensile strength and decreased WVP when coupled with TEMPO-ChNCS (TEMPO-oxidized chitin nanocrystal). The composite of epigallocatechingallate (EGCG) and KGM/carboxymethyl chitosan (KGM/CMCS) also shown beneficial qualities, including exceptional antioxidant and antibacterial activities that are crucial for food packaging. According to Sun *et al.* [47], EGCG improved the water vapor barrier qualities as well as the thermal stability, tensile strength, and UV-light barrier capabilities. Additionally, to reduce postharvest loss and increase the shelf life of tomatoes, a de-oiled crude green algal ethanolic extract (CAEE) was added to chitosan films [48]. Their mechanical and physical characteristics are enhanced by chitosan composite. Figure 10.4 shows the pictorial representation of tomato: (a) control, (b) nCS, (c) CS/CAEE of 2%, and (d) CS/CAEE of 4%, at a specific storage day.

10.3.3 Essential Oil-Chitosan–Based Composite for Food Applications

Natural plant essential oil extracts have showed promise as edible film food preservatives that are also friendly to the environment. Natural essential oil extracts from plants are added to the film, increasing its antifungal and antibacterial activity [14].

The essential oil in the chitosan-based film controls the pace at which antimicrobial agents are released to maintain the concentration of active compounds over an extended length of time. Between 85% and 99% of essential oils are made up of volatile compounds such terpenoids, terpenes, alpha-aromatic compounds, and aromatic compounds; the remaining 1% to 15% are non-volatile compounds, which are often used in the food sector [49].

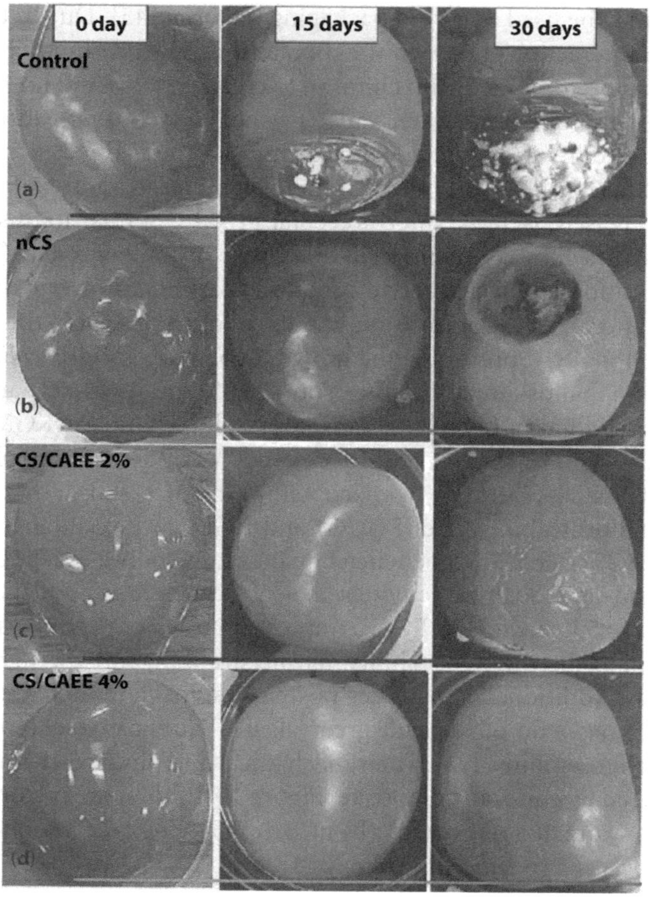

Figure 10.4 Pictorial representation of tomato: (a) control, (b) nCS, (c) CS/CAEE of 2%, and (d) CS/CAEE of 4%, at a specific storage day [48].

Chitosan and fish sarcoplasmic protein (FSP) film's physiochemical properties were enhanced by ginger essential oil, which was derived from ginger. All of the composite films made of chitosan, FSP, and ginger essential oil exhibit outstanding UV-barrier qualities from 200 nm to 280 nm, a drop in tensile strength, and a greater increase in elongation at break.

Through the use of composite film, fillet shelf life is extended by 2 days, and the antibacterial properties of chitosan film are also improved. In order to improve preservation, Ginger essential oil was also slowly released into the surface of the fish [50]. Another chitosan-based composite film that contains potato peel extract and PVA also contains fennel essential oil, making it to contribute significantly in the enhancement of the film's antibacterial and antioxidant characteristics.

The mechanical qualities of the NCH potato peel extract-PVA-fennel essential oil composites film were improved, extending the food's shelf life. *Pseudomonas putida* completely damaged the composite film in 2 days, but the chitosan control film took nearly a week [51]. Additionally, chitosan and honeysuckle flower extract (HFE) composite films improved antioxidant and antibacterial properties and decreased WVP. The DPPH scavenging activity and *E. coli* inhibition zone enhanced with the addition of HFE (30%). Mechanical testing, however, revealed that the addition of HFE decreased the tensile strength and elongation at break. According to Wang *et al.* [52], the composite films show enormous promise as antimicrobial packaging for extremely perishable commodities.

Extensive focus has been dedicated to the antibacterial properties of the essential oil derived from cinnamon. The enhancement of emulsion capacity in octenyl succinic anhydride (OSA) and Gum Arabic (GA)-based film was achieved by stabilizing OSA and GA with cinnamon essential oil (CEO). The CEO emulsions improved the antibacterial activity against *S. aureus* and *E. coli* in the chitosan films. It diminishes the film's structural integrity, yet, by employing OSA-GA stabilized CEO emulsions, it is possible to enhance the durability and safeguard the edible films [53]. The thermal, mechanical, and bactericidal properties of chitosan films enhanced by thymine essential oil are similar. Furthermore, by retaining increased total soluble solids, titrable acidity, and lower weight loss, the mixture of zinc oxide (ZnO), polyethylene glycol (PEG), calcium chloride, and nanoclay (NC) in the chitosan-based film retained the quality of delicious cherries (Zehra *et al.*, 2022) [88].

10.3.4 Gelatin-Chitosan–Based Composite for Food Packaging Applications

Natural biopolymer components are becoming more prevalent in food packaging applications in order to meet rising consumer demand and alleviate environmental problems brought on by petroleum-based plastic packaging in recent years [54]. The usage of tailored edible films made of edible biopolymers for food packaging that is safe and multifunctional is growing significantly. Chitosan and gelatin are common edible biopolymers used to create high-performance films [55–57]. Chitosan has a positive charge and is soluble in solutions with pH values lower than 6.5 as a result of the protonation of the amino groups. It can take on many different forms, including films, gels, and nanoparticles [58, 59]. The great film-forming potential of this polysaccharide makes it one of the most studied polysaccharides for packaging film. However, the weak mechanical and water barrier properties of chitosan films prevent them from being used as food packaging [59].

Gelatin, an odorless, colorless, and water-soluble protein, is created when collagen is partially or completely hydrolyzed. Gelatin is separated into two groups based on the method of extraction. Type A gelatin is produced using an acid extraction method, whereas type B gelatin is produced using an alkaline extraction method. The main sources of gelatin are pig skin (46%), bovine hides (29.4%), pig and cow bones (23.1%), and fish skin (1.5%). However, due to its great availability and biocompatibility, gelatin is excellent for usage in food packaging and other industries [60]. Different types of animal by-product gelatin have good mechanical and barrier properties. The triple helix structure of the gelatin protein gives it physical strength, and the different amino acids that make up the gelatin structure protect packaged goods from oxidative damage. Gelatin is a great material for packaging films because of these characteristics. Due to its functional features, such as its capacity to bind water, create films and foam, and have an emulsifying tendency, gelatin is a good substitute for food packaging in the food industry [61].

Polymer nanocomposite has garnered more attention than other modern nanotechnologies in the area of food packaging. It possesses superior barrier properties, moisture permeability, oxygen permeability, solvent resistance, thermal stability, anti-microbial properties, and mechanical toughness. Films for packaging composed of biopolymers have been marketed as alternatives to the often-used plastic packaging produced from petroleum. Disposing of traditional petroleum-based polymeric materials can potentially result in adverse consequences for both the environment and human well-being. It is necessary to use a different kind of packaging material. Polysaccharide-based biopolymers are highly valued by the packaging industry because they are biodegradable, biocompatible, cost-effective, non-toxic, renewable, and edible. When the ideal moisture, temperature, and oxygen availability conditions are met, they degrade and lessen the creation of toxic leftovers that may pollute the environment [62]. Gelatin (GE), a biopolymer generated from collagen, has the ability to thicken, texturize, and form gels. The food industry is able to use GE in a larger range of food products because of its emulsifying capabilities. Food has been extensively protected from the outside environment, including light, air, and pathogens, using GE's opulent film-forming capability [63]. Consequently, it creates high-quality food packaging material, contributing to the preservation of food ingredients' freshness. In order to quickly absorb UV rays and prevent light from reaching the food contents, GE contains a chain of aromatic amino acids. However, due to its poor mechanical, elasticity, and thermal qualities, it cannot be used in a wide range of applications. A range of additives, including synthetic and biopolymers like chitosan (CH), starch, or pectin, can be mixed to work around these limitations. Chitosan (CH) is an organic and cationic biopolymer with intrinsic biocompatibility, biodegradability, and antibacterial characteristics. In-depth research is done on it in a wide range of disciplines, including those involving powder, films, membranes, and sheets. CH is used increasingly commonly in food packaging applications because it is naturally antibacterial and prevents bacteria from entering and growing on food goods [64].

The preservation of the products' quality and appearance is the major objective of food packaging. This is achieved by reducing lipid oxidation, restricting microbial development, and lengthening food products' shelf lives. Petroleum is the primary source of the majority of the polymers used in conventional packaging. The international output of plastic has expanded over the past few decades, and 40% of that plastic is utilized for packaging. Due to its low cost, light weight, ease of form-molding, great mechanical strength, and thermal sealing, plastic is a beneficial material for packing; however, its widespread usage in packaging may have adverse environmental implications. Because plastic packaging degrades slowly and is not reused or recycled frequently, it has a detrimental impact on the environment. A lot of plastic could, therefore, pollute and contaminate the world [64].

Edible and eco-friendly packaging applications have recently been investigated as an alternative to conventional or synthetic packaging due to the increasing interest of consumers in natural, safe, and healthy food as well as researchers' efforts to satisfy these needs and produce biodegradable and non-toxic films and coatings for the food industry. Consequently, a vast array of biopolymers are being carefully examined as potential materials for food packaging. For use in packaging, proteins, lipids, polysaccharides, and any conceivable combinations of these have all been researched as edible biopolymers. These biopolymers or their mixes have occasionally included antioxidants, antimicrobials, fragrances, and colors to enhance the qualities of the films [64].

Collagen is a protein present in animal connective tissue that is partially destroyed to produce gelatin. It is a solid that is translucent, shiny, and white or yellowish. The ability to bind water, create gels, operate as a water vapor barrier, form films and foams, and have a propensity to emulsify are just a few of the numerous beneficial qualities of gelatin. Gelatin can be used in the food, pharmaceutical, photography, and cosmetic sectors thanks to these qualities. Gelatin has great swelling and water absorption properties, yet it has low mechanical resistance and is permeable to water vapor molecules. Gelatin's use as a packing material is restricted by its weak water vapor barrier property. Gelatin can be improved by mixing it with additional beneficial elements and active ingredients [65].

Edible films and coatings can be produced using a variety of techniques, including solution casting, extrusion, coating, layer-by-layer (LBL) assembly, and others. Solution casting is a widely used method for making edible films and coatings because it is simple, affordable, and effective. The polymer structure is produced during this step by intermolecular electrostatic and hydrogen bond formation. Although extrusion technology is a potential approach for producing films and is commonly chosen over solution casting techniques because of its faster processing times and lower energy consumption, there are not many studies on the application of extrusion technology to gelatin film. Fresh foods like fruits, vegetables, seafood, meat, and other goods are routinely coated in order to increase their shelf life. To develop edible active packaging films and coatings, active agents can be included in the structure of a single polyelectrolyte or between layers via LBL self-assembly [65].

10.3.4.1 Application of Gelatin-Based Composites for Packaging Various Food Items

Modern consumers have a preference for "green" consumerism; they typically desire a greater variety of naturally occurring foods with little to no processing, fewer artificial food additives, and food packaging that is easily biodegradable, recyclable, and environmentally friendly. To achieve these goals, various biodegradable packaging materials other than plastic have been explored [66].

10.3.4.1.1 Gelatin-Based Composite for Fruit Packaging and Preservation

Due to their high moisture content and oxygen permeability, fruit lipids and vitamins can oxidize, which can also cause Maillard reactions, enzymatic browning, physiological aging, biochemical changes, and microbiological degradation [67]. These fruits' nutrient content, microbiological safety, sensory quality, and shelf life may all be enhanced by coating them in gelatin-based films. In order to increase the shelf life of strawberries and facilitate storage between harvest and consumption, Fakhouri *et al.* [68] employed gelatin CNCs to produce biofilms. In a recent study, strawberries were coated with a filmogenic solution (G-CNC) for 1 min before being dried for 24 h at 15°C. After that, the strawberries were maintained in the refrigerator, and their water content, weight, ascorbic acid content, and titratable acidity were all monitored on a regular basis. It was shown that strawberries coated with GEL or CNC had a significantly longer shelf life by reducing microbial degradation. Additionally, after 8 days of storage, the coating reduced weight loss from 65% to 31%. A slower fall in ascorbic acid content was seen in samples coated in gelatin than in samples of fruits that were left exposed, according to the study. Soradech *et al.* [69] also used a shellac-gelatin composite film coating to extend the shelf life of banana fruits. In this study, bananas were covered using a dipping method and stored at 25°C for 30 days. The study showed that a composite film comprised 40% gelatin and 60% shellac provided a strong barrier against the ripening process and retained the fruit's quality for more than 30 days.

10.3.4.1.2 Gelatin-Based Composite for Vegetable Packaging and Preservation

Numerous studies have demonstrated the value of gelatin-based films in fresh-cut goods. Halim *et al.* [70] developed a gelatin-based film comprising CS, methylcellulose, and tannic acid to preserve cherry tomatoes and grapes. The cherry tomatoes were wrapped in gelatin (22.1%–15.5%), CS (21.3%–19.6%), methylcellulose (26.2%–20.5%), and tannic acid (15%). They monitored the cherry tomatoes and grapes' physical, chemical, and microbiological characteristics over time in order to determine how well the complex covering preserved them by reducing weight loss and browning index values. Gelatin-coated tomatoes were also resistant to microbial deterioration due to their resistance to *E. coli* and *S. aureus*. Cui *et al.* [71] also developed a gelatin-based film combined with CO-loaded CS nanoparticles (CO@CNPs) to further increase antibacterial activity against *E. coli* O157:H7 biofilms on cucumbers. *In vitro*, the population of *E. coli* O157:H7 biofilms was reduced by around 99.99% when cucumber samples were treated with gelatin (9 mg/mL) and CO@CNPs for 24 h. *E. coli* O157:H7 biofilm in the population also decreased from 4.28 to 4.97 log10 points. The findings of the sensory evaluation showed that the gelatin-CO@CNP nanofiber coating maintained the color and flavor of cucumbers for more than 4 days.

10.3.4.1.3 Gelatin-Based Composite for Fish and Meat Packaging and Preservation

Gelatin coatings that stop microbial growth and lipid oxidation can be used to improve the shelf life of fish and meat products when kept chilled. Sun *et al.* [72] developed a coating consisting of gelatin and curcumin/

cyclodextrin (CUR/CD) to shield grass carp (*Ctenopharyngodon idella*) fillets when they were stored at 4°C. According to the study, the gelatin-based coating stopped lipid oxidation and proteolysis in GCFs during chilled storage at 4°C. The coating decreased oxidative damage, chemical deterioration, and microbiological deterioration while extending the shelf life of GCFs by about 6 days. Mohebi and Shahbazi [73] developed a chitosan-gelatin active packaging film consisting of cellulose nanoparticles, Ziziphora clinopodioides EO, and pomegranate peel extract. The gelatin-based composite film displayed high antibacterial action by inhibiting the growth of numerous foodborne pathogenic bacteria, including *Pseudomonas fluorescens*, *Shewanella putrefaciens*, *Pseudomonas* sp., *Listeria monocytogenes*, and lactic acid bacteria in chilled shrimp. The sensory and organoleptic qualities of the shrimp, such as color, texture, and flavor, were also found to be retained for at least 10 days in cold storage using chitosan-conjugated gelatin films. Cardoso *et al.* [74] talked about the effects of monolayer and bilayer chitosan-gelatin coatings on beef steaks. In this study, the weight loss, rate of lipid oxidation, appearance, and shelf life of gelatin-coated beef steaks were examined after they had been kept in the fridge for 10 days. The study found that gelatin-CS coatings efficiently prevented the development of metmyoglobin on the surface of beef as well as the growth of microbes such as yeast, mold, and psychrotrophic bacteria. Chitosan-gelatin coatings prolonged the shelf life and display life of the beef during the course of the 6–10 days of retail storage.

10.3.5 Clay-Chitosan–Based Composite for Food Packaging

Biopolymer-based nanocomposites have superior thermal, mechanical, and biodegradable qualities, and these have attracted a lot of attention recently. To make polysaccharide-clay nanocomposites, inorganic nanoscale clay is frequently utilized as a filler in polysaccharide matrixes. Polysaccharide-clay nanocomposites, when employed as packaging materials or when combined with antimicrobial substances, have the potential to be transformed into active packaging solutions that showcase antimicrobial properties suitable for diverse food safety applications [75]. Through the incorporation of nanoclays, chitosan can undergo physical enhancements, resulting in enhanced properties suitable for practical applications. Within the chitosan framework, nanoclays, a category of inorganic minerals, exhibit the ability to permeate and generate three distinct structures: tactoid, intercalation, and exfoliation. Nanoclays are typically regarded as traditional fillers, as their layers remain unseparated, and the complete clay particle disperses uniformly within the chitosan matrix in a tactoid configuration.

The most effective clay nanocomposites with chitosan are characterized by intercalation or exfoliation structures. Intercalation involves the insertion of 1.0%–3.0% of chitosan chains between the clay layers, resulting in an expansion of the interlayer spacing while maintaining the parallel orientation of the nanoclays in an orderly layered structure. Various techniques, such as agitation, ultrasonication, and homogenization, are employed to facilitate the integration of chitosan and nanoclays during exfoliation, leading to the dispersion of individual clay layers throughout the chitosan matrix and their departure from parallel alignment [75].

10.3.5.1 Food Packaging-Related Applications of Chitosan-Clay Nanocomposites

An efficient way to create reasonably priced, ecologically friendly, and fiercely competitive packaging materials is to use chitosan-clay-based nanocomposites. Because chitosan and clay are both natural substances—a biopolymer and a naturally occurring inorganic mineral, respectively—biodegradability can also be preserved, especially because the degradation of the nanocomposites does not release any harmful or artificial chemicals into the environment [76]. These biocompatible and biodegradable materials can thus replace traditional polymers made from fossil fuels because they have usable structural and functional characteristics. It has been proven that incorporating clay into chitosan can give pure chitosan films and coatings unique physical and chemical properties at a loading level of 1 to 5 weight percent. These property improvements include improved mechanical properties, improved barrier, antioxidant, and antibacterial capabilities, as well as increased heat stability [75].

10.3.6 Polyphenolics-Chitosan-Based Composite for Food Packaging

The polyphenol family is one of the most prevalent and comprehensive groups of natural antioxidants from plant sources, such as herbs and spices. According to Talón *et al.* [35], the ability of polyphenols to chelate metals, restrict the activity of the lipoxygenase enzyme, and scavenge free radicals is related to their antioxidant capabilities.

Polyphenols are secondary plant metabolites that are essential for a plant's development, reproduction, pigmentation, and defense against UV radiation and disease. According to the number of phenolic rings that they contain and the different compounds that link these rings together, polyphenols are divided into four classes: phenolic acids, flavonoids, stilbenes, and lignans. Their wide range of biological activities, including antioxidant, antiallergic, antibacterial, anti-inflammatory, anti-cancer, and antiviral, have all been well investigated [77]. They can frequently be discovered in fruits, vegetables, teas, and coffees.

In addition to their nutritional advantages for human health, plant polyphenols have a wide range of positive functional attributes, including exceptional antibacterial and antioxidant activities [78, 79]. Polyphenols can be found in fruits, vegetables, green and black tea, red wine, coffee, chocolate, and extra virgin olive oil. According to Zhang and Jiang [80], the majority of plant polyphenols are thought to be safe and can be used directly as food additives. Numerous studies have lately focused on adding plant polyphenols to biodegradable food packaging films in an effort to boost the antioxidant and antibacterial properties of active films and thus improve food preservation. The various plant polyphenols include tea polyphenols (TP), which are natural polyphenols obtained from tea (*Camellia sinensis*). Its beneficial functional features, which include antioxidant, antibacterial, anti-inflammatory, anti-tumor, and anti-cancer properties, have been extensively utilized by the food and pharmaceutical industries. The creation of biodegradable food packaging sheets has gained attention recently. A few of the active films that TP has shown substantial potential for boosting are alginate, gelatin, chitosan, pectin-konjac glucomannan, and starch [80].

Active films that incorporate TP are now achieving excellent results in food applications and have the potential to greatly increase the application efficiency of active films for food preservation. The use of active films in food preservation can be used as a film for food packaging as well as an edible coating. The edible coating contains active films, which are extremely small. A thin layer of a little thickness is directly created on the surface of the food as a fresh-keeping technique. To develop the edible coating, the meal is either immediately immersed in the film-forming solution or has the solution sprayed on its surface. Although an independent film cannot be created with the film-forming solution, active films can [81].

The strong antioxidant and antibacterial properties of chitosan films, including TP, showed great potential for increasing the shelf life of beef samples [82]. Compared to unpackaged and pure chitosan film-packaged beef samples, chitosan film containing TP extended the shelf life of beef samples from 3 and 4 days to 6 days, along with fewer bacterial colonies and a degree of lipid oxidation. Additionally, when the concentration of TP in the chitosan film increased, the pH level of the beef sample decreased. This may be due to lactic acid generation by lactic acid bacteria as well as the synergistic antibacterial action of TP and chitosan [82]. The use of pectin-chitosan, curdlan-chitosan, and distiller-dried grain active films incorporating TP significantly extended the shelf life of fresh meat.

Fresh fruits continue to rapidly respire in postharvest storage, resulting in water loss and quality degradation [80]. Fresh produce's quality is maintained throughout storage by lowering water loss and gas exchange thanks to the active packaging film or coating's high barrier properties. When compared to unpackaged and polylactic acid/chitosan-packaged cherry fruits, the cherry fruits packaged with polylactic acid/chitosan film containing TP demonstrated a lower decay rate and weight loss while maintaining better firmness and ascorbic acid content [83]. Oil storage quickly degrades when exposed to light and oxygen, which contributes to deterioration. Studies have shown that the addition of TP might improve the efficiency of the light and oxygen barriers created by active films; this might therefore prevent fatty foods from losing quality while being stored. Wu *et al.* [84] pioneered the creation of a biologically active, ingestible packaging material derived from pomelo peel powder, enriched with TP, with the goal of reducing the reliance on oil. In contrast to both un-encased samples and oil samples wrapped in pomelo peel flour film, their findings revealed that the inclusion of TP in the pomelo peel flour film significantly lowered the peroxide value throughout the storage period. Moreover, the flour film crafted from pomelo peel combined with TP showcased a remarkable capacity to block the infiltration of oil. The rationale behind the enhanced efficacy of employing active films or edible coatings for food preservation when TP is incorporated can be attributed to the subsequent factors: The interaction between TP and the film's framework can enhance the protective properties of active films or edible coatings, thereby retarding the food's water loss and gas exchange; TP can impart UV-resistant attributes to active films or edible coatings, thus postponing food oxidation caused by UV radiation; and TP can provide active films or edible coatings with exceptional antioxidant and antimicrobial properties, thereby delaying food spoilage [81].

10.3.7 Polyvinyl Alcohol-Chitosan–Based Composite for Food Packaging

Food packaging is essential for protecting food from environmental hazards and maintaining it throughout storage and transit. Additionally, it provides information to retailers and consumers regarding food producers, expiration dates, storage needs, and chemical composition. It significantly contributes to boosting product sales by utilizing enticing colors and designs.

According to Oun *et al.* [85], materials with the following qualities are most in demand for food packaging films: good mechanical and thermal properties; desirable optical properties; biodegradability; a high barrier to water vapor, aroma, and gases; and functional properties like antibacterial, antioxidant, UV protection, detection, and tracking properties.

PVA is permitted by the United States Department of Agricuture (USDA) for use in the packaging of meat and poultry products and has been acknowledged by the FAO/WHO Joint Committee of Experts on Food Additives as safe [85].

Polymerizing one of the hydrophilic synthetic biodegradable polymers, PVA, is created by polymerizing vinyl acetate monomer to create polyvinyl acetate (PVAc), which is then hydrolyzed to create PVA. PVA is made commercially available in a number of forms by controlling the alkali process known as saponification, which happens when the ester groups in vinyl acetate are partially replaced with OH groups (as a monomer) in the presence of a NaOH medium. As a result, the saponification reaction's conditions have an impact on how much PVA is hydrolyzed. PVA's mechanical, physical, and barrier properties are typically determined by its molecular weight (200,000–400,000 g/mol) and degree of hydrolysis (80%–99%, or 84.2%–89% partial hydrolysis, 92.2%%–96.5% moderate hydrolysis, and 98%–99% complete hydrolysis). PVA polymers with higher molecular weights and levels of hydrolysis, according to Tang and Alavi [86], offer improved tensile strength, water resistance, and adhesive strength. PVA has been employed in a number of industries due to its compatibility, safety, capacity to produce high-quality hydrogels, and high swelling characteristics. These include packaging, optical and electronic sensors, construction and building materials, packaging, and biomedical and pharmaceutical applications [87]. Application might vary so depending on which one, PVA can be found in hydrogels, beads, fibers, films, membranes, and scaffolds [85].

10.4 Conclusion

Replacing disposable synthetic shopping bags made from plastic to environmentally friendly options such as chitosan-based composites is an important step toward minimizing the environmental dangers posed by plastics that are not biodegradable. Chitosan, obtained from natural sources, improves food packaging by providing antibacterial properties, biodegradability, and safety. The further incorporation of nanotechnology and the coupling of chitosan with other chemicals have resulted in materials with superior qualities, guaranteeing longer shelf life for perishable goods and lowering our dependency on these toxic plastics. The increased usage of biopolymers such as chitosan and gelatin, as well as improvements in nanocomposites and the use of environmentally friendly polymers such as PVA, demonstrates an obligation to eco-friendly food packaging. These solutions address consumer preferences as well as the critical need to reduce plastic pollution and environmental effect in the food companies. The incorporation of nano-metals and GO to chitosan improves its antibacterial and storage properties, strengthening its status as a flexible material for innovative food packaging solutions. These findings point to a viable route for more sustainable, efficient, and environmentally responsible food packaging practices.

References

1. Cesa-Luna, C., Alatorre-Cruz, J.M., Carreño-López, R., Quintero-Hernández, V., Baez, A., Emerging Applications of Bacteriocins as Antimicrobials, Anticancer Drugs, and Modulators of The Gastrointestinal Microbiota. *Polym. J. Microbiol.*, 70, 143–159, 2021.
2. Jiang, A., Patel, R., Padhan, B., Palimkar, S., Galgali, P., Adhikari, A., Varga, I., Patel, M., Chitosan Based Biodegradable Composite for Antibacterial Food Packaging Application. *Polymers*, 15, 2235, 2023. https://doi.org/10.3390/polym15102235.
3. Kontominas, M.G., Use of Alginates as Food Packaging Materials. *Foods*, 9, 1440, 2020.
4. Flórez, M., Guerra-Rodríguez, E., Cazón, P., Vázquez, M., Chitosan for food packaging: Recent advances in active and intelligent films. *Food Hydrocoll.*, 124, 107328, 2022.
5. Castillo, B.E., Prokhorov, E., Luna-Bárcenas, G., Kovalenko, Y., Potential Use of Chitosan-TiO2 Nanocomposites for the Electroanalytical Detection of Imidacloprid. *Polymers*, 14, 1686, 2022.
6. Omerovic´, N., Djisalov, M., Živojevic´, K., Mladenovic´, M., Vunduk, J., Milenkovic´, I., Kneževic´, NZ., Gadjanski, I., Vidic´, J., Antimicrobial nanoparticles and biodegradable polymer composites for active food packaging applications. *Compr. Rev. Food Sci. Food Saf.*, 20, 2428–2454, 2021.
7. Famiglietti, M., Savastano, A., Gaglione, R., Arciello, A., Naviglio, D., Mariniello, L., Edible Films Made of Dried Olive Leaf Extract and Chitosan: Characterization and Applications. *Foods*, 11, 2078, 2022.
8. Tao, W., Wang, G., Pei, X., Sun, W., Wang, M., Chitosan Oligosaccharide Attenuates Lipopolysaccharide-Induced Intestinal Barrier Dysfunction through Suppressing the Inflammatory Response and Oxidative Stress in Mice. *Antioxidants*, 11, 1384, 2022.
9. Chen, M.Y., Alexiev, A., McKenzie, V.J., Bacterial Biofilm Thickness and Fungal Inhibitory Bacterial Richness Both Prevent Establishment of the

Amphibian Fungal Pathogen Batrachochytrium dendrobatidis. *Appl. Environ. Microbiol.*, 88, e01604–e01621, 2022.

10. Lu, N. and Liu, Y., Structural, physicochemical, and functional (antioxidant-antimicrobial) properties of 2-O-methyl-β-cyclodextrin inclusion with hexahydro-β-acids in chitosan films. *Colloids Surf. B Biointerfaces*, 191, 111002, 2020.
11. Farmoudeh, A., Shokoohi, A., Ebrahimnejad, P., Preparation and Evaluation of the Antibacterial Effect of Chitosan Nanoparticles Containing Ginger Extract Tailored by Central Composite Design. *Adv. Pharmeutical Bull.*, 11, 643–650, 2021.
12. Chouljenko, A., Chotiko, A., Reyes, V., Alfaro, L., Liu, C., Dzandu, B., Sathivel, S., Application of water-soluble chitosan to shrimp for quality retention. *LWT*, 74, 571–579, 2016.
13. Kumari, P., Barman, K., Patel, V.B., Siddiqui, M.W., Kole, B., Reducing post-harvest pericarp browning and preserving health promoting compounds of litchi fruit by combination treatment of salicylic acid and chitosan. *Sci. Hortic.*, 197, 555–563, 2015.
14. Giannakas, A., Vlacha, M., Salmas, C., Leontiou, A., Katapodis, P., Stamatis, H., Barkoula, N.M., Ladavos, A., Preparation, characterization, mechanical, barrier and antimicrobial properties of chitosan/PVOH/clay nanocomposites. *Carbohydr. Polym.*, 140, 408–415, 2016.
15. Shahabi-Ghahfarrokhi, I. and Babaei-Ghazvini, A., Using photo-modification to compatibilize nano-ZnO in the development of starch-kefiran-ZnO green nanocomposite as food packaging material. *Int. J. Biol. Macromol.*, 124, 922–930, 2019.
16. Shahbazi, Y. and Shavisi, N., Current advancements in applications of chitosan based nano-metal oxides as food preservative materials. *Nanomedical Resour. J.*, 4, 3, 122–129, 2019.
17. Babaei-Ghazvini, A., Acharya, B., Korber, D.R., Antimicrobial Biodegradable Food Packaging Based on Chitosan and Metal/Metal-Oxide Bio-Nanocomposites: A Review. *Polymers*, 13, 2790, 2021.
18. Jiménez, M., Ignacio Maldonado, M., Rodríguez, E.M., Hernández-Ramírez, A., Saggioro, E., Carra, I., Supported TiO2 solar photocatalysis at semi-pilot scale: degradation of pesticides found in citrus processing industry wastewater, reactivity and influence of photogenerated species. *J. Chem. Technol. Biotechnol.*, 90, 1, 149–57, 2014.
19. Kumara, S., Mukherjee, A., Dutta, J., Chitosan based nanocomposite films and coatings: Emerging antimicrobial food packaging alternatives. *Trends Food Sci. Technol.*, 97, 196–209, 2020.
20. Lin, B., Luo, Y., Teng, Z., Zhang, B., Zhou, B., Wang, Q., Development of silver/titanium dioxide/chitosan adipate nanocomposite as an antibacterial coating for fruit storage. *LWT Food Sci. Technol.*, 63, 2, 1206–13, 2015.
21. Zhang, X., Xiao, G., Wang, Y., Zhao, Y., Su, H., Tan, T., Preparation of chitosan-TiO2 composite film with efficient antimicrobial activities under

visible light for food packaging applications. *Carbohydr. Polym.*, 169, 101–107, 2017.
22. Galatage, T.S., Hebalkar, S.A., Dhobale, S.V., Mali, R.O., Kumbhar, S.P., Nikade, V.S., Killedar, S.G., Silver Nanoparticles: Properties, Synthesis, Characterization, Applications and Future Trends. *Intech Open*, 3, 1–12, doi: DOI:10.5772/intechopen.99173.
23. Kowsalya, E., Mosachristas, K., Balashanmugam, P., Tamil, S.A., Biocompatible silver nanoparticles/poly (vinyl alcohol) electrospun nanofibers for potential antimicrobial food packaging applications. *Food Packag. Shelf Life*, 21, 100379, 2019.
24. Alavi, M. and Kennedy, J.F., Recent advances of fabricated and modified Ag, Cu, CuO and ZnO nanoparticles by herbal secondary metabolites, cellulose and pectin polymers for antimicrobial applications. *Cellulose*, 28, 3297–3310, 2021.
25. Qin, Y., Liu, Y., Yuan, L., Yong, H., Liu, J., Preparation and characterization of antioxidant, antimicrobial and pH-sensitive films based on chitosan, silver nanoparticles and purple corn extract. *Food Hydrocoll.*, 96, 102–111, 2019.
26. Pandey, V.K., Upadhyay, S.N., Niranjan, K., Mishra, P.K., Antimicrobial biodegradable chitosan-based composite Nano-layers for food packaging. *Int. J. Biol. Macromol.*, 157, 212–219, 2020.
27. Wang, H., Ding, F., Ma, L., Zhang, Y., Edible films from chitosan-gelatin: Physical properties and food packaging application. *Food Biosci.*, 40, 100871, 2021.
28. Al-Tayyar, N.A., Youssef, A.M., Al-hindi, R., Antimicrobial food packaging based on sustainable Bio-based materials for reducing foodborne Pathogens: A review. *Food Chem.*, 310, 125915, 2020.
29. Bingren, T., Dan, X., Jianhua, C., Yumei, L., Chitosan-silica with hops β-acids added films as prospective food packaging materials: Preparation, characterization, and properties. *Carbohydr. Polym.*, 272, 118457, 2021.
30. Vincent, M., Hartemann, P., Engels-Deutsch, M., Antimicrobial applications of copper. *Int. J. Hyg. Environ. Health*, 219, 7, 585–591, 2016.
31. Ali, A.I., Dandago, M.A., Ali, F.I., Immanuel, G., Naskar, J., Oxide-Based Nanocomposites for Food Packaging Application: A Review. *Food Scientech J.*, 4, 2, 145–174, 2022.
32. Mirtalebi, S.S., Almasi, H., Khaledabad, A.M., Physical, morphological, antimicrobial and release properties of novel MgO-bacterial cellulose nanohybrids prepared by *in-situ* and *ex-situ* methods. *Int. J. Biol. Macromol.*, 128, 848–857, 2019.
33. Baloch, H., Siddiqua, A., Nawaz, A., Latif, S.M., Zahra, Q.S., Alomar, Y.S., Ahmad, N., Elsayed, T.M., Synthesis and Characterization of sulfur nanoparticles of citrus limon Extract Embedded in Nanohydrogel Formulation: *In Vitro* and *In vivo* Studies. *Gels*, 9, 4, 284, 2023.
34. Sekhar, C.R., Application and Uses of Graphene Oxide and Reduced Graphene Oxide, vol. 2, pp. 40–55, 2015.

35. Talón, E., Trifkovic, K.T., Nedovic, V.A., Bugarski, B.M., Vargas, M., Chiralt, A., González-Martínez, C., Antioxidant edible films based on chitosan and starch containing polyphenols from thyme extracts. *Carbohydr. Polym.*, 157, 1153–1161, 2017.
36. Bhardwaj, A., Sharma, N., Sharma, V., Alam, T., Sahu, J.K., Hamid, H., Assessing the consumer acceptance and storability of chitosan and beeswax coated cellulose packaging for whole wheat bread. *Food Control*, 133, 108682, 2022.
37. Costa, S.M., Ferreira, D.P., Teixeira, P., Ballesteros, L.F., Teixeira, J.A., Fangueiro, R., Active natural-based films for food packaging applications: The combined effect of chitosan and nanocellulose. *Int. J. Biol. Macromol.*, 177, 241–251, 2021.
38. Pinem, M.P., Wardhono, E.Y., Nadaud, F., Clausse, D., Saleh, K., Guénin, E., Nanofluid to Nanocomposite Film: Chitosan and Cellulose-Based Edible Packaging. *Nanomaterials*, 10, 660, 2020.
39. Jiménez-Regalado, E.J., Caicedo, C., Fonseca-García, A., Rivera-Vallejo, C.C., Aguirre-Loredo, R.Y., Preparation and Physicochemical Properties of Modified Corn Starch-Chitosan Biodegradable Films. *Polymers*, 13, 4431, 2021.
40. Castillo, L.A., Farenzena, S., Pintos, E., Rodríguez, M.S., Villar, M.A., García, M.A., López, O.V., Active films based on thermoplastic corn starch and chitosan oligomer for food packaging applications. *Food Packag. Shelf Life*, 14, 128–136, 2017.
41. Llanos, J.H.R., Avezum, L., Dacanal, G.C., Tadini, C.C., Increase in the physical performance of nanostructured starch/chitosan blends with montmorillonite. *Colloidal Polym. Sci.*, 299, 1901–1915, 2021.
42. Istiqomah, A., Utami, M.R., Firdaus, M., Suryanti, V., Kusumaningsih, T., Antibacterial chitosan-Dioscorea alata starch film enriched with essential oils optimally prepared by following response surface methodology. *Food Biosci.*, 46, 101603, 2022.
43. Othman, S.H., Kechik, N.R.A., Shapi'i, R.A., Talib, R.A., Tawakkal, I.S.M.A., Water Sorption and Mechanical Properties of Starch/Chitosan Nanoparticle Films. *J. Nanomater.*, 2019, 3843949, 2019.
44. Shapi'i, R.A., Othman, S.H., Nordin, N., Kadir Basha, R., Nazli Naim, M., Antimicrobial properties of starch films incorporated with chitosan nanoparticles: *In vitro* and *in vivo* evaluation. *Carbohydr. Polym.*, 230, 115602, 2020.
45. Zhao, Y., Teixeira, J.S., Gänzle, M.M., Saldaña, M.D.A., Development of antimicrobial films based on cassava starch, chitosan and gallic acid using subcritical water technology. *J. Supercrit. Fluids*, 137, 101–110, 2018.
46. Gao, H.-X., He, Z., Sun, Q., He, Q., Zeng, W.-C., Functional polysaccharide film forming by pectin, chitosan, and tea polyphenols. *Carbohydr. Polym.*, 215, 1–7, 2019.

47. Sun, J., Jiang, H., Li, M., Lu, Y., Du, Y., Tong, C., Pang, J., Wu, C., Preparation and characterization of multifunctional konjac glucomannan/carboxymethyl chitosan biocomposite films incorporated with epigallocatechin gallate. *Food Hydrocoll.*, 105, 105756, 2020.
48. Mondal, K., Goud, V.V., Katiyar, V., Effect of Waste Green Algal Biomass Extract Incorporated Chitosan-Based Edible Coating on the Shelf Life and Quality Attributes of Tomato. *ACS Food Sci. Technol.*, 2, 1151–1165, 2022.
49. Meenu, M., Padhan, B., Patel, M., Patel, R., Xu, B., Antibacterial activity of essential oils from different parts of plants against Salmonella and Listeria spp. *Food Chem.*, 404, 134723, 2023.
50. Cai, L., Wang, Y., Cao, A., The physiochemical and preservation properties of fish sarcoplasmic protein/chitosan composite films containing ginger essential oil emulsions. *J. Food Process. Eng.*, 43, e13495, 2020.
51. Sadadekar, A.S., Shruthy, R., Preetha, R., Kumar, N., Pande, K.R., Nagamaniammai, G., Enhanced antimicrobial and antioxidant properties of Nano chitosan and pectin based biodegradable active packaging films incorporated with fennel (Foeniculum vulgare) essential oil and potato (Solanum tuberosum) peel extracts. *J. Food Sci. Technol.*, 60, 938–946, 2022.
52. Wang, L., Wang, Q., Tong, J., Zhou, J., Physicochemical Properties of Chitosan Films Incorporated with Honeysuckle Flower Extract for Active Food Packaging. *J. Food Process. Eng.*, 40, e12305, 2017.
53. Xu, K., Dai, Q., Dong, K., Wei, N., Qin, Z., Double noncovalent network chitosan/hyperbranched polyethylenimine/Fe^{3+} films with high toughness and good antibacterial activity. *RSC Adv.*, 12, 5255–5264, 2022.
54. Wang, L., Ning, C., Pan, T., Cai, K., Role of Silica Nano particles in Abiotic and Biotic stress tolerance in plants: A Review. *Int. J. Mol. Sci.*, 23, 4, 1947, 2021.
55. Ding, F., Hu, B., Lan, S., Wang, H., Flexographic and screen printing of carboxymethyl chitosan based edible inks for food packaging applications. *Food Packag. Shelf Life*, 26, 100559, 2020.
56. Nair, M.S., Tomar, M., Punia, S., Kukula-Koch, W., Kumar, M., Enhancing the functionality of chitosan- and alginate-based active edible coatings/films for the preservation of fruits and vegetables: A review. *Int. J. Biol. Macromol.*, 164, 304–320, 2020.
57. Priyadarshi, R. and Rhim, J.-W., Chitosan-based biodegradable functional films for food packaging applications. *Innov. Food Sci. Emerg. Technol.*, 62, 102346, 2020.
58. Koc, B., Akyuz, L., Cakmak, Y.S., Sargin, I., Salaberria, A.M., Labidi, J., Kaya, M., Production and characterization of chitosan-fungal extract films. *Food Biosci.*, 35, 100545, 2020.
59. Wang, H., Li, B., Ding, F., Ma, T., Improvement of properties of smart ink via chitin nanofiber and application as freshness indicator. *Prog. Org. Coat.*, 149, 105921, 2020.

60. Ciannamea, E.M., Castillo, L.A., Barbosa, S.E., Angelis, M.G.D., Barrier properties and mechanical strength of bio-renewable, heat-sealable films based on gelatin, glycerol and soybean oil for sustainable food packaging. *React. Funct. Polym.*, 125, 29–36, 2018.
61. Luo, Q., Hossen, M.A., Zeng, Y., Dai, J., Li, S., Qin, W., Liu, Y., Gelatin-based composite films and their application in food packaging: A review. *J. Food Eng.*, 313, 110762, 2022.
62. Ramakrishnan, R.K., Wacławek, S., Cerňík, M., Padil, V.V.T., Biomacromolecule assembly based on gum kondagogu-sodium alginate composites and their expediency in flexible packaging films. *Int. J. Biol. Macromol.*, 177, 526–534, 2021.
63. Sancakli, A., Basaran, B., Arican, F., Polat, O., Effects of bovine gelatin viscosity on gelatin-based edible film mechanical, physical and morphological properties. *SN Appl. Sci.*, 3, 1, 8, 2021.
64. Ştefănescu, B.E., Socaciu, C., Vodnar, D.C., Recent progress in functional edible food packaging based on gelatin and chitosan. *Coatings*, 12, 12, 1815, 2022.
65. Lu, Y., Luo, Q., Chu, Y., Tao, N., Deng, S., Wang, L., Li, L., Application of gelatin in food packaging: A review. *Polymers*, 14, 3, 436, 2022.
66. Cristofoli, N.L., Lima, A.R., Tchonkouan, G., Quintino AC, R.D.,., Vieira, M.C., Advances in the Food Packaging Production from Agri-Food Waste and By-Products: Market Trends for a Sustainable Development. *Sustainability*, 15, 6153, 2023.
67. Bell, L., Moisture effects on food's chemical stability, in: *Water activity in foods*, pp. 227–253, 2020.
68. Fakhouri, F.M., Casari, A.C.A., Mariano, M., Yamashita, F., Mei, L.H.I., Soldi, V., Martelli, S.M., Effect of a gelatin-based edible coating containing cellulose nanocrystals (CNC) on the quality and nutrient retention of fresh strawberries during storage. *IOP Conf. Ser. Mater. Sci. Eng.*, 64, 12–24, 2014.
69. Soradech, S., Nunthanid, J., Limmatvapirat, S., Luangtana-anan, M., Utilization of shellac and gelatin composite film for coating to extend the shelf life of banana. *Food Control*, 73, 1310–1317, 2017.
70. Halim, A.L.A., Kamari, A., Phillip, E., Chitosan, gelatin and methylcellulose films incorporated with tannic acid for food packaging. *Int. J. Biol. Macromololecules*, 120, 1119–1126, 2018.
71. Cui, H., Bai, M., Rashed, M.M.A., Lin, L., The antibacterial activity of clove oil/ chitosan nanoparticles embedded gelatin nanofibers against Escherichia coli O157: H7 biofilms on cucumber. *Int. J. Food Microbiol.*, 266, 69–78, 2018.
72. Sun, X., Guo, X., Ji, M., Wu, J., Zhu, W., Wang, J., Cheng, C., Chen, L., Zhang, Q., Preservative effects of fish gelatin coating enriched with CUR/βCD emulsion on grass carp (Ctenopharyngodon idellus) fillets during storage at 4 °C. *Food Chem.*, 272, 643–652, 2019.

73. Mohebi, E. and Shahbazi, Y., Application of chitosan and gelatin based active packaging films for peeled shrimp preservation: a novel functional wrapping design. *LWT Food Sci. Technol.*, 76, 108–116, 2017.
74. Cardoso, G.P., Andrade, M.P.D., Rodrigues, L.M., Massingue, A.A., Fontes, P.R., Ramos A de, L.S., Ramos, E.M., Retail display of beef steaks coated with monolayer and bilayer chitosan-gelatin composites. *Meat Sci.*, 152, 20–30, 2019.
75. Qu, B. and Luo, Y., A review on the preparation and characterization of chitosan-clay nanocomposite films and coatings for food packaging applications. *Carbohydr. Polym. Technol. Appl.*, 2, 100102, 2021.
76. Laaraibi, A., Moughaoui, F., Damiri, F., Ouakit, A., Charhouf, I., Hamdouch, S., Bennamara, A., Chitosan-clay based (CS-NaBNT) biodegradable nanocomposite films for potential utility in food and environment. Chitin-Chitosan Myriad Functionalities in Science and Technology, 2, 45–69, 2018.
77. Hu, Q. and Luo, Y., Polyphenol-chitosan conjugates: Synthesis, characterization, and applications. *Carbohydr. Polym.*, 151, 624–639, 2016.
78. Marquardt, K.C. and Watson, R.R., Polyphenols and public health, in: *Polyphenols in human health and disease*, vol. 9–15, 2014.
79. Li, D., Zhang, X., Li, L., Aghdam, M.S., Wei, X., Liu, J., Xu, Y., Luo, Z., Elevated CO2 delayed the chlorophyll degradation and anthocyanin accumulation in postharvest strawberry fruit. *Food Chem.*, 285, 163–70, 2019.
80. Zhang, W., Li, X., Jiang, W., Development of antioxidant chitosan film with banana peels extract and its application as coating in maintaining the storage quality of apple. *Int. J. Biol. Macromol.*, 154, 1205–14, 2020.
81. Zhang, W., Jiang, H., Rhim, J.-W., Cao, J., Jiang, W., Tea polyphenols (TP): a promising natural additive for the manufacture of multifunctional active food packaging films. *Crit. Rev. Food Sci. Nutr.*, 63, 2, 288–301, 2022.
82. Ashrafi, A., Jokar, M., Nafchi, A.M., Preparation and characterization of biocomposite film based on chitosan and kombucha tea as active food packaging. *Int. J. Biol. Macromol.*, 108, 444–54, 2018.
83. Ye, J., Wang, S., Lan, W., Qin, W., Liu, Y., Preparation and properties of polylactic acid-tea polyphenol-chitosan composite membranes. *Int. J. Biol. Macromol.*, 117, 632–9, 2018.
84. Wu, H., Lei, Y., Zhu, R., Zhao, M., Lu, J., Xiao, D., Jiao, C., Zhang, Z., Shen, G., Li, S., Preparation and characterization of bioactive edible packaging films based on pomelo peel flours incorporating tea polyphenol. *Food Hydrocoll.*, 90, 41–9, 2019.
85. Oun, A.A., Shin, G.H., Rhim, J.W., Kim, J.T., Recent advances in polyvinyl alcohol-based composite films and their applications in food packaging. *Food Packag. Shelf Life*, 34, 100991, 2022.
86. Tang, X. and Alavi, S., Recent advances in starch, polyvinyl alcohol based polymer blends, nanocomposites and their biodegradability. *Carbohydr. Polym.*, 85, 1, 7–16, 2011.

87. Dhineshbabu, N.R., Jakku, S.K.D., Mada, S.P., Nathi, V.S., *Analysis the Linear Optical Constant of Lanthanum Doped Nickel Ferrites Nanoparticles from Absorption Spectra*, vol. 13, pp. 426–435, 2021.
88 Zehra, A., Wani, S.M., Jan, N., Bhat, T.A., Rather, S.A., Malik, A.R., Hussain, S.Z., Development of chitosan-based biodegradable films enriched with thyme essential oil and additives for potential applications in packaging of fresh collard greens. *Sci. Rep.*, 12, 1, 16923, 2022.

11

Application of Nanochitosan as Biofertilizers for Sustainable Agriculture

Ibrahim Dawud[1*], Ahmad Muhammad Abubakar[2], Mujahid Hussaini[2], Sherifat Enejiyon Ozavize[3], Oluwafemi Adebayo Oyewole[3], Charles Oluwaseun Adetunji[4], Eniola K. I. T.[5] and Mohammed Bello Yerima[6]

[1]*Department of Applied Biology, Federal University of Technology Babura, Jigawa State, Nigeria*
[2]*Department of Microbiology, Umaru Musa Yar'adua University Katsina, Katsina State, Nigeria*
[3]*Department of Microbiology, Federal University of Technology, Minna, Niger State, Nigeria*
[4]*Department of Microbiology, Edo State University, Uzairue, Edo State, Nigeria*
[5]*Department of Biological Sciences, Joseph Ayo Babalola University, Ikeji Arakeji, Osun State, Nigeria*
[6]*Department of Microbiology, Sokoto State University, Sokoto, Sokoto State, Nigeria*

Abstract
Nanofertilizer particles are designed in a nanoscale so as to enhance the delivery of nutrient to the plant and improve crop production in order to meets the demands of the ever increasing global population for sustainable future. Nanofertilizers are a set of new-generation fertilizers that provide an efficient and sustainable method of improving crop yields and soil fertility through utilizing advanced nanotechnology and are classified on the basis of nutritional constituent, mode of action, and consistency. Biofertilizers are natural fertilizers produced from organic particulate of matter such as living microbial inoculants of bacteria, fungi, and algae or materials from plant and other animal sources. In the present context, agricultural significance of biofertilizers is overwhelming due to the increased cost of chemical fertilizers and their hazardous effects on soil health, particularly the soil texture, greenhouse effect, pollution, and deterioration of microbial communities. Chitosans are group of a linear hydrophilic biopolymeric compound with a

**Corresponding author*: idawud.bio@futb.edu.ng

Charles Oluwaseun Adetunji, Maulin P. Shah, Yerima Mohammed Bello, Daniel Hefft, Jay Singh, Shyam S. Pandey and Ravindra Pratap Singh (eds.) Nanochitosan Applications for Enhanced Crop Production and Food Security, (249–280) © 2025 Scrivener Publishing LLC

random structural distribution of β-(1→4)-linked D-glucosamine and *N*-acetyl-D-glucosamine. They are abundant in nature and have a lot of agricultural, medical, and industrial potentials. In agriculture, nanochitosans are among the group of action-based nanofertilizers with numerous potential in enhancing and ensuring proper utilization of the nutrients. Several studies have described the used of chitosan and its nanoparticles in enhancing the nutrient uptake by plants and reducing the cost of and negative environmental impacts associated with the use of conventional fertilizers. This study provides an insight in the use of chitosan and its nanoparticles as a substitute for conventional fertilizers, in ensuring eco-friendly and in enhancing agricultural productivity for sustainable future.

***Keywords*:** Chitosan, biofertilizers, sustainable, eco-friendly, biopolymer, agriculture, nanofertilizer

11.1 Introduction to Nanoparticle and Chitosan

Nanoparticles synthesized from chitosan have been the center of attention recently due to their biodegradability, antimicrobial potentials, and toxic-free nature. In agriculture, the ultimate goal for every farmer is to record a bumper harvest every year. The vast and diverse potentials of nanosynthesized substances cannot be overlooked in the present world; as a result of the agricultural products, demand created by the ever-increasing populace of the world increased, and the space for agricultural cultivation remains the same, thereby posing a significant threat to the sustainable future of agricultural activities. Therefore, an alternative route must be invented such that the agricultural industry finds means to increase yield from the farms in order to satisfy the agricultural needs of the population and ensure future sustainability. Nanochitosan's potentialities can proffer a magnificent panacea to that effect as they have proven to pose a broad spectrum effect against pathogens that may alter plant growth, thereby promoting plant growth, and also its non-toxicity implores its application in the coating of vegetables and fruits [1]. The growth-promoting activities of nanochitosan are a result of their soil fertility improvement property. The use of fertilizer to boost crop yields and maintain soil fertility has become a crucial practice for agricultural sustainability for the world's expanding population. Common fertilizers generated chemically include nitrogen, urea, phosphorus, potassium, diammonium phosphate, and monoammonium phosphate [2, 3]. These fertilizers put up with leaching which results in low nutrient utilization and decreases soil fertility. Furthermore, these chemically synthesized fertilizers used currently in improving soil fertility and other agricultural activities are a significant contributor to environmental contamination because of their methane releases during runoff, cultivable land degradation, and longtime uses of these chemicals that can cause imbalance in the ecosystem food chain distribution and loss of natural resources [4]. Therefore, an alternative to agricultural chemicals containing pollutants is imperative, and the harm done to the environment by the uncontrollable release of excess nutrients necessitates the development of more eco-friendly and efficient fertilizers [5, 6]. Proper utilization of chitosan nanoparticles can be a substitute for chemical fertilizer. Thus, producing biofertilizer from the nanoparticles can serve as a launching pad for promoting safe and sustainable agriculture.

11.2 Nanofertilizers

In agriculture, nontechnology is primarily deployed to increase production yields and minimize losses through enhancing the plant defense mechanisms against insects, pests, and other environmental distress. Nanofertilizers are a set of new-generation fertilizers that provide an efficient and sustainable method of improving crop yields and soil fertility through utilizing advanced nanotechnology [7, 8]. Nanofertilizers have emerged as a promising option in curtailing the global challenges encountered in the use of common or conventional fertilizers, as they have offered to reduce environmental impact, thereby providing higher yield efficiency. Nanofertilizers are designed to deliver nutrients to plants in a gradual and controlled manner, thereby ensuring a steady supply of essential nutritional elements to plants over an extended period. Nanofertilizers are made to release nutrients to plants gradually and under regulated conditions, ensuring a consistent supply of vital nutrients for plants over an extended period of time. Because the nanofertilizers contain one or more essential nutrients for plant growth, more than half of the particles have the least minimal-sized particles of less than 100 nm. The smaller nature of the nanomaterials makes the particles have a high surface area–to–volume ratio and have coating ability with various materials; these make them ideal for holding and releasing nutrients in an efficient controlled system, thereby reducing the need for the frequent application of fertilizer and its significant quantity [9, 10]. Nanofertilizers can improve agricultural sustainability in order to meet the increasing demand for food by the ever-growing population, through controlling and reducing the nutrient leaching and reducing the general cost of fertilizer application during the farming activities.

11.2.1 Types of Nanofertilizers

Categorically, nanofertilizers can be classified on the basis of their nutritional constituent, mode of action, and consistency.

11.2.1.1 Classification Based on Their Modes of Action

Nanofertilizers can be grouped on the basis of their modes of action into controlled release, targeted delivery, plant growth stimulating, and water and nutrient loss controlling.

11.2.1.1.1 Controlled Release Nanofertilzers

These groups of nanofertilizers bring about an array of benefits to sustainable agricultural activities, which include proper delivery of the target nutrient, enhancing the plant growth, optimum utilization of nutrients, controlling the release of nutrient, reducing the loss of nutrient, and decreasing the contamination of water bodies their by protecting the aquatic ecosystem. In this formulation, the nutrients are encapsulated and carried within nanoscale carrier materials, which may be composed of lipids, polymers, or some inorganic substances [11]. Thus, the activities of these groups of nanofertilizers can be affected by some environmental conditions such as pH, temperature, and moisture or stimuli-responsive mechanisms such as enzyme-mediated degradation and other biodegradation mechanisms

11.2.1.1.2 Target Delivery

Small molecules made up of either oligonucleotides or peptides known as aptamers are used in modifying the surface of the nanofertilizers. Aptamers are a sequence of artificially synthesized deoxyribonucleic acid (DNA), ribonucleic acid (RNA), xeno nucleic acid (XNA), or peptides that bind to specific target molecules. Nanoaptamers are synthesized to target some specific molecules in the soil and deliver nutrients and other elements directly to the plant. The nanoaptamers can efficiently and effectively bind with the plant hormones and other growth enzymes to stimulate and enhance nutrient uptake from the soil. The surface modification of the fertilizers by the aptamers allows the release and activation of nutrients within the nanostructure by the signals arising from the rhizosphere [12].

The fertilizers are delivered by the aptamer through attachment of the fertilizer components to the incorporated nanoparticles such as liposomes in order to safeguard it from degradation and enable its targeted delivery to the intended plant cells [13]. The nanoaptamers can also be employed in regulating fertilizer dissemination by forming a connection between soil microorganisms and the root system, which aids in precise administration of the desired fertilizer quantity. Not only essential nutrients can be delivered through the nano target delivery system but also pesticides and other modified genetic materials can be delivered successfully through the system, and this can enhance agricultural productivity and future sustainability.

11.2.1.1.3 Plant Growth-Stimulating Nanofertilizers

The categories of nanofertilizers function by stimulating plant growth through increasing interaction between some elemental nutrients such as carbon with some indigenous soil nutrients and the plant root system by boosting hormone synthesis [14]. This kind of fertilizer helps in absorbing and releasing nutrients, increases water retention, enhances plant growth, and improves soil structure. They also assist in sustaining the plant flora for a very long period due to their ability to penetrate deep into the soil. The lengthy diffusion of the nutrient within the soil inhibits the manifestation of over-fertilization of the plant. Plant growth-stimulating nanofertilizers help in preventing erosion and augmenting the nutrient storage capacity of the soil, which leads to the downsizing of the amount of conventional fertilizers needed for the farming period. These fertilizers have proven to have significant tensile strength ability, making them one the smallest known fibers, which also can be used as a protective coat on seeds, and to prevent pest activities on the fruits and, at the same time, enhance the seed capacity to absorb nutrients and water and also to make the entire plant materials to resist collapsing or breaking as a result of some extreme environmental conditions and, most importantly, to improve the plant yield for sustainable farming [15, 16].

11.2.1.1.4 Water and Nutrient Loss–Controlling Fertilizer

These are the category of action-based nanofertilizers whose nanoparticles can coordinate and control the release of nutrients and water into the soil. Some of the techniques used in formulating these fertilizers include encapsulation and surface modification. The encapsulation method is the act by which the nanofertilizer becomes incorporated in a porous matrix, which then allows the gradual release of nutrients over time, whereas, the surface modification is the process of modifying the surface of the nanofertilizers to become hydrophilic in order to increase their water retention capacity and reduce the loss of water through evaporation. One of the traditional fertilizers that may be suited for water and nutrient loss control is urea fertilizer. The urea-synthesized nanoparticles can be coated with some elements such as sulfur, calcium, iron oxide, magnesium, copper, ammonium sulfate, zinc, potassium, boron, and molybdenum to generate nanofertilizer with the ability to control nutrient and water loss in soil [17–19]. The two prominent types of soil water and nutrient loss–controlling nanofertilizers are nanoemulsions and nanobeads. Also, water and nutrient loss–controlling fertilizer allows farmers to use smaller amounts of fertilizer while sustaining the same crop outputs.

11.2.1.2 Classification Based on Nutrient Constituents

These are types of nanofertilizers designed to improve nutrient availability for effective uptake and utilization within the plant. Nutrient-based nanofertilizers are grouped into inorganic-based fertilizers, organic-based fertilizers, hybrid nanofertilizers, and nutrient-loaded nanofertilizers.

11.2.1.2.1 Inorganic Nanofertilizers

These are nanofertilizers designed to increase the production yield for sustainable agriculture by enhancing efficient nutrient uptake. These categories of nutrient-based nanofertilizers can provide vital nutrients to the plants such as potassium, nitrogen, and phosphorus [20]. Their entire fertilizer can be synthesized from metalloids, metals or nonmetallic nanoparticles, and the choice of material is determined by the desired nutrient needed. Thus, they can be specifically customized based on nutrient requirements targeted by a particular plant and the nutrient can be sub-grouped into macro and micronutrients based inorganic nanofertilizers.

11.2.1.2.1.1 PLANT MACRONUTRIENTS

Plant macronutrients are essential nutrients required by plants in large quantities for proper growth and development. Nanofertilizers synthesized with macronutrients proffer numerous trump cards for plant growth and environmental safety for sustainable agriculture [20–22]. Some examples of macronutrients and their functions are as follows:

- Phosphorous-based micronutrient nanofertilizers: These nanofertilizers aid in plant nutrient absorption and play a great role in the repossession of the soil.
- Nitrogen-based micronutrient nanofertilizers: They strengthen the efficacy of nitrogen utilization and reduce the acidification

of the aquatic ecosystem also reducing the rate of greenhouse gas emission.

- Potassium-based micronutrient nanofertilizers: These nanofertilizers are resistant to leaching with very high nutrient absorption and can assist in enhancing soil properties.
- Sulfur-based micronutrient nanofertilizers: They ensure a slow and steady nutrient supply to the plant, thereby reducing the risk of soil acidification.
- Magnesium-based micronutrient nanofertilizers: The set of nanofertilizers plays a vital role in improving the quality and quantity of crop growth and increasing the resistance of the plant to pests and diseases.
- Calcium-based micronutrients nanofertilizers: They also play a great role in increasing disease resistivity of the plant, enhancing the quality of vegetables and fruits, and contributing to higher yield realization.

11.2.1.2.1.2 Plant Micronutrients

Plant micronutrients are categories of nutrients required by plants in a trace quantity for the enchantment of their growth and development. Plant micronutrients play a crucial role in improving plant health. Nanoparticles synthesized from plant nutrient deliver essential micronutrients more effectively than conventional fertilizers [23].

Some common example of micronutrients and their functions includes the following:

- Nickel-based micronutrient nanofertilizers: These are a category of micronutrient nanofertilizers that aid in ensuring proper metabolic functions such as photosynthesis, hormone and enzyme production within the plant system. These fertilizers also ensure proper growth of the plant by preventing the plant from stunted growth.
- Boron-based micronutrient nanofertilizer: Boron is a macronutrient required for optimal growth of the plant, hence lacking in the soil. Boron-based macronutrient nanofertilizers assist in providing sufficient baron to the soil for the uptake by the plants. Boron-based nanofertilizers have a promising future in enhancing the quality and quantity of pomegranate fruits.

- Zinc-based micronutrient nanofertilizers: These groups of nanofertilizers play an active role in accelerating the metabolic activities of the plant. They are also found to decrease leaching and environmental contamination, so also enhance crop yields.
- Copper-based micronutrients nanofertilizer: These nanofertilizers are very good candidates for fertilizers that promise a sustainable agricultural future. They are effective in delivering nutrients and other essential elements directly to the plant root system. They can penetrate the plant cells and are resistant to leaching.
- Titanium-based micronutrient nanofertilizer: These nanofertilizers serve by increasing or boosting the plant immune system, thereby providing long-term immunity against microorganisms, pests, and other diseases. They also play a role in increasing the plant root and shoot length.
- Iron-based micronutrients nanofertilizers: Iron is also one of the essential elements that requires plant growth that is often lacking in the soil. These nanofertilizers are designed to increase the bioavailability of iron within the soil for effective utilization by the plant. The fertilizers can restore iron deficiency in the soil, improve the quality of water, and enhance the crop yield.

11.2.1.2.2 Organic Nanofertilizer

Organic nanofertilizers are a group of nutrient-based nanofertilizers that are synthesized from organic materials delineated to supply nutrients to the soil in a gradual manner. Their natural source origin makes the fertilizers eco-friendly and plays a role in maintaining soil moisture, altering the soil pH and ensuring adequate and efficient nutrient uptake and utilization. Their molecular or atomic size enables them to be processed into various forms of nanoparticles such as liposomes, polymer conjugates, capsules, and others. The fertilizers can be prepared from natural polymers such as seed polysaccharides (*Mimosa pudica* seed, *Leucaena leucocephala* seed, and tamarind seed), bio-derived materials (carrageenan, chitosan, and pectin), and gums (mucuna gum, xanthan gum, and guar gum). The gradual nature of releasing nutrients by these nanofertilizers ensures more effective and efficient nutrient absorption and utilization, leading to enhanced growth of the plant and increased yield of the crops [24, 25]. Furthermore, they have a significant impact on the structure of the soil through increasing aeration rate, infiltration of water porosity, and root penetration. The natural organic maters can serve as pivotal sources of nutrients to the soil microbiota by stimulating and improving their activities, which, in turn, assist in cycling and enhancing nutrient availability within the soil. Thus, nanofertilizers can be prepared from organic waste such as compost, plant waste, and manure.

11.2.1.2.3 Hybrid Nanofertilizer

These are engineered forms of conventional fertilizers incorporated with desirable nutrients or elements in a nanoparticulate size to increase nutrient availability within the plant and strengthen its efficacy. Hybridized nanofertilizers provide a slow and controlled release of nutrient material and reduce the negative impact of conventional fertilizer within the environment [26]. These fertilizers performed much better than the various traditional fertilizers due to their enhancement in the following properties: (i) mechanical properties such as such as increased pH tolerance, resistance to heat distortion, and improved stability in storage conditions; (ii) physical properties such as strength, toughness, and rigidity; (iii) thermal and electrical conductivity; and (iv) enhanced the moisture and gas barrier within the soil.

Hybridized nanofertilizers offer various benefits that support agricultural sustainability such as cost saving in fertilizer usage improving and controlling the release of nutrients, reducing nutrient wastage, enhancing nutrient uptake, and enhancing nanofertigation.

11.2.1.2.4 Nutrient-Loaded Nanofertilizers

These are groups of nutrient-based nanofertilizers that are modified and loaded with particular nutrients of interest to provide an efficient and effective nutrient to the crops. Example of nutrient-loaded nanofertilizers includes nanoporous zeolites. Structurally, these fertilizers are honeycomb in shape due to formation from a mixture of alkaline lake water with volcanic ash. The zeolite nanofertilizer is a very good candidate for soil fertilization and amendments as it tends to facilitate the absorption and retention of nutrients and moisture within the soil and also provides a high degree of porosity in the soil environment [27], thus making the soil suitable for crop growth, reducing the need for irrigation, and controlling the rate of erosion.

11.2.1.3 Consistency-Based Nanofertilizers

These are nanofertilizers designed to maintain an adequate and proportionate supply of nutrients for a sustainable future in agriculture. These nanofertilizers provide solutions to nutrient uptake and utilization, nutrient wastage, and environmental impact in various plants [28]. They are grouped into four main groups, *viz*, surface-coated nanofertilizers, synthetic polymers, biological product coatings, and nanocarrier-based nanofertilizers.

11.2.1.3.1 Surface-Coated Nanofertilizers

These are a group of consistency-based nanofertilizers designed with nanoparticles of iron, silicon, nitrogen, phosphorous, or other plant nutrients with additives or some coating particles such as carbon, titanium, gold, and silver [29–31]. This particle aids the surface interaction of the fertilizer particles with the plant cellular materials, thereby enhancing the absorption rate and utilization of the nutrient-coated fertilizer and improving plant growth.

11.2.1.3.2 Synthetic Polymer–Coated Nanofertilizer

These are consistency-based nanofertilizers coated with synthetic polymers to protect the fertilizer materials from environmental degradation and ensure better handling and efficient and even distribution of nutrients throughout the soil and plant materials. Fertilizers coated with polymer are often applied for high-value applications due to their role in preventing nutrient loss and also have a very complex nitrogen-releasing and distribution system. The coated polymer matrix diffuses continuously and consistently for a long time, which regulates the speed of the nutrient release and dispersion [32]. Furthermore, the temperature and humidity of the root zone control the rate by which nutrient is released from the synthesized fertilizer. Some examples of these forms are polyvinyl chloride–coated zeolite nanofertilizers and polyethylene-coated urea nanofertilizers.

11.2.1.3.3 Biological Product–Coated Nanofertilizers

These are groups of nutrient consistency–based nanofertilizers coated mainly with biological materials. These forms of fertilizers are grouped into forms, namely, organic compound–coated consistency-based nanofertilizers and microbial-coated consistency-based nanobiofertilizers.

(i) **Organic compound consistency–based nanofertilizers:** These are biological product–coated nanofertilizers synthesized with organic materials such as humic acid and oligo-alginate with nanoparticles [33]. Coating of nanofertilizers with organic compounds helps to enhance the

availability, uptake, and transportation of essential nutrients within the soil, thereby ensuring proper fertilization of soil. Application of this form of fertilizer provides various advantages more than the conventional fertilizers; some of the advantages are the following:
(a) boosting soil fertility;
(b) enhancing the water retention capacity of the soil;
(c) activating and enhancing the friendly soil microbiota;
(d) regulating the growth and development of the plants;
(e) enhance the plant tolerance to biotic stress.

(ii) **Microbial-coated nanobiofertilizers:** These are biological product–coated nanofertilizers synthesized with soil-beneficial microorganisms such as *Bacillus subtilis*, *Pseudomonas fluorescens*, *Paenibacillus elgii*, and *Pseudomonas putida* with fertilizer nanoparticles. The miniature size and structure of these nanofertilizers coupled with their high surface area ratio help them to transport nutrients directly to the plant roots and tissues [34, 35]. Some of the importance of these nanofertilizers include the following:
(a) the microbial cell helps in mineralization, solubilization, and fixation of nutrients within the soil;
(b) replenishing soil nutrients;
(c) ensuring the consistent release of nutrient;
(d) sustaining the bioavailability of nutrients within the plant;
(e) ensuring proper fertilization of the soil.

11.2.1.3.4 Nanocarrier-Based Nanofertilizers

Nanocarrier nanofertilizers are transport-based fertilizer that transport and delivers nutrients to the soil and the plants. These fertilizers can be engineered to ensure nutrient biofortification, harmful environmental mitigation, and effective dispersion of nutrients within a control system [36]. They also play a role in enhancing nutrient uptake, leaching control, efficiency in nutrient utilization, and improving crop yield.

11.2.2 Application of Nanofertilizers

There are three main techniques used for the effective application of nanofertilizer, they are soil treatment, foliar spray, and seed nanopriming.

11.2.2.1 Soil Treatment

This is the process of direct embodying of nanofertilizers into the soil by either banding, localizing placement, or broadcasting. This technique warrants slow and controlled release of nutrient. This method is best applied in the soil with retention capacity and congruous climate precipitation pattern. Therefore, this technique must be strictly managed to avoid nutrient imbalance and environmental pollution.

Merits of Soil Treatment

(i) It facilitates the plant root and soil interaction.
(ii) It ensures a steady supply of essential elements.
(iii) It controls the release of nutrients.
(iv) It enhances the growth productivity of the plant.
(v) It increases the crop productive yield.

11.2.2.2 Foliar Spray

This is a technique of application where the fertilizer is prepared in liquid form and used as a spray. The method involves direct spraying of the nanofertilizer on the leaf surface of the plant; it allows rapid absorption of the nutrient through leaves to the other part of the plant. It is the most preferred, effective, and faster nutrient-supplying route for plants growing in low fertile soil [37]. Although the efficacy in the uptake of nutrients in this technique can be affected by wind, water, temperature, humidity, and other environmental abiotic conditions, it is characterized by the following advantages:

(i) accurate and rapid transfer of nutrients to other parts of the plant;
(ii) it can serve as a route for delivering other essential items to the plants, such as pesticides, fungicides and others;
(iii) significant reduction of fertilizer leaching;
(iv) very quick in response;
(v) improve utilization of nutrient;
(vi) promote healthy growth of the plant;
(vii) increase the yields of the crops.

11.2.2.3 Seed Nanopriming

This is a technique of nanofertilizer application by which the seeds are impregnated and soaked in the nanofertilizer before planting. This makes the seed stronger and quick germination and allows for proper nutrient uptake throughout the plants across various cells and tissues. It can also be applied in nutrient-poor soil, and the concentration level of the fertilizer must be determined to avoid phytotoxicity in the crop [38].

Advantages of seed nanopriming

(i) It ensures faster germination of the seed.
(ii) It helps to regulate signaling and metabolic cascade.
(iii) It inhibits reacting oxygen species.
(iv) It stimulates multiple gene expression.
(v) It mitigates the plant stress.

11.3 Biofertilizers

Biofertilizers are groups of fertilizers synthesized from natural sources such as microorganisms, organic materials (like plants), or other forms of living organism and their product that can be directly applied to soil, seeds, or plant surfaces. They can settle in the soil surrounding the root system of the plants or the intramural part of the plants to increase bioavailability and supply of nutrient materials to the plant when applied. These fertilizers help to enhance the fertility level of the soil, improve plant growth, enhance crop yields, and protect the other biotic components of the environment [39, 40].

11.3.1 Types of Biofertilizers

Various microbial cells have been exploited in the production of biofertilizers; therefore, biofertilizers can be classified on the basis of the type of organisms, nature, and functions.

11.3.1.1 Microbial Biofertilizers

These are biofertilizers produced from selected efficient living microbial inoculum, which can be applied to the seed, soil or plant surface. These fertilizers are cheaply available to farmers more than the conventional chemical fertilizers.

When these biofertilizers are applied, they tend to colonize the rhizosphere or the interior part of the host plant, thereby increasing the availability, supply, and utilization of nutrients that are critical in promoting plant growth [41]. The microbial inoculum used in preparing these fertilizers is exploited on the basis of their symbiotic relationships with plants, and these include bacteria, fungi, and blue-green algae. Functionally, these bacteria aid in the supply of nitrogen and phosphorus to the plants.

11.3.1.1.1 Nitrogen-Fixing Bacteria Biofertilizers

Nitrogen is one of the key elements in human activities and biological, physical, chemical, and geologic processes. It is highly abundant in the air but is limited in the soil, and it is one of the essential macronutrients required by plants for proper development and function but yet insufficient due to the difficulty of its fixation and uptake by the plants. Elemental nitrogen is relatively inert; it does not easily react with other chemicals to form new compounds; therefore, it needs to be fixed for easy absorption and assimilation by the plant. Fixation of nitrogen can be natural and essential for agriculture in the fertilizer manufacturing process manufacture of fertilizer. Certain species of bacteria can fix nitrogen within the soil and increase its bioavailability and abundance within the rhizosphere for easy absorption, assimilation, and utilization by the plants [42]. Producing biofertilizer with these microbes improves the efficiency of nitrogen uptake and reduces losses by leaching, denitrification, and volatilization. Nitrogen-fixing bacteria are sub-grouped into three categories, namely:

a. **Free-living bacteria:** These are groups of bacteria mostly found to be dominant inhabitant in arable soil and are capable of fixing nitrogen in the soil. Furthermore, these bacteria help in soil aggregation due to their ability to produce a vast amount of slime particles. Example of these bacteria includes *Azotobacter*, *Bacillus polymyxa*, *Anabaena*, *Clostridium A. chroococcum*, *Tolypothrix*, *Beijerinkia*, *Derxia*, *Aulosira*, *Klebsiella*, *Cylindrospermum*, *Desulfovibrio*, and *Rhodopseudomonas*.
b. **Symbiotic bacteria:** These are groups of soil-inhabitant bacteria that can fix nitrogen in the soil and are mostly found symbiotically living within the noodles of leguminous plants. These bacteria are one of the most efficient biofertilizers due to the amount of nitrogen; they fix symbiotically with the roots of the plants, although the morphology,

physiology, and living conditions of these bacteria vary from other species of bacteria found within the noodles of the plant [42]. The prominent examples of symbiotic bacteria of biofertilizer importance are *Anabaena azollae*, *Trichodesmium*, *Frankia*, and *Rhizobium* with their group of genera (*Mesorhizobium*, *Bradyrhizobium*, *Allorhizobium*, *Azorhizobium*, and *Sinorhizobium*).

c. **Associative bacteria:** These are groups of symbiotic bacteria without much endophytic association. When compared with the symbiotic nitrogen-fixing bacteria, associative bacteria have a less intimate relationship with the plant noodles. Furthermore, apart from nitrogen fixation, they tend to improve disease resistivity and drought tolerance in the plant. Example of associative bacteria includes *Acetobacter diazotrophicus*, *Azospirillum* spp. (*A. brasilense*, *A. lipoferum*, *A. amazonense*, *A. halopraeferens*, *and A. irakense*) *Azoarcus* spp., *Pseudomonas*, *Klebsiella*, *Alcaligenes*, *Bacillus*, *Enterobacter*, and *Herbaspirillum* spp.

11.3.1.1.2 Phosphate-Solubilizing Microbial Biofertilizers

Phosphorous is an essential element required for plant growth; it is highly concentrated in the soil but unavailable or limited for plant utilization. Some certain species of bacteria help to solubilize phosphorous so that it will be readily available for plant uptake and utilization, these bacteria are called phosphate-solubilizing microorganisms. Biofertilizers can be essentially produced with higher concentrations of these organisms to mobilize and enhance phosphorous availability for easy uptake and utilization by the plant [43]. solubilization of phosphorous within the soil can be carried out by the soil bacteria and fungi such as *Bacillus subtilis*, *Bacillus polymyxa*, *Bacillus circulans*, *Bacillus megaterium* var. *phosphaticum*, *Pseudomonas striata*, *Erwinia*, *Agrobacterium*, *Achromobacter*, *Flavobacterium Microccocus*, *Aereobacter*, *Penicillium nonatum*, *Penicillium chrysogenum*, *Aspergillus awamori*, *Trichoderma*, *Rhizoctonia solani*, *Rhizobium*, and *Burkholderia*. These organisms help secrete organic acids in the soil, thereby lowering the soil pH that brings about the dissolution of the bounded phosphate in the soil. They have the capability of solubilizing the insoluble phosphate from organic and inorganic phosphate sources. Phosphate-solubilizing microbes supply more phosphorous to the soil than chemical fertilizers and also save the soil from environmental pollution that might be caused as a result of phosphate solubilization by chemical fertilizers.

11.3.1.1.3 Algae Biofertilizer

These are biofertilizers produced from blue-green algae such as the genera of cyanobacteria. Blue-green algae are photosynthetic prokaryotic plants with distinct physiological and morphological features like bacteria. These organisms help in fixing atmospheric nitrogen and are suitable for farming in both the upper and lower land conditions. Algal biofertilizers serve as a perpetual source of nutrients to plants and partake in depleting resources and contaminating the groundwater. Examples of these organisms include *Tolypothrix*, *Nostoc*, *Aulosira*, and *Anabaena*. The potentiality of blue algae in increasing soil fertility and fixing nitrogen was discovered in 1939 in a rice field in India. The rice field supplies the algae with some favorable growth conditions such as water, light, moderate to high temperature, and some nutrients for easy thriving and metabolic activities [44]. The algae live symbiotically with the rice plant, thereby helping the plant in fixing the atmospheric nitrogen, excretion of some substances that help in promoting plant growth, secretion of some organic matter to the soil, and solubilizing insoluble form of phosphorous that helps in modifying the physical and chemical properties of the soil [45]. Many people from different countries have substituted the use of chemical fertilizers with algae as farm manure. For instance, the Irish people generated Fucus in large quantities and used it as farm manure, and people from the coastal district used seaweeds to break down clay in the farm and also as manure and improve crop yields. Those in tropical countries use dried pond bottom debris with abundant blue-green algae as field manure for agricultural activities. Hitherto, it was reported that mixing blue-green algae and seaweed may serve as an ideal fertilizer and function more than chemical fertilizers.

11.3.1.1.4 Fungi Biofertilizers

These are biofertilizers produced from fungi culture; basically, the fungi mostly play a role in the mobilization of phosphorous, which is carried out by symbiotic association between the fungi and plant root by forming mycorrhiza association. The mycorrhizal fungi established functional mutualistic symbioses with more than 80% of the pant on land; thus, the plant supplies energy via photosynthesis to the fungi and in return, and the fungi provide abundance benefits to the host plants, such as phosphate mobilization. The mycorrhiza can serve as a critical linkage between plant roots and soil, which helps in mobilizing or absorbing the solubilized form phosphorous, thereby increasing its bioavailability for easy uptake and utilization by the plant. More so, the fungi use mycelia to expand its mycelia through the root system, thereby increasing the surface area volume for more efficient uptakes of other nutrient materials such as zinc, calcium, and copper [46–48]. The mycorrhiza also improves the soil quality, aeration and water retention capacity, drought, and heavy metal tolerance, which increases the plant's resistivity to pests and other pathogens. The mycorrhizal fungi are divided into groups and genera, namely:

(i) **Ectomycorrhiza:** This is mantle formed from the outer hyphae of both outer and inner and intercellular spaces of the epidermis and cortex of the root system. Examples of the fungal genera in this group are *Pisolithus* spp., *Amanita* spp., *Laccarias* pp., and *Boletus* spp.
(ii) **Endomycorrhiza:** This group of fungi does not form any external mantle but survives within the root system; they are produced by an internal hyphae network and link between cortical cells, which extends out into the soil and helps it to absorb water and mineral salt. Example of endomycorrhiza genera includes *Scutellospora* sp., *Sclerocystis* sp., Arbuscular mycorrhiza- *Glomus*sp., *Gigaspora* sp., and *Acaulospora* sp.

11.3.1.2 Compost Biofertilizers

These are biofertilizers formed as a result of the decomposition of some organic matter such as plant and animal debris. Compost biofertilizers can be produced from various biodegradable materials such as plant leaves, cattle and other animals' shed bedding, biogas plant slurry, fruit and vegetable wastes, industrial wastes, city garbage, sewage sludge, and factory waste. The decomposing debris forms a symbiotic food web within the soil, which aids in the transformation of phosphorus nitrogen and potassium as a result of metabolic activities from earthworms, microorganisms, and the beetles present in the dung [49]. The decomposed materials can serve as organic biofertilizers with the ability to sufficiently aerates, buffer, and aggregates and maintain soil moisture; this helps in maintaining healthy growth pattern of the plants to improve crop yields and to enhance the soil microbial diversity, which ensure future agricultural sustainability.

Compost biofertilizers can be grouped into two based on the nature and type of decomposing microbial species. Some composts are solely decomposed by bacteria and fungi such as *Aspergillus niger*, *Trichoderma viridae*, *Aspergillus terreus*, *Bacillus* spp., and numerous Gram-negative bacteria such as *Klebsiella*, *Pseudomonas*, *Enterobacter*, and *Serratia* spp. Some of the most important nature of this group of compost biofertilizers is their proteolytic and antimicrobial ability against certain parasites and disease-causing microorganisms and also their potentialities in plant cell wall degradation and recycling of the nutritional components of the plants due to their cellulolytic and lignolytic activities.

The second group of compost biofertilizers is the Vermicompost; this group of biofertilizers is produced from dung, earthworm cocoons, some microbial culture such as actinomycetes, and fungi with various organic matter. The vermicompost supplies more macronutrients such as phosphorous, nitrogen, and potassium with several other micronutrients to plants [20]. They also partake in the efficient recycling of agricultural residues and animal and industrial wastes in a very friendly cost and energy-effective manner.

11.3.2 Application of Biofertilizers

Biofertilizers are relatively cheap, locally available, and easier to produce, which are conventional carrier-based inoculants. It is common sense that biofertilizer carrier materials must be nontoxic, very cheap, and locally available with easy processing producers. They must also be biodegradable a high water retention capacity that will support microbial growth and survival for a long time. Examples of some materials used as a carrier for the production of biofertilizers are charcoal, farm yard manure, press mud, soil mixture, neutralized peat soil, or lignite and vermiculite. Thus, the majority of these materials is prone to contamination and can be affected by some conditions such as temperature, humidity, and light intensity, by affecting their shelf life and functions.

11.3.2.1 Methods of Biofertilizer Application

Generally, there are three methods for the effective application of biofertilizers, these are seedling root dipping, seed treatment, and soil application

(i) **Seedling root dipping:** This method is mostly used for transplanted plants. This is the process by which the seedling roots of the plants are dipped in prepared biofertilizers coated water suspension. This allows for proper uptake and utilization of nitrogen and other nutrient within the root system. This method is commonly applied to crops such as fruits, vegetables, trees, cereals, sugarcane, cotton, grapes tobacco, and banana. Examples of biofertilizers applied using the seedling root dipping include phosphorous-solubilizing and nitrogen-fixing Azotobacter or Azospirillum biofertilizers [50].

(ii) **Seed treatment:** In this method, the seeds are uniformly coated in slurry and then air-dried before sowing within 24 h. Seed treatment techniques are the most effective economical and most adopted method for all types of inoculants. Also, this method allowed for easy assimilation and proper utilization of the nutrients. Depending on the type and nature of the plants, this method is commonly applied to numerous crops such as Azotobacter (cereals, mustard, seasum, linseeds, sunflower, castor, forage crops, and grasses), Rhizobium (chickpea, pea, groundnut, soybean, beans, lentil, green gram, black gram, pigeon pea, and cowpea), and Azospirillum or phosphorus-solubilizing bacteria (rice, maize, and sorghum) [50].

(iii) **Direct soil application:** This method in the prepared biofertilizers is applied to the farm field directly without any application medium; this biofertilizer is mostly applied during sowing or planting; for example, direct application of the cow dung compost onto the farmland, a mixture of phosphorous-solubilizing microbial culture, or an overnight shade processed phosphate stone biofertilizers. Examples of biofertilizers applied using the direct soil application techniques include Rhizobium (leguminous plants or trees) and Azotobacter (tea, coffee, rubber, coconuts, and various forest trees) [50].

11.3.3 General Characteristics of Biofertilizers

Biofertilizers play vital roles in ensuring agricultural sustainability; some of their general characteristics include the following:

i. They are good candidates for improving soil fertility.
ii. They are one of the easier means of fixing nitrogen deficiencies in the soil.
iii. They are environmentally friendly due to the role they play in maintaining the microbial stability in the soil.
iv. They are relatively simple and cheaper than conventional fertilizers.
v. They are mostly made up of microbial inoculants.
vi. They are naturally friendly across all habitats.

vii. The microorganisms present in the fertilizer associate symbiotically with the plant root system.
viii. The microbial cells help in converting complex organic material into simple compounds for easy absorption and assimilation by the plant.
ix. They help stimulate plant growth and enhance crop yield.
x. They help in replacing the chemical nitrogen and phosphorus with about 30%.
xi. They protect the plant from desiccation, pests, and some soil-borne pathogens.
xii. They play a role in ensuring healthy soil for noble agricultural sustainability.
xiii. They can be applied easily and do not require extra proficiency.
xiv. They liberate growth-promoting substances such as vitamins.

11.4 Application of Nanochitosan for Sustainable Agricultural Activities

Chitin is one of the major biopolymers found naturally around the globe and is primarily exploited by various marine animals such as shrimp, crabs, crawfish lobster arthropods, and some plant bodies. Chitosan occurs as crystalline microfibrils and appears in three polymeric forms such as α chitin, β-chitin, and γ-chitin. Partial deacetylations of the chitins lead to the formation of natural polycationic polysaccharides from the polymeric s β-chitin forms known as chitosan. The unique properties of chitosan prepare them to be extensively used as a bioagent in numerous biological and biomedical processes such as wound healing, water treatment, scaffolding for tissue engineering, drug carriers, and various aspects of agricultural activities [51].

Chitosans are structurally cationic biopolymers that can easily interact with negatively charged materials. The structural and biopolymeric properties of chitosan give it advantageous proceeds in forming an interactive complex with various fertilizer molecules, which helps in enhancing the availability and uptake of various nutrient materials to plants [52]. Chitosan is one of the leading preferences in the agricultural sector. Having a look at the agricultural yards, nanoparticles synthesized from chitosan molecules can act as portent antimicrobial agents against a wide range of plant pathogens ranging from parasites, bacteria, and fungi, which help to prevent the plants from attack by various microbes dwelling in the soil and can further be used as nanocarriers for existing agrochemicals called chitosan based agro-nanochemicals [53, 54]. They can also be utilized as a substitute the chemically potent biocide agents because studies have demonstrated the use of chitosan in plant pathogen control and disease management. For instance, chitosan has been found to possess numerous fungicidal properties against various plants such as head blight and Fusarium wilt in chickpeas and wheat, leaf blast in rice, blast in finger millet, and leaf spot in maize [55]. Generally, chitosan is very valuable due to its nontoxic nature, biodegradability, biocompatibility, and low allergenicity. Furthermore, chitosan can act as a growth-promoting substance and stimulate both primary and secondary metabolic activities. The combination of chitosan and oligosaccharides amplifies the activities of various enzymes of carbon and nitrogen metabolism, which enhances photosynthetic activities at both light and dark reaction phases, thereby regulating the primary photochemistry of the plant, amplifying the efficiency of carbon fixation in the dark reaction, promoting carbohydrate synthesis, preventing pathogens proliferation, and preserving yield [56].

Synthesis of nanochitosan is paramount in ensuring cheap, eco-friendly, and reliable agricultural activities for a sustainable future. Chitosan nanoparticles are a natural source of eco-friendly nano-nitrogen fertilizers; exploring the use of nanochitosan in the agricultural sector as fertilizers, pesticides, or carries of some farming agrochemicals is immeasurable and plays a vital role in reducing the burden of global food shortage. Nanoparticles derived from chitosan are relatively simple and can serve as a vehicle for the controlled release of some of the conventional fertilizers such as the nitrogen fertilizers (NPK) in order to enhance their availability within the plants; to improve growth-stimulating and growth-regulating hormones; and to play role as insecticides, fertilizers, and herbicides [57, 58]. Nano-nitrogen fertilizers can serve as an alternative to conventional nitrogen fertilizers due to their potential to reduce the application of mineral nitrogen fertilizers and increase the quality and quantity of yield. Chitosan nanoparticles are relatively simple, and the simple nature of their molecules allows for their modification into simpler forms of molecule. Using nano-chitosan as a carrier and control release matrix for the spatial distribution of nutrients and other farming-enhancing agents helps in controlling the fluctuation of some biomolecules such as light, pH, and extreme temperatures. Chitosan-coated plant nutrients are absorbed gradually by preventing the hypersecretion of growth homes to plants and uttermost cell lyses [59, 60].

More so, the ability of the chitosan to form complexes with metal ions helps to reduce the toxicity level of the soil and enhance nutrient utilization. Several studies show potassium nutrient plays a role in improving the physical properties of the soil such as water conductivity, friability, and porosity. Enhancements of these physical properties allow the plant to uptake and utilize high-quality and quantity nutrients, improve the interaction and metabolism of soil microbiota, strengthen the bounding between soil particles, reduce leaching and fertilizer requirements, and increase plant productivity.

In terms of yield enhancement, several studies have shown the application of chitosan, resulting in the enhancement of various plant yields for instance; foliar application of oligo-chitosan and oligo-chitosan nano silica have shown to increase the yield of that soybean seed yield increased 10.5% and 17.0% [61, 62]. Comprehensive analysis of the use of chitosan nanoparticles (1–100 μg/mL) showed proper absorption of the surface of wheat seeds and, also at 5 μg/mL of chitosan nanoparticles, induced auxin-related gene expression [63]. The nanoparticles of chitosan when combined with other nanoparticles were found to be a good preservative when applied to act as preservative to fresh fruits such as Jujube, strawberries, longan, and loquat during storage [64]. Chitosan-based nanomaterials play an important role in managing abiotic stress in plants and water purification for agricultural activities.

11.4.1 Applications of Nanochitosan in Regulating Abiotic Stress in Plants

11.4.1.1 Salinity Control

One of the major significant abiotic factors that affect the growth of plants is the salinity level of the agricultural area. More than 20% of the entire global agricultural fields have high salinity levels that can exert a negative impact on the proper growth of the plant. It is forecasted that, by the year 2050, more farming areas (50%) will be polluted by high salinity levels; this can cause a great setback in food production for the larger global population and a threat to future agricultural sustainability. Therefore, an alternative mechanism has to be developed in order to save future generations from hunger and ensure secured agricultural sustainability. Some studies reveal the potentialities of nanochitosan to enhance the germination of *Phaseolus vulgaris* L. (Bean plant) when treated with 0.1%, 0.2%, and 0.3% nanochitosan at 100 Mm salt concentration [65]. Nitric oxide–releasing chitosan nanoparticles are more effective in combating salt stress in *Zea maize* than other free donors of nitric oxides. More so, the nitric oxides delivered from nanochitosan have proven to improve the activities of the plant photosystem II and chlorophyll levels through the increase of the bioavailability of the nitric oxides. Priming the solid matrix of mung bean seedlings with nanochitosan has been shown to reduce the negative effect of salinity stress and enhance the growth, metabolism, chlorophyll, and protein levels of the plants [66, 67].

Applying chitosan coated with polyvinyl alcohol hydrogels and copper nanoparticles to tomato plants under salt stress has shown to promote plant growth and elevate the expression of genes necessary for detoxification such as superoxide dismutase (SOD) and jasmonic acid (JA) [68]. Some research has also shown the ability of chitosan in large amounts to lessen the effect of salt stress in lentils, wheat, ajowan seeds, and chickpeas [69]. Neutralizing the salinity stress activities on the crops was made possible due to the high surface-to-volume ratio of nanochitosan particles, therefore resulting in higher penetrability and the ability to form more interactions.

11.4.1.2 Drought Stress Control

Water is the number one key component for the survival of plants and virtually every living thing on earth. Water deficiency affects the growth and survival of plants negatively to the extent of extension. Plant exposed to drought stress experiences chloroplast destruction, slowing the rate of the Calvin cycle of photosynthesis due to the reduced enzyme activities, so the anatomical, physiological, and biochemical structure of the plants become negatively affected thereby leading to lower yields [70, 71].

Furthermore, drought stress causes the closure of the plant stomata, which blocks the intake of CO2 by the leaves, leading to the reduction of photosynthesis and plant growth.

Chitosan nanoparticles have been demonstrated to alleviate the harmful effects of late-season drought stress related to relative water content, plant growth, and yield in an experiment conducted with barley plants. Nanochitosan emulsion applied using folia methods under drought stress has been shown to improve the water status of pearl millet plants by decreasing stomatal conductance and transpiration [72]. In sugarcane plants, nanochitosan encapsulating S-nitrosoglutathione was shown to reduce the effect of drought on the sugarcane due to the increase in the photosynthetic activities more than the plants treated with free S-nitrosoglutathione [73]. At a concentration of 90 ppm on water-deficient wheat plants, the application of nanochitosan on water-deficient wheat plants through soil or leaves was shown to enhance both the biochemical and physiological characteristics of the plants [74]. Furthermore, applying chitosan in large quantities has been seen to elevate the growth of apple plants under drought stress induced by agar at a concentration of 40 mg/L. More so, bulk application of chitosan using foliar techniques to wheat plants under water stress has been shown to lower the deleterious effects on plant growth and yield [70]. In the same vein, N-succinyl chitosan and N,O-dicarboxymethylated chitosan derivatives had increased the tolerance of water stress in hybrid maize species that are sensitive to drought stress. These chitosan derivatives were shown to induce the antioxidant defense system of the plants and increase the production of phenolic compounds and osmoregulators, and crop yield also promotes gas exchange in the leaves [75]. Thus, all these potentialities associated with nanochitosan have proven to exert a positive effect on the growth of these plants under drought stress.

11.4.1.3 Temperature and Heavy Metal Stress

High- and low-temperature stress coupled with contamination of heavy metals in soil is among the common abiotic factors exerting a negative effect on agricultural land. However, there are limited data available resources (data) to back the use of nanochitosan in alleviating stress caused by heat and heavy metals. However, there is a study that shows the use of bulk chitosan with zinc compound to reduce the heat stress on late-sowing *Phaseolus vulgaris* (dry bean). Maize seed priming with chitosan at a low temperature of 15°C has been shown to increase the germination index, root length, dry root weight, and shoot height of the plant and also to reduce the time taken for germination [76, 77]. Due to the complex formation ability of chitosan, it has been proven to effectively form a complex with metal ions of Cu(II) and Ag(I) in soil which coexist with other ionic substances such as Cl^-, K^+, and $NO3^-$.

11.4.1.4 Water Purification for Agricultural Activities

As a result of rapid industrial development and drought in some regions around the globe, searching for alternative routes of clean water for domestic and agricultural activities is of paramount importance. Several investigations have been carried out to ensure the use of nontoxic, cost-effective organic compounds with sufficient energy such as chitosan, activated carbon, cellulose, and activated carbon, in purifying seawater and non-conventional water, for agricultural irrigation and domestic activities [78–80]. So also to ensure and maintain crop health and to avoid soil damage that might arise as a result of direct use of wastewater in agricultural activities, sufficient treatment of water is essential in order to adequately meet the requirement for crop growth.

Chitosan nanocomposites have widely been selected and used as an absorbent to remove various inorganic and organic pollutants due to the abundant cross-linked nature of hydroxyl and amino groups present in the chitosan structure. The results demonstrate the ability of the chitosan nanocomposites to possess high and efficient adsorption capacity for dyes, phosphate, pesticides, and heavy metal ions such as Pb(II), Hg(II), Cd(II), Cr(III), Cr(VI), Cu(II), Co(II), Ni(II), and some rare earth metals [81–83]. In some advanced treatment options, chitosan nanoparticles can be personified into nanofiltration membranes and ultrafiltration to reduce the concentration of color, metal ions, and rate chemical oxygen demands and to improve antifouling properties [84, 85]. Furthermore, the economic importance of water treatment and purification in our lives cannot be overemphasized, as such chitosan nanoparticles made coated with Fe_3O_4 demonstrated a promising activity in the recycling of adsorbent under a magnetic field, thereby reducing the operational and treatment cost for easy availability to the farmers [86, 87]. Nanoparticles synthesized from chitosan can also be used as flocculants in settling down pollutant particles during the flocculation step in a series of water treatment processes [88].

11.5 Conclusion

Nanochitosan is a modified form of chitosan, which is derived from chitin, a natural biopolymer found in the shells of crustaceans like shrimp and crab. Nanochitosan has gained significant attention in the field of agriculture due to its potential applications as a biofertilizer and its ability to enhance sustainable agriculture practices. Nanochitosan as a biofertilizer for sustainable agriculture offers several benefits of improved nutrient delivery, nutrient uptake, biological nitrogen fixation, disease resistance, soil health, environmental benefit, biodegradation, stress tolerance, and crop yield and quality.

References

1. Al-Tawaha, A.R.M. and Al-ghzawi, A.L.A., effect of chitosan coating on seed germination and salt tolerance of lentil (Lens culinaris L.). *Res. Crops*, 14, 489–491, 2013.

2. Sharma, S.B., Sayyed, R.Z., Trivedi, M.H., Gobi, T.A., Phosphate solubilizing microbes: sustainable approach for managing phosphorus deficiency in agricultural soils. *Springer plus*, 2, 587–590, 2013.
3. Khordadi, V.J., Sinaki, J.M., Rezvan, S., Damavandi, A., Physiological response of sesame (Sesamum indicum L.) to application of chitosan and magnesium-nano fertilizers under irrigation cut-o_ in a sustainable agriculture system. *Plant Physiol.*, 9, 2629–2639, 2018.
4. Kalia, A. and Sharma, S.P., Nanomaterials and vegetable crops: Realizing the concept of sustainable production, in: *Nanoscience for Sustainable Agriculture*, pp. 323–353, Springer, Berlin/Heidelberg, Germany, 2019.
5. Kanmani, P., Aravind, J., Kamaraj, M., Sureshbabu, P., Karthikeyan, S., Jeyaseelan, A., Sivashanmugam, K., Environmental applications of chitosan and cellulosic biopolymers: A comprehensive outlook. *Bioresour. Technol.*, 242, 295–303, 2017.
6. Liu, P., Sehaqui, H., Tingaut, P., Wichser, A., Oksman, K., Mathew, A.P., Cellulose and chitin nanomaterials for capturing silver ions (Ag+) from water via surface adsorption. *Cellulose*, 21, 449–461, 2013.
7. Adisa, I.O., Pullagurala, V.L.R., Peralta-Videa, J.R., Dimkpa, C.O., Elmer, W.H., Gardea-Torresdey, J.L., White, J.C., Recent advances in nano-enabled fertilizers and pesticides: A critical review of mechanisms of action. *Environ. Sci. Nano*, 6, 2002–2030, 2019.
8. Mali, S.C., Raj, S., Trivedi, R., Nanotechnology a novel approach to enhance crop productivity. *Biochem. Biophys. Rep.*, 24, 100821, 2020.
9. Sivarethinamohan, R. and Sujatha, S., Unlocking the potentials of using nanotechnology to stabilize agriculture and food production, in: *Proceedings of the AIP Conference Proceedings*, vol. 12–13, p. 020022, 2021.
10. Zhang, X., Cao, H., Wang, H., Zhao, J., Gao, K., Qiao, J., Li, J., Ge, S., The effects of graphene-family nanomaterials on plant growth: A review. *Nanomaterials*, 12, 936, 2022.
11. Ghormade, V., Deshpande, M.V., Paknikar, K.M., Perspectives for nanobiotechnology enabled protection and nutrition of plants. *Biotechnol. Adv.*, 29, 792–803, 2011.
12. Okey-Onyesolu, C.F., Hassanisaadi, M., Bilal, M., Barani, M., Rahdar, A., Iqbal, J., Kyzas, G.Z., Nanomaterials as nanofertilizers and nanopesticides: An overview. *Chem. Select*, 6, 8645–8663, 2021.
13. Kamari, A., Pulford, I., Hargreaves, J.S., Chitosan as a potential amendment to remediate metal contaminated soil. A characterisation study. *Colloids Surf. B Biointerfaces*, 82, 71–80, 2011.
14. Singh, A., Singh, N., Hussain, I., Singh, H., Singh, S., Plant-nanoparticle interaction: An approach to improve agricultural practices and plant productivity. *Int. J. Pharm. Sci. Invent.*, 4, 25–40, 2015.
15. Liu, R. and Lal, R., Potentials of engineered nanoparticles as fertilizers for increasing agronomic productions. *Sci. Total Environ.*, 514, 131–139, 2015.

16. Jakhar, A.M., Aziz, I., Kaleri, A.R., Hasnain, M., Haider, G., Ma, J., Abideen, Z., Nano-fertilizers: A sustainable technology for improving crop nutrition and food security. *NanoImpact*, 27, 100411, 2022.
17. Eshghi, S., Hashemi, M., Mohammadi, A., Badii, F., Mohammadhoseini, Z., Ahmadi, K., Effect of Nanochitosan-Based Coating With and Without Copper Loaded on Physicochemical and Bioactive Components of Fresh Strawberry Fruit (Fragaria x ananassa Duchesne) During Storage. *Food Bioprocess Technol.*, 7, 2397–2409, 2014.
18. Song, H., Yuan, W., Jin, P., Wang, W., Wang, X., Yang, L., Zhang, Y., Effects of chitosan/nano-silica on postharvest quality and antioxidant capacity of loquat fruit during cold storage. *Postharvest Boil. Technol.*, 119, 41–48, 2016.
19. Krsti´c, V., Uroševi´c, T., Pešovski, B., A review on adsorbents for treatment of water and wastewaters containing copper ions. *Chem. Eng. Sci.*, 192, 273–287, 2018.
20. Carmona, F.J., Guagliardi, A., Masciocchi, N., Nanosized calcium phosphates as novel macronutrient nano-fertilizers. *Nanomaterials*, 12, 2709, 2022.
21. Ramirez-Rodriguez, G.B., Dal Sasso, G., Carmona, F.J., Miguel-Rojas, C., Perez-de Luque, A., Masciocchi, N., Guagliardi, A., Delgado-Lopez, J.M., Engineering biomimetic calcium phosphate nanoparticles: A green synthesis of slow-release multinutrient (NPK) nanofertilizers. *ACS Appl. Bio Mater.*, 3, 1344–1353, 2020.
22. Mejias, J., Salazar, F., Pérez Amaro, L., Hube, S., Rodriguez, M., Alfaro, M., Nanofertilizers: A cutting-edge approach to increase nitrogen use efficiency in grasslands. *Front. Environ. Sci.*, 9, 52, 2021.
23. Hernández-Hernández, H., Juárez-Maldonado, A., Benavides-Mendoza, A., Ortega-Ortiz, H., Cadenas-Pliego, G., Sánchez-Aspeytia, D., González-Morales, S., Chitosan-PVA and Copper Nanoparticles Improve Growth and Overexpress the SOD and JA Genes in Tomato Plants under Salt Stress. *Agronomy*, 8, 175, 2018.
24. Jariwala, D., Sangwan, V.K., Lauhon, L.J., Marks, T.J., Hersam, M.C., Carbon nanomaterials for electronics, optoelectronics, photovoltaics, and sensing. *Chem. Soc. Rev.*, 42, 2824–2860, 2013.
25. Shanmuganathan, R., Edison, T.N.J.I., LewisOscar, F., Kumar, P., Shanmugam, S., Pugazhendhi, A., Ponnuchamy, K., Chitosan nanopolymers: An overview of drug delivery against cancer. *Int. J. Boil. Macromol.*, 130, 727–736, 2019.
26. Jiang, D., Huang, D., Lai, C., Xu, P., Niu, C.-G., Wan, J., Tang, L., Dong, H., Huang, B., Hu, T., Difunctional chitosan-stabilized Fe/Cu bimetallic nanoparticles for removal of hexavalent chromium wastewater. *Sci. Total Environ.*, 644, 1181–1189, 2018.
27. Khodakovskaya, M., Dervishi, E., Mahmood, M., Xu, Y., Li, Z., Watanabe, F., Biris, A.S., Carbon nanotubes are able to penetrate plant seed coat and dramatically affect seed germination and plant growth. *ACS Nano*, 3, 3221–3227, 2009.

28. Kubavat, D., Trivedi, K., Vaghela, P., Prasad, K., Vijay Anand, G.K., Trivedi, H., Patidar, R., Chaudhari, J., Andhariya, B., Ghosh, A., *Characterization of a chitosan-based sustained release nanofertilizer formulation used as a soil conditioner while simultaneously*, 2020.
29. Janmohammadi, M., Amanzadeh, T., Sabaghnia, N., Ion, V., Effect of nano-silicon foliar application on safflower growth under organic and inorganic fertilizer regimes. *Bot. Lith.*, 22, 53–64, 2016.
30. Safdar, M., Park, S., Gwon, Y., Kim, Y.O., Kim, J., Engineering plants with carbon nanotubes: A sustainable agriculture approach. *J. Nanobiotechnol.*, 20, 1–30, 2022.
31. Zareie, C., Eshkalak, S.K., Najafpour-Darzi, G., Baei, M.S., Younesi, H., Seeram, R., Uptake of Pb(II) Ions from Simulated Aqueous Solution via Nanochitosan. *Coatings*, 9, 862, 2019.
32. Kah, M., Kookana, R.S., Gogos, A., Bucheli, T.D., A critical evaluation of nanopesticides and nanofertilizers against their conventional analogues. *Nat. Nanotechnol.*, 13, 677–684, 2018.
33. Kumaraswamy, R.V., Kumari, S., Choudhary, R.C., Pal, A., Raliya, R., Biswas, P., Saharan, V., Engineered chitosan based nanomaterials: Bioactivities, mechanisms and perspectives in plant protection and growth. *Int. J. Boil. Macromol.*, 113, 494–506, 2018.
34. Timmusk, S., Muthoni, J., Muraya, A., Aronsson, A., Perspectives and challenges of microbial application for crop improvement. *Front. Plant Sci.*, 8, 49–53, 2017.
35. Krishnani, K.K., Boddu, V.M., Chadha, N.K., Chakraborty, P., Kumar, J., Krishna, G., Pathak, H., Metallic and non-metallic nanoparticles from plant, animal, and fisheries wastes: Potential and valorization for application in agriculture. *Environ. Sci. Pollut. Res. Int.*, 29, 81130–81165, 2019.
36. Kolangare, I.M., Isloor, A.M., Karim, Z.A., Kulal, A., Ismail, A.F., Asiri, A.M., Antibiofouling hollow-fiber membranes for dye rejection by embedding chitosan and silver-loaded chitosan nanoparticles. *Environ. Chem. Lett.*, 17, 581–587, 2018.
37. Ahmed, S. and Fahmy, A., Environmental Impact for applications of neem cake coated urea and nano iron foliar on rationalization of chemical nitrogen fertilizers and wheat yield. *J. Soil Sci. Agric. Eng.*, 8, 613–620, 2017.
38. do Espirito Santo Pereira, A., Caixeta Oliveira, H., Fernandes Fraceto, L., Santaella, C., Nanotechnology potential in seed priming for sustainable agriculture. *Nanomaterials*, 11, 267, 2021.
39. Singh, S., Singh, B.K., Yadav, S.M., Gupta, A.K., Potential of biofertilizers in crop production in Indian agriculture. *Am. J. Plant Nutr. Fertil. Technol.*, 4, 33–40, 2014.
40. Thakur, P., Singh, I., Biocontrol of Soilborne Root Pathogens: An Overview. In: Giri, B., Prasad, R., Varma, A. (eds). *Root Biology. Soil Biology*, vol 52. Cham, Springer, 2018, https://doi.org/10.1007/978-3-319-75910-4_7.

41. Meena, V.S., Maurya, B.R., Verma, J.P., Does a rhizospheric microorganism enhance K+ availability in agricultural soils? *Microbiol. Res.*, 169, 337–347, 2014.
42. Cakmakci, R., Dönmez, M.F., Erdoğan, U., The effect of plant growth promoting rhizobacteria on barley seedling growth, nutrient uptake, some soil properties, and bacterial counts. *Turk J. Agric.*, 31, 189–199, 2007.
43. Etesami, H., Emami, S., Alikhani, H.A., Potassium solubilizing bacteria (KSB): mechanisms, promotion of plant growth, and future prospects, a review. *J. Soil Sci. Plant Nutr.*, 17, 897–911, 2017.
44. Ramachandran, S.K. and Gangasalam, A., Reduction of chemical oxygen demand and color from the rice mill wastewater by chitosan/2(5H)-furanone-incorporated ultrafiltration membrane system. *Sep. Sci. Technol.*, 54, 409–425, 2018.
45. Saraiva, R., Ferreira, Q., Rodrigues, G.C., Oliveira, M., Phosphorous nanofertilizers for precise application in rice cultivation as an adaptation to climate change. *Climate*, 10, 183, 2022.
46. Tayel, A.A., Ibrahim, S.I., Al-Saman, M.A., Moussa, S.H., Production of fungal chitosan from date wastes and its application as a biopreservative for minced meat. *Int. J. Boil. Macromol.*, 69, 471–475, 2014.
47. Singh, I. and Giri, B., Arbuscular mycorrhiza mediated control of plant pathogens, in: *Mycorrhiza Nutrient uptake, biocontrol, ecorestoration*, pp. 131–160, Springer, Cham, 2017.
48. Jones, M., Weiland, K., Kujundzic, M., Theiner, J., Kählig, H., Kontturi, E., John, S., Bismarck, A., Mautner, A.Waste-Derived Low-Cost Mycelium Nanopapers with Tunable Mechanical and Surface Properties. *Biomacromolecules*, 20, 3513–3523, 2019.
49. Basavegowda, N. and Baek, K.H., Current and future perspectives on the use of nanofertilizers for sustainable agriculture: The case of phosphorus nanofertilizer. *Biotech*, 11, 357, 2021.
50. Kumar, R., Kumawat, N., Sahu, Y.K., Role of Biofertilizers in Agriculture. *Popular Kheti*, 5, 4, 1–5, 2017.
51. Pourmortazavi, S.M., Sahebi, H., Zandavar, H., Hajimirsadeghi, S., Fabrication of Fe3O4 nanoparticles coated by extracted shrimp peels chitosan as sustainable adsorbents for removal of chromium contaminates from wastewater: The design of experiment. *Compos. Part B Eng.*, 175, 107130, 2019.
52. Divya, K. and Jisha, M.S., Chitosan nanoparticles preparation and applications. *Environ. Chem. Lett.*, 16, 101–112, 2017.
53. Cho, K.Y., Yoo, C.H., Won, Y.J., Hong, D.Y., Chang, J.S., Choi, J.W., Lee, J.H., Lee, J.S., Surface-concentrated chitosan-doped MIL-100(Fe) nanofiller-containing PVDF composites for enhanced antibacterial activity. *Eur. Polym. J.*, 120, 109221, 2019.
54. Nadaroglu, H., Mosber, G., Gungor, A.A., Adıguzel, G., Adiguzel, A., Biodegradation of some azo dyes from wastewater with laccase fromWeissella

viridescens LB37 immobilized on magnetic chitosan nanoparticles. *J. Water Process Eng.*, 31, 100866, 2019.
55. Khati, P., Chaudhary, P., Gangola, S., Bhatt, P., Sharma, A., Nanochitosan supports growth of Zea mays and also maintains soil health following growth. *Biotech*, 7, 81, 2017.
56. Huang, Y., Lee, X., Grattieri, M., Macazo, F.C., Cai, R., Minteer, S.D., A sustainable adsorbent for phosphate removal: Modifying multi-walled carbon nanotubes with chitosan. *J. Mater. Sci.*, 53, 12641–12649, 2018.
57. Corradini, E., A preliminary study of the incorparation of NPK fertilizer into chitosan nanoparticles. *eXPRESS Polym. Lett.*, 4, 509–515, 2010.
58. Mustafa, I.F. and Hussein, M.Z., Synthesis and technology of nanoemulsion-based pesticide formulation. *Nanomaterials*, 10, 1608, 2020.
59. DeRosa, M.C., Monreal, C., Schnitzer, M., Walsh, R., Sultan, Y., Nanotechnology in fertilizers. *Nat. Nanotechnol.*, 5, 91, 2010.
60. Oliveira, H.C., Gomes, B.C., Pelegrino, M.T., Seabra, A.B., Nitric oxide-releasing chitosan nanoparticles alleviate the effects of salt stress in maize plants. *Nitric Oxide*, 61, 10–19, 2016.
61. Wang, J., Zhou, Y., Shao, Y., He, F., Wu, M., Ni, H., Zheng, Y., Sun, Y., Chitosan–silica nanoparticles catalyst (M@CS–SiO2) for the degradation of 1,1-dimethylhydrazine. *Res. Chem. Intermed.*, 45, 1721–1735, 2018.
62. Rabêlo, V.M., Magalhães, P.C., Bressanin, L.A., Carvalho, D.T., Dos Reis, C.O., Karam, D., Doriguetto, A.C., Dos Santos, M.H., Filho, P.R.D.S.S., De Souza, T.C., The foliar application of a mixture of semisynthetic chitosan derivatives induces tolerance to water deficit in maize, improving the antioxidant system and increasing photosynthesis and grain yield. *Sci. Rep.*, 9, 8164, 2019.
63. Ibrahim, E.A. and Ramadan, W.A., Effect of zinc foliar spray alone and combined with humic acid or/and chitosan on growth, nutrient elements content and yield of dry bean (Phaseolus vulgaris L.) plants sown at different dates. *Sci. Hortic.*, 184, 101–105, 2015.
64. Hong, J., Wang, C., Wagner, D.C., Gardea-Torresdey, J.L., He, F., Rico, C.M., Foliar application of nanoparticles: Mechanisms of absorption, transfer, and multiple impacts. *Environ. Sci. Nano*, 8, 1196–1210, 2021.
65. Mahdavi, B., Safari, H., Modarres Sanavy, S.A.M., Effect Of Chitosan Seed Priming On Germination, Ion Relations and Biochemical Characteristics of Chickpea Under Salinity Stress. *Plant Prod. Technol. (Agric. Res.)*, 15, 163–172, 2015.
66. Mahdavi, B. and Rahimi, A., Seed priming with chitosan improves the germination and growth performance of ajowan. *Eurasian J. Biosci.*, 7, 69–76, 2013.
67. Sen, S.K., Chouhan, D., Das, D., Ghosh, R., Mandal, P., Improvisation of salinity stress response in mung bean through solid matrix priming with normal and nano-sized chitosan. *Int. J. Boil. Macromol.*, 145, 108–123, 2019.

68. Sun, C., Fu, D., Jin, L., Chen, M., Zheng, X., Yu, T., Chitin isolated from yeast cell wall induces the resistance of tomato fruit to Botrytis cinerea. *Carbohydr. Polym.*, 199, 341–352, 2018.
69. Ma, L., Li, Y., Yu, C., Wang, Y., Li, X., Li, N., Chen, Q., Bu, N., Alleviation of exogenous oligochitosan on wheat seedlings growth under salt stress. *Protoplasma.*, 249, 393–399, 2011.
70. Behboudi, F., Tahmasebi Sarvestani, Z., Kassaee, M.Z., Modares Sanavi, S.A.M., Sorooshzadeh, A., Ahmadi, S.B., Evaluation of Chitosan Nanoparticles Effects on Yield and Yield Components of Barley (Hordeum vulgare L.) under Late Season Drought Stress. *J. Water Environ. Nanotechnol.*, 3, 22–39, 2018.
71. Priyaadharshini, M., Sritharan, N., Senthil, A., Marimuthu, S., Physiological studies on effect of chitosan nanoemulsion in pearl millet under drought condition. *J. Pharmacogn. Phytochem.*, 8, 3304–3307, 2019.
72. Iriti, M., Picchi, V., Rossoni, M., Gomarasca, S., Ludwig, N., Gargano, M., Faoro, F., Chitosan antitranspirant activity is due to abscisic acid-dependent stomatal closure. *Environ. Exp. Bot.*, 66, 493–500, 2009.
73. Silveira, N.M., Seabra, A.B., Marcos, F.C., Pelegrino, M.T., Machado, E.C., Ribeiro, R.V., Encapsulation of S-nitrosoglutathione into chitosan nanoparticles improves drought tolerance of sugarcane plants. *Nitric Oxide*, 84, 38–44, 2019.
74. Masjedi, M.H., Roozbahani, A., Baghi, M., Assessment Effect of Chitosan Foliar Application on Total Chlorophyll and Seed Yield of Wheat (Triticum aestivum L.) Under Water Stress Conditions. *J. Crop Nutr. Sci.*, 3, 14–26, 2017.
75. Avestan, S., Naseri, L., Barker, A.V., Evaluation of nanosilicon dioxide and chitosan on tissue culture of apple under agar-induced osmotic stress. *J. Plant Nutr.*, 40, 2797–2807, 2017.
76. Guan, Y.J., Hu, J., Wang, X.J., Shao, C.X., Seed priming with chitosan improves maize germination and seedling growth in relation to physiological changes under low temperature stress. *J. Zhejiang Univ.-Sci. B.* 10, 6, 427–433, 2009.
77. Fan, H., Zhou, S., Jiao, W.-Z., Qi, G.-S., Liu, Y.-Z., Removal of heavy metal ions by magnetic chitosan nanoparticles prepared continuously via high-gravity reactive precipitation method. *Carbohydr. Polym.*, 174, 1192–1200, 2017.
78. Olivera, S., Muralidhara, H.B., Venkatesh, K., Guna, V.K., Gopalakrishna, K., Kumer, Y., Potential applications of cellulose and chitosan nanoparticles/composites in wastewater treatment: A review. *Carbohydr. Polym.*, 153, 600–618, 2016.
79. Alarcón-Payán, D.A., Koyani, R.D., Vazquez-Duhalt, R., Chitosan-based biocatalytic nanoparticles for pollutant removal from wastewater. *Enzym. Microb. Technol.*, 100, 71–78, 2017.
80. Namasivayam, S.K.R., Kumar, S.N., Kamil, T.M., Ravi, T., Biopolymer-Mediated Coating Influence on Wastewater Treatment Efficacy of Silver

Nanoparticles Synthesized from Fungal Consortium. *Natl. Acad. Sci. Lett.*, 43, 6, 1–5, 2020.

81. Altinisik, A. and Yurdakoc, K., Chitosan-/PVA-coated magnetic nanoparticles for Cu (II) ions adsorption. *Desalin. Water Treat.*, 57, 18463–18474, 2016.
82. Eladlani, N., Dahmane, E.M., Ouahrouch, A., Rhazi, M., Moha, T., Recovery of Chromium(III) from Tannery Wastewater by Nanoparticles and Whiskers of Chitosan. *J. Polym. Environ.*, 26, 152–157, 2017.
83. Khalil, T.E., Elhusseiny, A.F., El-Dissouky, A., Ibrahim, N.M., Functionalized chitosan nanocomposites for removal of toxic Cr (VI) from aqueous solution. *React. Funct. Polym.*, 146, 104407, 2020.
84. Bratina, B., Šorgo, A., Kramberger, J., Ajdnik, U., Zemljič, L.F., Ekart, J., From municipal/industrial wastewater sludge and FOG to fertilizer: A proposal for economic sustainable sludge management. *J. Environ. Manag.*, 183, 1009–1025, 2016.
85. Rahimi, Z., Zinatizadeh, A.A., Zinadini, S., Membrane bioreactors troubleshooting through the preparation of a high antifouling PVDF ultrafiltration mixed-matrix membrane blended with O-carboxymethyl chitosan-Fe3O4 nanoparticles. *Environ. Technol.*, 40, 3523–3533, 2018.
86. Cadogan, E.I., Lee, C.-H., Popuri, S.R., Lin, H.-Y., Efficiencies of chitosan nanoparticles and crab shell particles in europium uptake from aqueous solutions through biosorption: Synthesis and characterization. *Int. Biodeterior. Biodegrad.*, 95, 232–240, 2014.
87. Zhang, H., Tan, X., Qiu, T., Zhou, L., Li, R., Deng, Z., A novel and biocompatible Fe_3O_4 loaded chitosan polyelectrolyte nanoparticles for the removal of Cd2+ ion. *Int. J. Boil. Macromol.*, 141, 1165–1174, 2019.
88. Picos-Corrales, L.A., Sarmiento-Sánchez, J.I., Ruelas-Leyva, J.P., Crini, G., Hermosillo-Ochoa, E., Gutierrez-Montes, J.A., Environment-Friendly Approach toward the Treatment of Raw Agricultural Wastewater and River Water via Flocculation Using Chitosan and Bean Straw Flour as Bioflocculants. *ACS Omega*, 5, 3943–3951, 2020.

12

Application of Nanochitosan in Plant Growth and Crop Protection

Innocent Ojeba Musa[1*], Job Oloruntoba Samuel[2], Mustapha Adamu[2], Oluwafemi Adebayo Oyewole[3], Charles Oluwaseun Adetunji[4], Eniola K. I. T.[5] and Mohammed Bello Yerima[6]

[1]*Department of Microbiology, Skyline University Nigeria, Kano State, Nigeria*
[2]*Department of Microbiology, Federal University of Technology, Minna, Niger State, Nigeria*
[3]*Department of Microbiology, Federal University of Technology, Minna, Nigeria*
[4]*Department of Microbiology, Edo State University, Uzairue, Nigeria*
[5]*Department of Biological Sciences, Joseph Ayo Babalola University, Ikeji Arakeji, Osun State, Nigeria*
[6]*Department of Microbiology, Sokoto State University, Sokoto, Sokoto State, Nigeria*

Abstract
The escalation of worldwide food consumption, along with the growth of the global population, has generated a need for enhanced research on sustainable food production and agriculture. The integration of chitosan, a naturally occurring biopolymer, with nanotechnology has the potential to offer a viable and sustainable substitute for conventional agrochemicals, thereby promoting a safer agricultural industry. The potential of chitosan-based agronanochemicals as a workable and long-term crop protection against diseases, pests, and plant growth enhancers is discussed in this study. These endeavors offer improved alternatives by utilizing chitosan nanocarriers to encapsulate existing agricultural active ingredients. This results in the development of potent biocides that effectively combat plant pests and plant pathogens. The high bioavailability and controlled release capabilities of the nanoformulations prevent the loss and leaching of the active components often associated with agrochemicals. Additionally, the nanometer-scale dimensions of these formulations facilitate enhanced penetration through the plant cuticle and cell wall. Moreover, this article presents a concise summary of the mechanism by

**Corresponding author*: innocentmusa0011@gmail.com

Charles Oluwaseun Adetunji, Maulin P. Shah, Yerima Mohammed Bello, Daniel Hefft, Jay Singh, Shyam S. Pandey and Ravindra Pratap Singh (eds.) Nanochitosan Applications for Enhanced Crop Production and Food Security, (281–300) © 2025 Scrivener Publishing LLC

which chitosan acts against pathogens, along with its ability to induce plant immunity and activate defense responses in plants treated with chitosan.

Keywords: Agriculture, chitosan, plant, nanoformulations, agronanochemicals, pathogens

12.1 Introduction

The potential uses of nanochitosan, a modified form of chitosan produced from chitin, in agriculture, especially in promoting plant development and protecting crops, have garnered a lot of attention in recent years [1]. Benefits of chitosan, a biopolymer isolated from the shells of crustaceans like prawns and crabs [2], include its biodegradability, biocompatibility, and low toxicity.

The escalating global need for food has significantly contributed to the expanded utilization of agrochemicals. There has been a surge in research efforts investigating sustainable agricultural production systems due to the heightened attention given to food security and agriculture [3, 4]. The utilization of pesticides poses significant implications for both human health and the environment, hence presenting a substantial obstacle in the pursuit of attaining global food security. In the preceding 10 years, the yearly worldwide utilization of agrochemicals has consistently hovered at approximately 2 million metric tons. China was found to have purchased almost 1.8 million metric tons of goods on an annual basis, whereas the United States acquired 500,000 metric tons and Argentina obtained 236,000 metric tons. On the contrary, the yearly use in Malaysia amounted to a mere 49,199 metric tons, as reported by Sharma [5] and Peralta *et al.* [6]. There is a projected increase of 3.5 million metric tons in the global utilization of agrochemicals in the year 2020. Agrochemicals encompass a range of substances, like insecticides, fertilizers, and regulators of plant growth, as documented by Gomes *et al.* [7] and Ali *et al.* [3]. Agrochemicals are considered an essential component of contemporary crop management due to several factors, as highlighted by Hassan *et al.* [8]. The risks connected with agrochemicals are widely acknowledged, alongside the advantages that they offer. The potential impact of application losses resulting from degradation, volatilization, photolysis, and leaching on organisms in soil and rivers is a matter of concern. The efficacy of agrochemicals in efficiently targeting and eliminating pests for which they were specifically developed is less than 1% [9].

Agronanochemicals, also known as nano-enabled agrochemicals, provide the capacity to enhance integrated pest and disease management by virtue of their controlled release of active components and targeted delivery to specified locations [2, 10, 11]. Sharma *et al.* [5] and Wang *et al.* [12] have provided an exposition of the enduring approaches employed for the management of agricultural diseases and pests. According to Singl *et al.* [13], the utilization of agronanochemicals has the potential to mitigate the environmental impact associated with conventional agrochemicals. Zhang *et al.* [14] suggest that agronanochemicals have the potential to address a range of agrochemical challenges, including but not limited to toxicities, absorption, solubility, stability, volatilization, leaching, and runoff.

12.2 Chitosan-Based Agronanochemicals

Chitosan is cartegorized by the United States Food and Drug Administration to be non-toxic, biodegradable, and biocompatible substance [15]. According to Husen and Jawad [16], this substance possesses various capabilities such as antibacterial, anti-inflammatory, antioxidant, antiviral, bio-adhesion, antifungal, and adsorption enhancement. Chitosan dissolves easily in acidic environments because of the protonation of its amino group. Chitin is a biopolymer that occurs naturally and may be extracted from a wide range of organisms, examples are fungal cell walls, insect cuticles, and crab shells [3]. Bio-waste generated by industries dealing with crustaceans is put to good use in the chitosan production process. Southeast Asian countries are expected to contribute roughly 1.5 million tons annually to the global chitosan production total of 6–8 million tons [17]. The aim of Kapour *et al.* [18] is to create a "zero-waste" food sector that is good for business and the planet. The sustainability of agronanochemicals based on chitosan in crop management has generated scientific controversy. The production of chitosan uses a lot of water, but the benefits of turning trash into treasure, as well as the effects of synergistic, abilities of toxic-protection, and decreased agrochemical runoff and leaching of chitosan nanoformulations in crop management, make it worthwhile [19]. Environmental and human health concerns from pesticide accumulation and runoff may be mitigated by the use of controlled-release formulations with high bioavailability.

Additionally, these formulations can contribute to reducing labor costs in the agricultural industry. Due to their minimal toxicity, these substances pose no harm to farmers and applicators. According to Kapour *et al.* [18], empirical evidence supports the notion that chitosan nanoparticles possess the ability to enhance plant growth in agricultural environments while concurrently impeding the proliferation of detrimental bacteria and fungi. The utilization of chitosan-based agronanochemicals as nanocarriers for pre-existing agrochemicals has been suggested by both Peralta *et al.* [6] and Kadam *et al.* [9]. Techniques that can be employed to encapsulate agriculturally active chemicals within a nanodelivery system formulation, such as entrapment within a chitosan-based polymeric matrix or by the use of ionic or covalent inter/intramolecular interactions [6]. According to Kadam *et al.* [9], various procedures such as reverse micellar, precipitation, spray drying, emulsion cross-linking, ionic gelation, and screening have been used in the synthesis of chitosan-based agronanochemicals. The sieve method is the most basic and direct among these procedures. The process has been observed to generate nanoparticles with inconsistent morphology and dimensions. The simplicity and low cost of the ionic gelation approach make it a subject of extensive research in formulating chitosan nanoparticulate systems. The approach's minimal chemical usage reduces the likelihood of hazardous side effects. In acidic environments, polyanions such as tripolyphosphate are utilized to form bonds with the positively charged protonated amino group of chitosan [9, 15].

Stable nanoparticulate structures can be created using emulsion cross-linking. This approach, however, requires the use of cross-linking chemicals such glutaraldehyde, formaldehyde, alginate, and others and takes a considerable amount of time. It is possible that the active chemicals would not get along with these. The ultimate particle size is significantly impacted by the size of the droplets employed in the emulsion generation process. The size of these droplets is influenced by several factors, such as the quantity of cross-linking, the speed at which mixing occurs, the molecular weight of chitosan, and the type of surfactant used [20, 21].

Thermodynamically stable, uniformly sized chitosan nanoparticles can be made via the reverse micellar method. The process calls for a cross-linking agent of a surfactant solution like cetyl trimethylammonium bromide and organic solvent. However, this option is rarely chosen due to the lengthy and expensive processes required. To create coacervate nanoparticles, one technique of precipitation includes blowing the solution of chitosan through an air nozzle that is compressed. The nanoparticles used in this method have weak mechanical strength and irregular morphologies, which is a drawback. Dry granules, powders, and pellets made from chitosan have been produced in large quantities using spray drying [22]. Each method begins with dissolving chitosan in acetic acid and then adding the active components and cross-linking agent one after the other. A heated air stream is used to evaporate the fluid, leaving behind the nanoparticles.

With the use of chitosan and agrochemicals, nanoformulations aim to increase the former while reducing the latter. Chitosan encapsulation is a viable solution to the problem of many agrochemicals having poor water solubility because of their amphiphilic nature. Because of this, inert compounds can be used in place of more harmful ones in conventional agrochemicals [11, 14]. Because of chitosan's bio-adhesive properties, encapsulating agrochemicals are more likely to remain stable and be absorbed by the plant [10].

12.2.1 Controlled Release Formulations

The introduction of controlled release mechanisms for active components in agrochemicals therefore aims to control the issues caused as a result of the excessive use of agrochemicals in agricultural contexts by restricting the amount and frequency of application. There are two distinct stimuli that can cause the agrochemicals to be released from the matrix of chitosan. This release can first be activated by biotic stressors, such as those brought on by the presence of plant diseases such fungus, nematodes, insects, pests, and bacteria [13, 14]. A second possible cause of agrochemical release is abiotic stress, which can be caused by changes in pH, temperature, salinity, flooding, and drought [13, 14, 23].

Multiple processes, like capsule expansion, pore diffusion, surface desorption, and disintegration, are involved in the stimulus-response mechanism, as stated by Faizan *et al.* [24]. Surface desorption describes the deposition of the active component on the nanoformulation's surface, whereas the diffusion-controlled process requires a gradient in diffusion rates. The subsequent discharge of agrochemicals during hydration is contingent upon the expansion of the chitosan capsule. In addition, the capsule matrix may experience rupture or degradation due to enzymatic activities or other environmental factors. Consequently, the implementation of controlled release mechanisms in nanoformulations enables the precise and efficient delivery of agrochemicals to the desired target site. Specifically, a decline in pH from 3 to 1 led to an increase in Cu release from 21.5% to 44.1% [13, 25]. The observed phenomenon can be attributed to the process of protonation of the amino group present in chitosan. The release of Cu was significantly decreased at elevated pH values of 6 and 7 due to the deprotonation of the amino group in chitosan, resulting in reductions of 6.1% and 4.9%, respectively. Moreover, under a pH of 4.5, there was a continuous release of Cu observed for a duration of 96 h. An investigation of chitosan-Zn nanoparticles revealed the presence of a stimulus-response release mechanism. This mechanism was characterized by the release of Zn primarily by stomatal absorption, followed by subsequent diffusion and polymer swelling following water penetration. The intracellular media, which exhibits a somewhat acidic nature, is purported to facilitate the liberation of Zn from chitosan nanoparticles. According to Zhang [14], the release of fungicides from chitosan-hexaconazole nanoparticles and chitosan-dazomet nanoparticles at a pH of 5.5 followed a diffusion-controlled mechanism. The half release times (t1/2) for chitosan-dazomet and chitosan-hexaconazole nanoparticles were found to be 42 h and 11 h, respectively. In addition, the simultaneous release of dazomet and hexaconazole from nanoparticles composed of chitosan-hexaconazole-dazomet resulted in an extended release duration of 53 h and 15 h, correspondingly. The combination of methomyl with azidobenzaldehyde and carboxymethyl chitosan led to the release of the compound being regulated by diffusion at a pH of 6.0. The half-life of this release ranged from 36 h to 70 h, the release of the pesticide acetamiprid from chitosan-alginate nanocapsules is influenced by pH. The study found that the release of acetamiprid was similar at pH 10 after 24 h, compared to pH 7.0 and 4.0 after 36 h.

12.2.2 Plant Growth Promoter

Chitosan nanoformulations have garnered significant attention as potential enhancers of plant growth. Protonated chitosan exhibits a higher affinity for plant cell membranes due to its abundance of positive charges, resulting in increased reactivity within the plant system. Plants use chitosan, a macronutrient consisting of 9%–10% nitrogen [26]. Micronutrients (copper, manganese, nickel, zinc, boron, iron, and chlorine) and macronutrients (nitrogen, phosphorous, potassium, magnesium, calcium, and sulfur) can be combined with chitosan in plants. Nanochitosan and macro/micronutrient nanocarriers have been the focus of recent studies [14]. As a chemical-free method, chitosan nanoformulations are widely utilized in seed treatment to boost germination rates and biomass production. Increases in chlorophyll content and photosynthetic rate are two of the ways in which chitosan nanoformulations have been used to stimulate development [8, 27]. The data reveal that nanochitosan of all sizes is superior to chitosan oligomer in nutritional absorption. Researchers found that body size had no discernible impact on nutrition uptake. Hussen *et al.* [27] found that chlorophyll concentration and photosynthetic rate vary with organism size. The growth rate of seedlings subjected to nano-sized chitosan treatment was seen to be higher in comparison to seedlings treated with chitosan oligomer or those that were left untreated. In an independent investigation, it was observed that the application of nanochitosan had a positive impact on the ability of bean seedlings to withstand abiotic stress, namely, salt stress [14]. The wheat plants were treatment with nanoparticles of chitosan, poly(methacrylic acid), and nitrophytin. A comparative analysis was conducted to assess the effectiveness of nanoformulations in comparison to conventional bulk NPK fertilizers [8]. Plant height, main spike weight, crop yield, and harvest index were measured after treatment with nanoformulations containing nitrogen (N; 500 mg/mL), phosphorus (P; 60 mg/mL), and potassium (K; 400 mg/mL). The measurements recorded for each plant include height (41.29 inches), weight (0.178 g and 6.95 g), and yield (26.94 cm), listed in sequential order. In a recent study, it was shown that the use of nanoformulations of bulk NPK had a substantial positive impact on the growth of wheat plants and the overall yield of the crop. The observed results indicate that there was a significant rise in plant height, with a mean growth of 38.85 cm under the same dosage. Additionally, the main spike weight exhibited a slight but statistically significant increase of 0.136 g. Moreover, crop output demonstrated a noteworthy improvement, with an average increase of 6.13 gg per plant. Lastly, the harvest index experienced a substantial increase of 21.64. Xing *et al.* [28] examined the impact of bulk chitosan, copper sulfate ($CuSO_4$), and chitosan-Cu nanoparticles on the growth of maize seedlings. In contrast to bulk chitosan, the utilization of chitosan nanoformulations significantly impacted seedling growth and development, total protein content, as well as the activity of amylase and protease enzymes.

The notion postulated that nanoformulations, by making it simpler for seeds to penetrate, would boost metabolic rates in plants. However, bulk chitosan can coat the seed's exterior and prevent it from absorbing water and nutrients.

High-hormone bioavailability is achieved by a slow-release nanodelivery method created by encapsulating plant growth regulators into chitosan nanocarriers. Hormones in plants are chemical compounds that regulate and modify the growth and cellular development of plants. These include cytokinin, ethylene, abscisic acid, cytokinin, and the hormones gibberellin and auxin. Adding chitosan-polyglutamic acid-gibberellic acid nanoparticles to French bean seedlings stimulated the development of more lateral roots than did the application of free hormone. The advantages of using nanoparticulate systems are highlighted. Researchers Sharma *et al.* [29] found that the germination rate of chitosan-thiamine nanoparticle-treated chickpea seeds was 90%, compared to 84% and 75% for seeds treated with a mixture of chitosan-thiamine and control (water), respectively. The nanoparticulate technology also resulted in a 10-fold increase in auxin levels and increased levels of defense enzymes in treated seedlings compared to untreated seedlings.

When chitosan is processed into nanoscale particles (nanochitosan), its properties are further enhanced, leading to its utilization in sustainable agricultural practices.

1. **Seed Germination and Root Growth Enhancement:** Treatments with nanochitosan have been shown to increase seed germination rates and stimulate early seedling growth. Nanoparticles create a favorable microenvironment around the seed, enhancing water absorption and nutrient uptake [7]. Nanochitosan has a positive effect in challenging environments such as drought or low soil quality by aiding seeds in the establishment and growth of robust root systems [30, 31].
2. **Nutrient Uptake and Soil Amendment:** Nanochitosan has the ability to adhere to soil particles and retain water, thereby improving soil structure and increasing nutrient accessibility. Positively charged nanoparticles attract negatively charged nutrient ions, thereby reducing leaching and enhancing plant root uptake. This can lead to improved nutrient utilization, decreased need for excessive fertilization, and reduced environmental contamination [2].
3. **Disease Management:** Nanochitosan exhibits antifungal and antibacterial properties, making it a promising method for crop protection. Nanochitosan exhibits inhibitory effects on various plant diseases when applied as foliar sprays or incorporated into the soil. The application of a physical barrier on the surface of plants can prevent the entry of pathogens and also enhance the plant's innate defense mechanisms [2].
4. **Pest Control:** Nanochitosan formulations have demonstrated potential in pest control. Nanoparticles can impact insect feeding and development by either disrupting their digestive systems or serving as a physical barrier. Sharma [5] suggests that this alternative to chemical pesticides is environmentally friendly, reducing the risk of pesticide resistance and minimizing harm to non-target organisms.

Chitosan, regardless of its combination with macronutrients, has biocidal properties against pathogenic bacteria, fungi, and viruses for a prolonged duration. The significance of nanochitosan in plant maintenance has been emphasized by recent studies. In the management of a variety of plant diseases, including *Fusarium* head blight and wilt disease in wheat and chickpea, post-flowering stalk rot in maize, blast leaf in rice, blast disease in finger millet, and maize leaf spot, chitosan has shown promise as a viable, sustainable alternative to traditional fungicides [27, 29]. Soybean farmers could benefit greatly from using a nanoformulation of chitosan and polyacrylic acid to combat common pests like cotton aphids and beetles. The application of chitosan nanoformulations has been proven in multiple studies to enhance plant defense mechanisms by activating defense enzymes [14, 27]. *Fusarium culmorum* and *Gibberella zeae* are examples of chitosan-resistant fungi, whereas *Nigrospora sphaerica*, *Botryosphaeria dothidea*, *Nigrospora oryzae*, and *Alternaria tenuissima* are examples of chitosan-sensitive fungi [2, 22]. Nanochitosan's antibacterial properties were tested against three fungal pathogens: *Pyricularia grisea*, *Fusarium oxysporum*, and *Alternaria solani*. Nanochitosan was more effective than bulk chitosan at preventing mycelia from spreading. Nanochitosan is more durable and effective against the studied fungal infections due to its high permeability, small size, and increased zeta potential [2, 22].

Using the chitosan nanodelivery system, chitosan-agrochemical nanoparticles are made that are very strong and potent because the active ingredient can now reach the target cell or plant section more effectively and within a certain time frame [13]. When making chitosan-agrochemical nanoparticles, it is important to think about their active agent loading content, how well they are encapsulated, how they release, their shape, and their size.

Several publications and papers discuss about how these nanoformulations are made and how they are made. By combining carboxymethyl chitosan with 2-nitro benzyl (average high resolution transmission electron microscopy (HRTEM) diameter: 140 nm), a nanocarrier system was made that carried herbicides (diuron) that stopped photosynthesis from happening. Photo-controlled release was used to make the nanoformulations [13]. Zhang *et al.* [14] found that the smart mixture of chitosan-alginate nanocapsules (30–40 nm diameter of HRTEM) released 50% of the insecticide after 24 h at pH 10 and after 24 h at pH 7 and 4. This is different from how most insecticides only work for about 6 h at all pH levels. Carboxymethyl chitosan and ricinoleic acid were used to make an azadirachtin nanoemulsion that was between 200 nm and 500 nm in size from a hydrodynamic point of view. Due to the enhanced solubility and stability provided by the nanoformulations, the insecticides were able to be released slowly and steadily [20].

5. **Abiotic Stress Tolerance:** Nanochitosan treatment has been found to enhance plant tolerance to abiotic stresses, including drought, salinity, and extreme temperatures. This is achieved through the regulation of plant water balance, enhancement of antioxidant systems, and reduction of oxidative damage, ultimately enabling plants to withstand challenging environmental conditions [12].
6. **Improved Crop Yield and Quality:** Nanochitosan treatments can improve crop yields and quality due to their numerous benefits. Plants treated with nanochitosan frequently have increased biomass, larger fruits, and improved nutritional value, all of which contribute to more marketable produce [12].
7. **Environmental Sustainability:** The usage of nanochitosan is consistent with sustainable agriculture concepts. When compared to conventional agrochemicals, its biodegradability and low toxicity ensure minimum environmental impact. Furthermore, its ability to eliminate the need for synthetic fertilizers and pesticides can help to conserve soil and water [30].

Nanochitosan holds significant promise in revolutionizing contemporary agriculture through its ability to address challenges related to sustainable crop production, disease control, and environmental preservation. Nanochitosan's significance in promoting food security and environmentally sustainable farming is expected to grow with technological advancements [13].

12.3 The Mechanism of Actions of Chitosan Against the Pathogens

Several processes account for chitosan's antimicrobial effect against microorganisms (bacteria, fungus, and viruses): (1) Electrostatic interactions are made possible by the positive charge of protonated chitosan and the negatively charged surface of the pathogen; (2) the damage inflicted by pathogens on cells leads to an increase in membrane permeability, ultimately resulting in cell death; (3) chitosan subsequently forms complexes with crucial elements necessary for pathogen growth, such as metal ions, minerals, and nutrients, thereby impeding their availability for normal

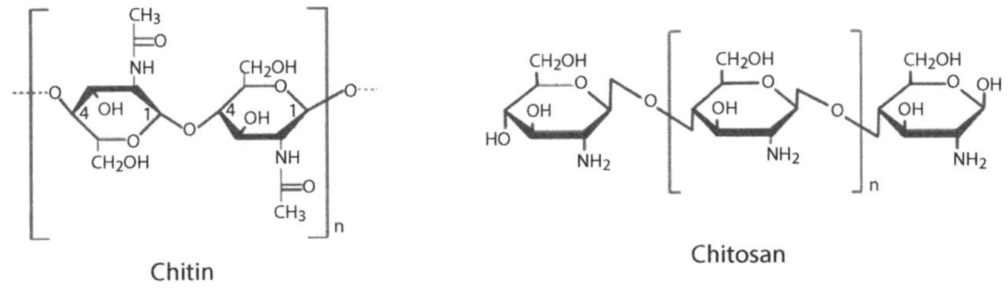

Figure 12.1 Chemical structure of chitin and chitosan.

pathogen growth; (4) the interaction between pathogen DNA/RNA and chitosan that has penetrated the cell inhibits the synthesis of messenger RNA (mRNA) and consequently hinders pathogen reproduction; and (5) when chitosan builds up on the surfaces of microbes, it works as a physical barrier that keeps extracellular interactions from happening.

The effectiveness of the mechanism in action is enhanced by the microscopic size of the chitosan nanoformulations, which increases the surface area in contact with the pathogens. Increased elicitations of plant immunological and defense response activities are possible due to the capacity of microscopic size to facilitate absorption and increase penetration of chitosan on the thick coat of seeds, plant tissues, and pathogen cell membranes [11].

Furthermore, chitosan increases defense-related enzymes such as phenylalanine ammonia-lyase (PAL), polyphenol-oxidase, catalase, and peroxidase, hence improving plant immunity and defense [14]. PAL catalyzes the conversion of L-phenylalanine to trans-cinnamic acid and ammonium, which is an important first step in plant metabolism [27]. Grapes treated with 1% chitosan reduced the occurrence of gray mold before and after harvest, resulting in a twofold increase in PAL [15]. Similar findings were obtained when wheat and rice were treated with chitosan, which increased PAL production and decreased illness rates [29]. During lignin synthesis, the enzyme polyphenol oxidase facilitates the catalysis of phenolic chemicals. Therefore, elevated levels of these enzymes are indicative of enhanced lignin synthesis, which helps in the development of the cell wall structure and acts as a defense mechanism against pathogen invasion [25].

When catalase (antioxidant enzyme) reacts with hydrogen peroxide (H_2O_2), it creates water (H_2O) and oxygen (O_2). Enzymes protect plant cells from reactive oxygen species (ROS), which can cause oxidative damage. Increased catalase activity, faster germination, and increased resistance to temperature stress have all been seen in seeds treated with chitosan [18]. Enzymes called peroxidases can be found in plants that have been subjected to both biotic and abiotic stress. ROS formation is aided by this procedure. In this way, catalase prevents damage to plant cells from reactive oxygen species (ROS), yet ROS generation ultimately results in the death of the pathogen [1, 2]. Similarly, chitosan-treated pears show significant inhibition of *Physalospora piricola* and *Alternaria kikuchian*a spore germination and mycelial growth due to increased peroxidase activity. Peroxidase-gene expression was significantly increased in chitosan-treated peaches compared to untreated peaches [2].

Secondary metabolites involved in plant defense, such as phenolic compounds, phytoalexins, callose and lignin, have been reported to accumulate in plants treated with chitosan. Phytoalexins are metabolite reactions against pathogen infection that produce antibacterial and antioxidant toxins. It was also found that chitosan-infused plants could produce phytoalexins. In 1980, it was observed that chitosan at a dosage of 0.9 g/mL strongly generated phytoalexins and provided resistance to attack for pea plants. Increasing phytoalexins and fostering resistance to *Fusarium* wilt in cotton seeds with the use of a chitosan derivative (carboxymethyl chitosan). The production of lignin, signal molecules (such as salicylic acid and flavonoids), and defense response chemicals (such as tannins and phytoalexins) are all ways in which phenolic compounds in plants help plants resist pathogenic illnesses [21].

Grapes had a higher total phenolic compound content after being treated with chitosan for 60 h. The polyphenol content of treated oregano was 38% and 29% higher, respectively, at 50 and 200 mg/mL of chitosan oligosaccharides. Polysaccharide callose is a normal component of plant cell walls. To protect plant cells from abiotic and biotic environmental stressors, it controls plasmodesmata permeability, phragmoplast development, and phloem pore formation. Plants treated with chitosan have been shown in multiple studies to produce callose in response to pathogen attacks [2].

Additionally, plants develop pathogenesis-related proteins after chitosan treatment, including chitinase and -1,3-glucanase. In response to pathogen attack, plants synthesize proteins and peptides with antibacterial properties. Chitosan, unexpectedly, can also function as an elicitor, producing proteins involved in pathogenesis and enhancing plant resistance to infections. Fungal cell wall chitin and insect exoskeleton -D-glucans are hydrolyzed by the enzymes chitinase and -1,3-glucanase, respectively. Thus, it inhibits fungal growth on the host plant by destroying the fungal cell wall [32].

12.3.1 Transportation, Translocation, and Uptake of Agronanochemicals in Plant

Absorption, translocation, and transport of agrochemicals are crucial to the efficacy of fungicides in combating fungal infections in crops. Inadequate disease control may occur if fungicides' active components are not delivered to the site where pathogenic fungi are growing. Understanding how nanoparticles affect plant metabolism and growth requires knowing how efficiently plants absorb nanoparticles through their leaves and roots. The great permeability of chitosan across plant surfaces was discovered by [34]. This includes the leaves, stems, and roots. Increased bioavailability of active agents in plant tissues has been linked to the use of chitosan-based agronanochemical systems [34].

Systemic and contact agrochemicals are the two main categories. By entering the leaf cuticle, systemic agrochemicals are transported via the phloem to the plant's tissues. Systemic agrochemicals are both medicinal and eradicative because of how well they get rid of infections that have spread throughout a plant's tissue. They also play a role in stopping the spread of disease within the plant. Benzimidazole, hexaconazole, avermectin, azoxystrobin, pyraclostrobin, and many others are only a few examples of the systemic agrochemicals commonly used in farming. In contrast, the treated leaf absorbs and retains contact agrochemicals, which do not move into the leaf's interior. As a preventative approach, contact agrochemicals eliminate pathogenic spores before they may establish and multiply within the plant [33]. Once a plant has been infected, it is no longer susceptible to treatment with contact agrochemicals like copper, sulfur, and fludioxonil [32].

Systemic agrochemicals are frequently selected as the active element in the creation of successful agronanochemical delivery systems. It is believed that agronanochemicals enter plant cells by endocytosis, ion channels, and aquaporin binding to carrier proteins. Exposure to agronanochemicals can occur either through the leaves (foliar uptake) or the roots (root uptake) [35]. Agronanochemicals can enter plant tissue through the stomatal or cuticular channels during foliar uptake. Agronanochemicals, being so tiny, are able to easily penetrate leaf tissue through the stomata, which typically have widths of 3–10 m and lengths of 25 m. Agronanochemicals' small size, between 0.6 nm and 4.8 nm, is one factor in the restricted movement of these substances via the cuticle pores [27].

Agronanochemicals may move from the soil to the root tissues via one of two possible routes: either root diffusion or the cuticular pathway. Root diffusion allows chemicals to flow from regions of higher concentration to regions of lower concentration inside the root. It is dependent on the concentration gradient that exists between the root and the surrounding soil. Agronanochemicals enter the body through the cell wall barrier during the absorption process. Several research using various approaches have found that the plant cell wall's pore diameter ranges from 5 nm to 50 nm. The usage of agronanochemicals, however, has been linked to the widening of pores, which facilitates the absorption of agronanochemicals [13].

Systemic phloem and/or systemic xylem refer to the pathways through which agronanochemicals that have entered the plant are transported to different parts of the plant. Systemic phloem migration in response to foliar application is driven by the symplastic pathway, which involves cytoplasmic movement. This migratory pattern includes both downward travel from the shoot to the root and upward travel from the root to the shoot. After receiving a foliar application, the symplastic and apoplastic channels carry systemic xylem flow upward. Some examples of agrochemicals that are truly systemic arefosetyl-aluminum, acibenzolar-S-methyl, harpin, and others [23]. Movement of chitosan-NPK nanoparticles, made up of nanochitosan-N (50 nm), nanochitosan-P (68 nm), and nanochitosan-K (45 nm), was seen in the xylem and phloem tissue of wheat leaves after 10 days of foliar application. This finding was made using HRTEM imaging of ultramicrotome-cut leaf samples. According to Singh *et al.* [13], the xylem and phloem receive nutrients via absorption through the stomata. The stomatal uptake mechanism of chitosan-Zn nanoparticles on wheat leaves was demonstrated by Deshpande *et al.* [36] using foliar application, whereas a study by Singh *et al.* [13]. found that the nanoparticles were only identifiable in phloem tissue and not in xylem tissue after 30 days. Kadam *et al.* [9] state that field emission scanning electron microscopy (FESEM) and fluorescence microscopy confirmed the localization of Zn within the stomata. The internalization of the nanoparticulate system was examined using confocal laser scanning microscopy, and the results showed a high concentration of zinc in the aleurone layer, embryo, and also endosperm of wheat grains.

12.3.2 Phytoprotection, Genoprotection, and Cytoprotection of Chitosan

Chitosan exhibits promising properties such as antioxidant activity, biocompatibility, bioadhesion, and free radical scavenging, making it a potential candidate for use as a protective agent against harmful agrochemicals. Agrochemicals have been recognized for their phytotoxic, cytotoxic, and genotoxic impacts on plants, cells, and DNA. Agrochemicals can be encapsulated within chitosan spherical nanocapsules, creating a protective barrier that prevents direct contact between the hazardous pesticide and cellular components such as DNA [13]. Our *in vitro* cell viability investigations employing MTT assays for cytotoxicity and greenhouse/nursery trials for phytotoxicity confirmed this hypothesis. Shahrajabian [15] showed that the comet assay may be used to assess genotoxicity.

All leaf mass, leaf area, fresh weight, root weight, and leaf area increased when the therapy was applied to chilli seeds. Control groups for the experiments used untreated bulk chitosan and chitosan nanoparticles. It was determined that chitosan does not have a phytotoxic effect and that its nano-sized particles can stimulate plant development in seedlings. Adding chitosan nanoparticles to wheat and barley plants corroborated the results [15]. Roots infected with Fusarium oxysporum were shown to have considerably less cell death when treated with chitosan-thiamine nanoparticles than when left untreated. That chitosan-thiamine nanoparticles are safe for plant cells to absorb is demonstrated here. Herbicide nanocarriers (imazapic and imazapyr) were made from chitosan-alginate and chitosan-tripolyphosphate. Cell modification damage caused by the herbicides was tested on onion roots and found to be reduced by 100% after being encapsulated. Whereas free herbicides were found to cause more DNA damage in the comet assay, nanoparticles were found to cause significantly less damage when tested on the hamster ovary cell line.

Husen [27] reports that the genotoxicity of chitosan-tripolyphosphate-herbicide nanoparticles was similar to that of the control group (untreated cells); compared to free hexaconazole and regular hexaconazole, chitosan-hexaconazole nanoparticles are safer for the Vero cell line. Also, chitosan-carbendazim nanoparticles were found to be safer than free carbendazim in cytotoxicity tests conducted on a human lens epithelial cell line. A human lens epithelial cell line was shown to be unaffected by chitosan nanoparticles, as found by Hassan *et al.* [8]. Furthermore, it was observed that both conventional and free pyraclostrobin exhibited acute toxicity toward zebrafish following a 24-h exposure period. However, when pyraclostrobin was encapsulated within chitosan-poly(2-dimethylamino-ethyl methacrylate) microcapsules, a notable decrease in the LC50 value was seen. This finding demonstrates the potential of chitosan in safeguarding plants, cells, and genes against the detrimental impacts of harmful pesticides. Hassan *et al.* [8] also discovered that chitosan nanocapsules and nanoemulsions containing triclabendazole exhibit reduced cytotoxicity toward intestinal absorptive cells than other formulations. The cytotoxicity and genotoxicity of chitosan-quinapyramine sulfate nanoparticles were found to be significantly lower compared to those of their conventional counterparts when tested on the HeLa cell line by Hassan *et al.* [8]. In addition, the therapeutic efficacy against trypanosomes was enhanced by these nanoformulations, and the survival time of infected rabbits was increased.

12.4 Agronanochemicals Pose Adverse Effects on Human Health and Environmental Welfare

Agrochemicals are of importance in the realm of crop management, and scholars have placed significant emphasis on mitigating the adverse impacts of these substances on both human well-being and the natural environment. The determination of risks or hazards to human health is contingent upon factors such as the toxicity levels of agrochemicals utilized, as well as the kind and extent of exposure resulting from their application or presence in food and drinking water [4]. In addition, conventional agrochemicals have significant environmental implications and provide a risk to both terrestrial and aquatic ecosystems because of their dispersal and leaching properties. Improper management and prolonged contact with agrochemicals in the agricultural sector have been found to result in adverse effects such as, dyspnea, emesis, nausea, headaches, and skin and ocular irritation [3].

As previously mentioned, the utilization of chitosan nanoparticles for the encapsulation of agrochemicals has been found to exhibit controlled release characteristics, hence contributing to the mitigation of agrochemical waste and leaching. The utilization of chitosan-based agronanochemicals has several advantages, including prolonged circulation duration and enhanced efficacy. Due to these inherent attributes, a diminished quantity of the active constituent can be administered, hence mitigating pollution stemming from phenomena such as runoff and the accumulation of agrochemicals. Furthermore, investigations on the composition and behavior of soil microbial communities in response to the application of agrochemicals provide valuable insights into the dynamics of nutrient cycling within the soil. The microbial communities under examination encompass a range of organisms, including bacteria, yeast, algae, protozoa, actinomycetes, and fungi [13]. Taking this into consideration, we conducted a comparative analysis between the impact of chitosan-based agronanochemicals and free agrochemicals on the soil microbiota. In the study conducted by Pileggi *et al.* [37] it was observed that the presence of free herbicide, specifically, paraquat, resulted in a considerable reduction in both soil enzyme activity and microbial application. The findings of Ali *et al.* [3] and de'Souza *et al.* [4] indicated that the application of chitosan-herbicide nanoparticles did not yield any significant impact in comparison to the control group. A study conducted by Maruyama *et al.* [38] shown that the application of chitosan-alginate-herbicide nanoparticles had a more pronounced impact on soil microbial populations compared to unbound herbicides (imazapic and imazapyr) after a period of 7 days. Based on the aforementioned evidence, it is evident that chitosan nanocarriers has the capacity to mitigate the adverse impacts of noxious agrochemicals on soil health.

12.5 Conclusion

Chitosan exhibits multifunctional properties, serving as a growth accelerator, antibacterial agent, and enhancer of plant immunity and defense mechanisms against many plant diseases in isolation. In an alternative approach, chitosan nanoparticles can serve as carriers for agrochemical active ingredients, resulting in the creation of a highly effective biocide. The encapsulation or loading of these active ingredients into chitosan nanoparticles allows for the formation of a protective barrier that safeguards plants against the harmful impacts of the agrochemicals. Moreover, the resulting chitosan nanoparticles exhibit non-toxic properties and are compatible with biological systems. In addition, the utilization of chitosan-based agronanochemicals is intended to enhance the effective delivery of active agents to the designated location, enabling a prolonged duration of action and thus enhancing the effectiveness of agrochemicals. In addition, the implemented solutions have the potential to enhance adoption rates while simultaneously mitigating the adverse effects of pesticide leaching and runoff, which pose risks to both human well-being and the ecosystem. Consequently, the utilization of chitosan-based agronanochemicals presents a potentially more sustainable option for the treatment of crop diseases when compared to conventional agrochemicals.

References

1. Hidangmayum, A. and Dwivedi, P., Chitosan based nanoformulation for sustainable agriculture with special reference to abiotic stress: a review. *J. Polym. Environ.*, 1–20, 2021, doi: 10.1007/S10924-021-02296-Y.
2. Hidangmayum, A., Dwivedi, P., Katiyar, D., Hemantaranjan, A., Application of chitosan on plant responses with special reference to abiotic stress. *Physiol. Mol. Biol. Plants*, 25, 2, 313e326, 2019.

3. Ali, E.F., El-Shehawi, A.M., Ibrahim, O.H.M., Abdul-Hafeez, E.Y., Moussa, M.M., Hassan, F.A.S., A vital role of chitosan nanoparticles in improvisation the drought stress tolerance in Catharanthus roseus (L.) through biochemical and gene expression modulation. *Plant Physiol. Biochem.*, 161, 166–175, 2021. https://doi.org/10.1016/j.plaphy.2021.02.008
4. de Souza Silva, P.T., de Souza, L.M., de Morais, M.B., de Moraes, M.M., da Camara, C.A.G., Ulisses, C., Effect of biotic elicitors on the physiology, redox system, and secondary metabolite composition of Lippia alba (Mill.) cultivated *in vitro*. *South Afr. J. Bot.*, 147, 415e424, 2022.
5. Sharma, P., Sangwan, S., Kaur, H., Process parameters for biosurfactant production using yeast Meyerozyma guilliermondii YK32. *Environ. Monit. Assess.*, 191, 9, 1e13, 2019.
6. Peralta-Ruiz, Y., Grande-Tovar, C.D., Navia Porras, D.P., Sinning-Mangonez, A., DelgadoOspina, J., Gonza´lez-Locarno, M., Chaves-Lo´pez, C., Packham's triumph pears (Pyrus communis L.) post-harvest treatment during cold storage based on chitosan and rue essential oil. *Molecules*, 26, 3, 725, 2021.
7. Gomes, D.G., Pelegrino, M.T., Ferreira, A.S., Bazzo, J.H., Zucareli, C., Seabra, A.B., Oliveira, H.C., Seed priming with copper-loaded chitosan nanoparticles promotes early growth and enzymatic antioxidant defense of maize (Zea mays L.) seedlings. *J. Chem. Technol. Biotechnol.*, 96, 8, 2176e2184, 2021.
8. Hassan, F.A.S., Ali, E., Gaber, A., Fetouh, M.I., Mazrou, R., Chitosan nanoparticles effectively combat salinity stress by enhancing antioxidant activity and alkaloid biosynthesis in Catharanthus roseus (L.) G. Don. *Plant Physiol. Biochem.*, 162, 291e300, 2021.
9. Kadam, P.M., Prajapati, D., Kumaraswamy, R.V., Kumari, S., Devi, K.A., Pal, A., Saharan, V., Physio-biochemical responses of wheat plant towards salicylic acid-chitosan nanoparticles. *Plant Physiol. Biochem.*, 162, 699e705, 2021.
10. Husen, A. and Jawaid, M., *Nanomaterials for Agriculture and Forestry Applications*, Elsevier Inc., 50 Hampshire St., 5th Floor, Cambridge, MA 02139, USA, 2020, https://doi.org/10.1016/C2018-0-02349-X.
11. Hussien, E.T., Dahab, M.I., Mohammed, K., Abd-Elatty, I.H.E.S., Farouk, H.M., Tarek, A., Abd-Elwahab, R.A., Gene Transformation by Chitosan Nanoparticle to Enhance Fatty Acid Production in Zea mays, *Rom. Biotechnol. Lett.*, 26, 5, 2971–2978, 2021. DOI: 10.25083/rbl/26.5/2971-2978.
12. Wang, W., Xue, C., Mao, X., Chitosan: structural modification, biological activity and application. *Int. J. Biol. Macromol.*, 164, 4532e4546, 2020.
13. Singh, S., Sangwan, S., Sharma, P., Devi, P., Moond, M., Nanotechnology for sustainable agriculture: an emerging perspective. *J. Nanosci. Nanotechnol.*, 21, 6, 3453e3465, 2021.
14. Zhang, J., Cheng, X., Chang, L., Zhang, L., Zhang, S., Combined treatments of chitosan and sodium silicate to inhibit Alternaria alternata pathogens of postharvest winter jujube. *Food Sci. Biotechnol.*, 30, 4, 589e597, 2021.

15. Shahrajabian, M.H., Chaski, C., Polyzos, N., Tzortzakis, N., Petropoulos, S.A., Sustainable agriculture systems in vegetable production using chitin and chitosan as plant biostimulants. *Biomolecules*, 11, 6, 819, 2021.
16. Husen, A., Carbon-based nanomaterials and their interactions with agricultural crops, in: *Nanomaterials for Agriculture and Forestry Applications*, A. Husen, and M. Jawaid, (Eds.), p. 199e218, Elsevier Inc, 50 Hampshire St., 5th Floor, Cambridge, MA 02139, USA, 2020b, https://doi.org/10.1016/B978-0-12-817852-2.00008-1.
17. Kumar, A., Choudhary, A., Kaur, H., Guha, S., Mehta, S., Husen, A., Potential applications of engineered nanoparticles in plant disease management: a critical update. *Chemosphere*, 295, 133798, 2022. https://doi.org/10.1016/j.chemosphere.2022.133798.
18. Kapoor, D., Sharma, P., Sharma, M.M.M., Kumari, A., Kumar, R., Microbes in pharmaceutical industry, in: *Microbial Diversity, Interventions and Scope*, p. 259e299, Springer, Singapore, 2020.
19. Maluin, F.N., Hussein, M.Z., Yusof, N.A., Fakurazi, S., Seman, I.A., Hilmi, N.H.Z., Daim, L.D.J., Enhanced fungicidal efficacy on Ganoderma boninense by simultaneous co-delivery of hexaconazole and dazomet from their chitosan nanoparticles. *RSC Adv.*, 9, 27083–27095, 2019.
20. Mosavikia, A.A., Mosavi, S.G., Seghatoleslami, M., Baradaran, R., Chitosan nanoparticle and pyridoxine seed priming improves tolerance to salinity in milk thistle seedling. *Not. Bot. Horti Agrobot. Cluj-Napoca*, 48, 1, 221e233, 2020.
21. Jogaiah, S., Satapute, P., De Britto, S., Konappa, N., Udayashankar, A.C., Exogenous priming of chitosan induces upregulation of phytohormones and resistance against cucumber powdery mildew disease is correlated with localized biosynthesis of defense enzymes. *Int. J. Biol. Macromol.*, 162, 1825e1838, 2020.
22. Saleem, M.S., Anjum, M.A., Naz, S., Ali, S., Hussain, S., Azam, M., Ejaz, S., Incorporation of ascorbic acid in chitosan-based edible coating improves postharvest quality and storability of strawberry fruits. *Int. J. Biol. Macromol.*, 189, 160e169, 2021.
23. Wang, A., Li, J., Al-Huqail, A.A., Al-Harbi, M.S., Ali, E.F., Wang, J., Eissa, M.A., Mechanisms of chitosan nanoparticles in the regulation of cold stress resistance in banana plants. *Nanomaterials*, 11, 10, 2670, 2021.
24. Faizan, M., Rajput, V.D., Al-Khuraif, A.A., Arshad, M., Minkina, T., Sushkova, S., Yu, F., Effect of foliar fertigation of chitosan nanoparticles on cadmium accumulation and toxicity in Solanum lycopersicum. *Biology*, 10, 7, 666, 2021.
25. Yu, J., Wang, D., Geetha, N., Khawar, K.M., Jogaiah, S., Mujtaba, M., Current trends and challenges in the synthesis and applications of chitosan-based nanocomposites for plants: a review. *Carbohydr. Polym.*, 261, 117904, 2021.
26. Sharma, M.M.M., Sharma, P., Kapoor, D., Beniwal, P., Mehta, S., Phytomicrobiome community: an Agrarian perspective towards resilient

agriculture, in: *Plant Performance under Environmental Stress*, p. 493e534, Springer, Cham, 2021a.

27. Husen, A., *Harsh Environment and Plant Resilience (Molecular and Functional Aspects)*, Springer Nature Switzerland AG, Gewerbestrasse 11, 6330 Cham, Switzerland, 2021a, https://doi.org/10.1007/978-3-030-65912-7.
28. Xing, K., Chen, X.G., Liu, C.S., Cha, D.S., Park, H.J., Oleoyl-chitosan nanoparticles inhibits Escherichia coli and Staphylococcus aureus by damaging the cell membrane and putative binding to extracellular or intracellular targets. *Int. J. Food Microbiol.*, 132, 127–133, 2017.
29. Sharma, P., Pandey, V., Sharma, M.M.M., Patra, A., Singh, B., Mehta, S., Husen, A., A review on biosensors and nanosensors application in agroecosystems. *Nanoscale Res. Lett.*, 16, 1, 1e24, 2021b.
30. Husen, A., *Plant Performance Under Environmental Stress (Hormones, Biostimulants and Sustainable Plant Growth Management)*, Springer Nature Switzerland AG, Gewerbestrasse 11, 6330 Cham, Switzerland, 2021b, https://doi.org/10.1007/978-3-030-78521-5.
31. Sharma, P., Sharma, M.M.M., Kapoor, D., Rani, K., Singh, D., Barkodia, M., Role of microbes for attaining enhanced food crop production, in: *Microbial Biotechnology: Basic Research and Applications*, p. 55e78, 2020b.
32. Adisa, I.O., Pullagurala, V.L.R., Peralta-Videa, J.R., Dimkpa, C.O., Elmer, W.H., Gardea-Torresdey, J.L., White, J.C., Recent advances in nano-enabled fertilizers and pesticides: A critical review of mechanisms of action. *Environ. Sci. Nanomaterial*, 6, 2002–2030, 2019.
33. Oliveira, H.C., Gomes, B.C., Pelegrino, M.T., Seabra, A.B., Nitric oxide-releasing chitosan nanoparticles alleviate the effects of salt stress in maize plants. *Nitric Oxide*, 61, 10–19, 2018.
34. Nguyen, T.K., Nguyen, N.D., Dang, V.P., Phan, D.T., Tran, T.H., Nguyen, Q.H., Synthesis of Platinium Nanoparticles by Gamma Co-60 Ray Irradiation Method Using Chitosan As Stabilizer. *Adv. Mater. Sci. Eng.*, 9624374, 1–5, 2019.
35. Niu, J., Liu, C., Huang, M. Liu, K., Yan, D., Effects of Foliar Fertilization: A Review of Current Status and Future Perspectives. *J. Soil Sci. Plant Nutr.*, 21, 104–118, 2021. https://doi.org/10.1007/s42729-020-00346-3
36. Deshpande, P., Dapkekar, A., Oak, M.J., Paknikar, K.M., Rajwade, J.M., Zinc complexed chitosan/TPP nanoparticles: A promising micronutrient nanocarrier suited for foliar application. *Carbohydr. Polym.*, 165, 1, 394–401, 2017. https://doi.org/10.1016/j.carbpol.2017.02.061
37. Pileggi, M., Pileggi, S. A. V., & Sadowsky, M. J., Herbicide bioremediation: from strains to bacterial communities. *Heliyon*, 6, 12, e05767, 2020. https://doi.org/10.1016/j.heliyon.2020.e05767
38. Maruyama, C., Guilger, M., Pascoli, M., Bileshy-José, N.,Abhilash, P.C., Fraceto, L.F., de Lima, R., Nanoparticles Based on Chitosan as Carriers for the Combined Herbicides Imazapic and Imazapyr. *Sci. Rep.*, 6, 19768, 2016. https://doi.org/10.1038/srep19768

13

Chitosan-Based Nanosystems: Antimicrobial Activity in Agrifood Sector

Sherifat Ozavize Enejiyon[1]*, Nana Hauwau Ahmad[2], Rukayat Abidemi Fasasi[1], Nasiru Usman Adabara[1], Oluwafemi Adebayo Oyewole[1], Charles Oluwaseun Adetunji[3], Eniola K. I. T.[4] and Mohammed Bello Yerima[5]

[1]*Department of Microbiology, School of Life Sciences, Federal University of Technology, Minna, Nigeria*
[2]*Newgate College of Health Technology, Minna, Niger State, Nigeria*
[3]*Department of Microbiology, Edo State University, Uzairue, Nigeria*
[4]*Department of Biological Sciences, Joseph Ayo Babalola University, Ikeji Arakeji, Osun State, Nigeria*
[5]*Department of Microbiology, Sokoto State University, Sokoto, Sokoto State, Nigeria*

Abstract

The agrifood sector is of paramount importance, providing sustenance to the global population. However, it faces constant challenges due to microbial contamination, leading to food spoilage, economic losses, and health hazards. Traditional methods of preserving food involve the use of chemical preservatives, which are increasingly facing scrutiny due to their potential health risks and environmental impacts. In response to these challenges, the search for safe and effective antimicrobial agents has led to the exploration of nanotechnology, with chitosan-based nanosystems emerging as a promising solution. This chapter aims to provide a comprehensive overview of chitosan-based nanosystems as antimicrobial agents in the agrifood sector. It delves into the properties of chitosan, its sources, and production methods. Furthermore, it discusses the significant challenges posed by microbial contamination in the agrifood industry and the need for innovative antimicrobial solutions. The essay explores the use of nanotechnology in addressing these challenges, highlighting the benefits of nanoscale materials. Specifically, it focuses on chitosan-based nanosystems, elucidating their various types, preparation methods, and characterization techniques.

**Corresponding author*: enejiyon.sherifat@futminna.edu.ng

Charles Oluwaseun Adetunji, Maulin P. Shah, Yerima Mohammed Bello, Daniel Hefft, Jay Singh, Shyam S. Pandey and Ravindra Pratap Singh (eds.) Nanochitosan Applications for Enhanced Crop Production and Food Security, (301–330) © 2025 Scrivener Publishing LLC

***Keywords*:** Agrifood, antimicrobial, chitosan, nanosystem, nanomaterials

13.1 Introduction

Antimicrobials have been utilized for a number of purposes in animal husbandry and aquaculture over the years, including promoting growth, enhancing feed efficiency, prevention, and the treatment of infectious diseases. The general health of farm animals and the cleanliness of agricultural surroundings are both improved by the use of antimicrobials from the perspective of animal welfare [1]. The agricultural food industry benefits from antimicrobials usage in crop protection and in the production of food animals. Antimicrobials, such as antibiotics and related semi- or synthetic medicines, have high antibacterial potency and selective toxicity, enabling their application as anti-infective treatments [2]. The bulk of antibiotics produced in the United States are used in animal husbandry [3]. Van Boeckel *et al.* [4] also predicted that, over the course of the globe, 1 kg of meat from hens, pigs, and cattle would result in the consumption of 148 mg, 172 mg, and 45 mg of antimicrobials, respectively.

The number of therapeutic alternatives available for veterinarians and clinicians has decreased since the introduction of antimicrobial resistance (AMR). Public health authorities have identified antibiotic resistance as a problem. According to the first economic analysis on the effects of AMR, the mortality rate resulting from AMR will increase from about 700,000 to 10,000,000 yearly by 2050 if nothing was done. In the healthcare sector, it would cost trillions of US dollars [5]. AMR has been chosen in contexts where food is produced as a result of the inappropriate application of antibiotics in conditions other than the diagnosis and treatment of illnesses. Through gene transfer (horizontal), bacteria can directly take on the resistance mechanisms of other bacteria or they can create new resistance when they are exposed to sub-inhibitory levels of antibiotics in the environment [6].

Despite the fact that the prevalence of AMR has mostly been attributed to the overuse and improper usage of antibiotics, recent evidence of co-selection for AMR among bacteria exposed to non-antibiotic chemicals used in the agrifood sector and agricultural production such as naturally occurring dense metals, additives, biocides, and disinfectants has raised concerns [7]. AMR development in bacteria that might enter the food chain is influenced both directly and indirectly by the use of antimicrobials, heavy metals, and biocides in food production and agriculture. The public outrage concerning the usage of antimicrobials in agricultural operations and the spread of multidrug resistance organisms has pressurized response from the agrifood industry. The animal food chain, which extends from farms through slaughterhouses and food processing facilities to packaging and retail facilities, is a significant area under investigation [8].

Numerous research studies have been conducted on the benefits of chitosan as antimicrobials with an extensive range of prospective applications against various types of organisms (yeasts, molds, bacteria, and algae). These studies looked explored chitosan in a variety of forms, including as composites, as films, and in solutions [9]. These investigations have encompassed both *in vivo* and *in vitro* trials. Kong *et al.* [9] postulated that chitosan may have antibacterial characteristics. Since then, the scientific community has paid considerable attention to them as a result of the widespread recognition of their commercial potential.

Investigations examining the organic occurrences connected to soil-borne and food-borne pathogenic fungus within the agriculture-food sectors started the investigation into chitosan's antimicrobial properties started over two decades ago [10]. Due to the intimate connection between these bacteria and human activity, research into chitosan's efficiency against microbes has become increasingly important.

The antibacterial efficacy of chitosan, its composites, and derivatives was commonly evaluated at this time using chemical, biochemical, microbiological, and medical assays, cellular and molecular level approaches were rarely used. Results in this phase showed that a variety of intrinsic and extrinsic parameters, including pH and molecular weight, affected how efficient chitosan and its derivatives were against microbes. In addition, several basic theories on the underlying principles of antimicrobial activity were put forth [10].

Numerous antibacterial agents centered on chitosan or its products have been developed as a result of these discoveries. The emergence of bacteria and fungi resistant to biocide, as well as increased public health concerns about harmful germs, all contributed to a greater desire for efficient and safe treatments that were unlikely to promote resistance development [11]. Furthermore, the advancement of nanotechnology coincided with significant advances in molecular biology, medications, cell biology technologies, and detection techniques. Nanotechnology, which played a particularly crucial role, provided the ability to expand antibacterial medicines to the submicroscopic level [12].

In order to better understand the antimicrobial capabilities of chitosan and its products, a variety of approaches have been explored. As a result, many systems of chitosan have developed, for used in agriculture and the food industry.

13.2 Chitosan

Chitin can be completely or partially deacetylated to produce the cationic biopolymer known as chitosan. It is the one of the most prevalent polymer on in the world, and it also is a linear polysaccharide made up of (1-4)-linked 2-amino-2-deoxy-D-glucopyranose that is produced during deacetylation of N-acetyl-D-glucosamine [13, 14]. Due to their polymerization, degree of deacetylation, molecular mass, acid dissociation constant or pKa, and viscosity, chitosan refers to a collection of co-polymers rather than a single chemical specifically [15]. Chitosan is recognized to be biodegradable, biocompatible, non-toxic, and changeable, among other qualities. These characteristics of chitosan allow its usage in extensive range of biological processes and applications, including the textile industry, waste water treatment, tissue engineering, medical, cosmetics pharmaceutics, agriculture, and food [16, 17].

The amino acids on the chitosan backbone demonstrate their distinctiveness by setting up its polycation in an acidic pH. This distinction is attributed to chitosan's muco-adhesiveness, permeability-enhancing impact, and antibacterial and antifungal activity. Antimicrobial characteristic of chitosan is the one that interest scientists the most among all the uses for it. Chitosan is employed as a preservative during the creation of beads, films, fibers, membranes, and hydrogels that have a variety of purposes because of these particular properties [18].

Chitosan has become increasingly popular because of its antibacterial, antioxidant, and chelated properties as well as the fact that it is biocompatible and has little to no harmful effects. Chitosan also possesses traits like adhesiveness, anti-inflammation, and anti-hypercholesterol because of its cationic polymer, which makes it easier to bind [19].

Chitosan is able to penetrate into the cell through its membrane or wall which makes it the perfect carrier for nanosystem [20]. The presence of amino and hydroxyl groups on chitosan allows for complexing, which results in more stable complexes with superior pharmacokinetic features [20]. Modification can also be done on chitosan in order to fit the research needs [19].

13.2.1 Source of Chitosan

Chitosan can be gotten from chitin (parent compound of chitosan) deacetylation. Chitin can be found in various sources ranging from exoskeleton of marine such as crustaceans (barnacles, crayfish, shrimps, krill, and lobster), mollusks (cuttlefish, clams, squids, octopus, and oysters snails), algae (diatoms and green and brown algae), insects (cockroaches, housefly, silkworms, ants, scorpions, beetles, brachiopods, and spiders) to the cell wall of fungi such as *Basidiomycetes, Ascomycetes, Phycomycetes (Mucor rouxii, Aspergillus niger, Trichoderma reesi, and Penicillium notatum*), and yeast species such as *Rhodosporidium paludigenum* and *Saccharomyces cerevisiae* [21, 22]. Chitosan was also synthesized from a pest named *Schistocerca piceifrons* while attacking crops. Demineralization, deproteinization, and decolorization (removal of mineral, protein, and color) occur to obtain chitin that is changed to chitosan by enzymatic deacetylation methods.

According to Abidin *et al.* [23], the quantity of chitin typically ranges from 5% to 25% in insect cuticles to 15% to 30% in cuticles of crabs, 20% to 30% in exoskeletons of crustaceans [24], 30% to 40% in shrimp cuticles, and 2% to 44% in fungal cell walls. The majority of the chitosan now employed in industrial applications comes from easily accessible crustaceans' exoskeletons obtained as waste from the food preparation sector. Conversely, it is also rapidly being made available as a byproduct of the extraction of insect proteins for the food and animal feed sectors, fungal fermentation, and the breeding of cocoons from the silk business.

Fish scales also comprised chitin, but, because of how little of them there is—just 1% of the weight of the fish—they are frequently ignored [25]. Because the underlying process could be modified to produce a clean and constituent product with desired specific features, chitosan generated from microbial sources is viewed as promising [26].

13.2.2 Antimicrobial Activity of Chitosan

One of the most studied function of chitosan is its antimicrobial activity. The antimicrobial activity of chitosan has uses ranging from biomedical to cosmetic to food and agricultural. Many research studies have been undertaken to date in order to explore the unique property of chitosan as well as its antimicrobial property, in order to develop self-preserving materials. This has resulted in the development of a wide range of chitosan-containing products such as fibers, beads, membranes, films, and hydrogels for a variety of applications [27].

Factors such as pH, polymerization degree, molecular weight, cross-linking, and concentration have great influence on the antibacterial characteristics of chitosan [28]. Chitosan solution is highly stable in neutral pH, and this stability is critical for its antimicrobial action against various foodborne pathogens [29].

Chitosan is efficient against both Gram-positive bacteria (GPB) and Gram-negative bacteria (GNB). The mechanisms of antibacterial action of chitosan are not clearly understood; however, the most frequently proposed mechanism encompasses attachment to the negatively charged bacterial cell wall, which distort the cell and disrupts the penetrability of the membrane, followed by binding to the cell DNA; this inhibits replication and eventually causes cell death. Chitosan functions as a chelating agent that attaches to trace metal components on an optional basis, producing toxins and preventing microbiological development [30]. The polycationic structure only arises in acidic settings, and the grafted groups of certain derivatives may change the chitosan's acidic strength and cause protonation at higher pH levels [31]. As the positive charge density of chitosan grows the chitosan's antibacterial capacity will increase proportionately [32]. Electrostatic interaction also depends on how many amino groups are connected to the C-2 on the chitosan backbones with protonation. Large amounts of amino groups can increase the antibacterial activity.

Chitosan's antifungal activity is thought to be fungistatic rather than fungicidal, much like bacteria, and both the host and the fungus can undergo regulatory changes as a result of chitosan. Generally speaking, it has been demonstrated that chitosan is highly effective in elongation of the germ tube, radial expansion, and inhibition of spore germination. Fungi such as vesicular arbuscular mycorrhizal fungi, entomopathogenic fungi, and necrotrophic and biotrophic mycoparasites can activate a variety of biological processes and production of chitinases in the presence of chitosan [33, 34].

The antiviral action of marine polysaccharides and oligosaccharides produced from chitosan, carrageenans, alginates, and fucans has been demonstrated [35]. Studies conducted on the antiviral action of chitosan in plants discovered that chitosan can prevent viral infections in a variety of plant. Chitosan has also been proven to impact the hypersensitive reaction to viral infections in plants as well as stopping the virus from spreading throughout the entire plant [36, 37]. According to several studies, chitosan does not directly fight viruses; rather, it boosts the plant's immune system and strengthens its overall defenses [38, 39]. Antiviral properties of chitosan derived from *Musca domestica* housefly larvae were investigated, and it was utilized to treat *Bombyx mori* that had been exposed to the occlusion-derived virus (ODV) associated with causing nucleopolyhedrovirus in *B. mori*. Chitosan exhibits antiviral activity by decreasing the rate of larval death as a result of ODV infection [40]. Additionally, they noted that chitosan also prevents nucleopolyhedro virus infections in Astographa California and *Botrytis* [40].

In a variety of biological systems, chitosan possesses antiviral properties, and it is effective in suppressing viral growth. Chitosan's antiviral capabilities are significantly influenced by degree of deacetylation, molecular weight, concentration, deamination, and chemical modification [36]. Chitosans have also been shown to suppress tobacco mosaic virus and potato virus X in treated potato and tomato plants [41].

13.2.3 Chitosan-Based Nanosytem

In the last decade, chitosan-nanoparticles (NPs) have attracted the interest of researchers and industries, and many applications for these materials have been developed. One of the major interest is the use of chitosan NPs as antimicrobial agent due to the fact that chitosan-based nanosystem presented an increased antimicrobial activity when compared to chitosan alone [42]. Nanoparticles (NPs) have significantly different physical characteristics due to their smaller size. Chitosan in combination with NPs has enhanced characteristics effect. Quantum size effects and enhanced surface area are all characteristics of chitosan-based nanosystems [30, 43, 44]. Several techniques, including spray drying [45], ionic gelation [46], emulsion droplet coalescence [47], microemulsion [48], and reverse micellar approach [49], have been used to create chitosan-based nanosystems.

The most successful technique for the production of chitosan-based nanosystem was found to be physical cross-linking by ionic gelation. The ionic gelation technique makes use of the electrostatic contact between the chitosan's positively charged group and tripolyphosphate (TPP) negatively charged group. The ratio of chitosan to TPP may be changed, in order to change the size and surface charge of NPs. Chitosan-based nanosystem has involved complexing chitosan with different types of metal-ions to increase their antimicrobial activity. These includes chitosan-silver NPs (CS-Ag^{+}NPs), chitosan-copper NPs (CS-Cu^{2+}NPs), chitosan-zinc NPs (CS-Zn^{2+}NPs), chitosan-manganese NPs (CS-Mn^{2+}NPs), and chitosan-iron NPs (CS-Fe^{2+}NPs) [50–53]. Furthermore, the antimicrobial efficacy of biomolecule compounds, such as protamine [54], lysozyme, oleic acid grafted chitosan oligosaccharide NPs [55], essential oil [56], and curcumin NPs, in hybrid with CS NPs has also been tested [57]. Polycationic chitosan NPs with increased surface charge density can interact with bacteria more effectively than chitosan alone. Chitosan NPs cause damage to the membrane and cell wall of bacterial cells causing intracellular chemical efflux and eventual death of the cell. The size of NP is influenced by a number of factors that include the kind of bacteria, development stage, zeta-potential, concentration, pH, molecular weight, and degree of acetylation, which, in turn, affects how effective they are against bacteria.

13.3 Nanotechnology in Agrifood Industry

One of the most crucial instruments of modern agriculture is nanotechnology, and it is predicted that agrifood nanotechnology will soon become a major economic force. Agriculture-produced foods, including crops for human consumption and animal feed, are the subject of agrifood themes that emphasize sustainability and protection [58]. Applications of nanotechnology in agriculture include production of nanoformulations of agrochemicals for crop improvement; the use of nanosensors and nanobiosensors in crop protection for the identification of diseases and residues of agrochemicals; usage of nanodevices for the genetic modification of plants; diagnosis of plant disease; and improvement of animal health, animal breeding, and poultry production.

According to several studies, nanotechnology has the ability to protect plants, monitor the plant growth, identify plant and animal diseases, boost global food production, improve food quality, and minimize waste for sustainable intensification [59–62]. However, the application of nanotechnology to the agriculture and food industries is still in its infancy, a lot still need to be done [63–66].

13.4 Various Forms of Chitosan-Based Nanosystems Used in Agrifood

Nanoparticles, microbeads, and microfibers made from chitosan have all been actively used for various agricultural applications [67, 68]. Chitosan is frequently utilized because of the functional groups in its chemical structure that make it easy to interact with other active chemicals [69]. In agriculture, chitosan is useful as an elicitor, fertilizer transporter, stress reliever, and plant growth regulator, among other things. Chitosan is potentially much more maneuverable compared to other biopolymers such as chitin, starch, gelatin, cellulose, and glucans without forfeiting its fundamental properties, such as antibacterial action, elicitor, affinity for cell membranes containing anionic components, and affinity for metals [70].

Chitosan can be combined with other biomaterials or active compounds to produce novel chitosan-based nanocomposites. These novel chitosan-based nanocompositescan improve the bioavailability and biological properties of chitosan as well as address problems like hydrophilicity, weak mechanical properties, low gas permeability, and low encapsulation efficiency, which are usually associated with chitosan. Biomaterials such as cellulose nanofibers have regularly been added to chitosan to boost its mechanical strength. Similarly, nanometals like iron (Fe), copper (Cu), silver (Ag), silicon (Si), zinc (Zn), zinc oxide (ZnO), and titanium dioxide (TiO2) when added to chitosan can increase its antibacterial properties [71–76].

Several production processes such as reverse micellar method [77], emulsion droplet coalescence [77], ionic gelation [46], spray drying, sieving, and precipitation [78] can be used to produce chitosan-based nanocomposites (ChNCs) [79].

13.5 Antimicrobial Activity of Chitosan Nanoparticles

13.5.1 Antifungal Activity of Chitosan-Based Nanocomposites

According to Rossman [80], fungi are among the most harmful pathogens for agricultural products. By creating mycotoxins, it seriously harms fruits, cereals, and other food products before and after harvest. Fungal infections cause a sizable amount of the world's agricultural products to waste in the current food scarcity times [81]. According to Nesic *et al.* [82], some three hundred metabolites of fungi have been identified as harmful to animals and humans. Infecting a variety of crops worldwide are fungi like *Rhizoctonia solani* and *Alternaria alternata*. Many conventional fungicides have been used extensively worldwide to address these issues, but most of them are exceedingly hazardous. Numerous crops have been explored using biopolymers such as chitin, chitosan, cellulose, and alginate as an alternative. Chitosan has stood out among them all thanks to its exceptional antifungal qualities. The molecular weight, amount of deacetylation, and pH of the chitosan solution that has been prepared, as well as the origin of the chitosan, all affects the antifungal activity of the substance. Inhibiting fungal growth at many growth phases, including sporulation, mycelial growth, germination, spore viability, and the synthesis of virulence factors, have been effectively evaluated for chitosan.

For improved antifungal properties, chitosan has also been used as a single ingredient or combined with additional components, particularly metal particles. In order to assess the efficacy of their formulation as an antifungal agent, Kaur *et al.* [83] combined chitosan with silver NPs and isolated strains of *Aspergillus flavus*, *Rhizoctonia solani*, and *Alternaria alternata*. Chitosan NPs, silver NPs, and composite silver-chitosan NPs were examined at various development stages. At all developmental stages, silver-chitosan NPs showed good antifungal activity against all three fungi pathogens.

13.5.2 Antibacterial Activity of Chitosan-Based Nanocomposites

Chitosan has been well-known over the past few decades as a potent antibacterial agent as a result of its polycationic structure. According to Wang *et al.* [75], the antibacterial properties of chitosan-based nanocomposites (ChNCs) are efficient against both GPB and GNB.

Direct anti-bacterial activity is one theory that could help to explain how chitosan might work against microbes. Chitosan directly interacts with the target pathogen by metal chelation or electrostatic interaction because the bacterial cells have certain strongly negatively charged molecules on their surface [84]. The plasma membrane is destroyed as a result of this direct connection, allowing the internal components of the cell to escape. Additionally, chitosan interferes with the target organisms' genetic processes, producing a malfunction of some crucial genes and finally leading to the pathogen's death [85]. Chitosan buildup on the pathogen's surface results in the blocking of nutrients and the movement of metabolic wastes, which is another sort of direct interaction.

In indirect bacterial eradication at the cellular level, chitosan improves plants' natural defenses. Here, chitosan functions as an elicitor, evoking the plants defense systems by increasing the production secondary metabolites, of pathogen-related proteins and strengthening the plants cell wall against pathogen infection [86]. In various crops for the elimination and inhibition of bacterial infections, ChNCs have been used. A harmful pathogen for tomatoes known to cause bacterial wilt is *Ralstonia solanacearum*. It adversely affects the productivity of the tomato crop and significantly reduces farmers' incomes, both in open fields and greenhouses. According to Dennis *et al.* [87], the application of a silica-nanocomposite significantly suppresses the growth of *Ralstonia solanacearum* while fostering the growth of tomato seedlings. Nanoformulation made of chitosan attaches to membranes and activates numerous pathways involved in the synthesis of defense-related proteins.

13.5.3 Antiviral Activity of Chitosan-Based Nanocomposites

There are not many reports on the use of NPs based on chitosan in plant sciences. The ChNCs' antiviral action is still unknown. Bean leaves have been treated with chitosan to stave off the tobacco necrosis virus. The outcomes showed that the chitosan application had a protective effect, which can be attributed to the induction of the plant's defensive mechanism. Chitosan's antiviral properties were first studied in bean leaves (*Phaseolus vulgaris*). By inducing systemic acquired resistance (SAR), the administration of chitosan considerably lowers the amount of necrotic lesions brought on by Alfalfa mosaic virus [88]. According to studies, Fabaceae had a greater level of Ch-NCs' antiviral activity than other botanical groups. Chitosan treatment was particularly responsive to beans and peas. The responses of potato, tomato, and tobacco to ChNCs were less even at higher doses. The recurrent infections caused by the radish mosaic virus (RaMV), turnip mosaic virus (TuMPV), and cauliflower mosaic virus (CaMV) in cabbage, are said to be resistant to chitosan [36].

13.6 Mechanisms of Antibacterial Activity of Chitosan-Based Nanocomposites

13.6.1 Disrupting the Cell Membrane/Cell Wall

The original and most widely accepted theory states that electrostatic interactions between chitosan and the anionic surfaces of GPB and GNB induce the breakdown of the cell membrane [88]. When negatively charged teichoic acid in peptidoglycans interacts electrostatically with positively charged chitosan in GPB, the cell membrane can be damaged, internal components can leak, and chitosan can enter the microbial cells [88, 89]. Protein and other intracellular components are leaked by chitosan, according to a prior study Kong *et al.*, [90]. The measurement of the bacterial mixture's electrical conductivity [91] and the release of cytoplasmic beta-galactosidase activity from *E. coli* into the culture medium [92] are both examples of how the hydrolysis of peptidoglycans can improve electrostatic interaction. High negative charges from lipopolysaccharide (LPS) can be balanced by positive charges from chitosan in GNB, disrupting the outer membrane and allowing chitosan to enter the cell membrane and cause bacterial cell death [93].

In addition to the finding that the outer membrane's permeability is enhanced, studies on the effect of chitosan on *E. coli* revealed that the inner cell membrane is also disrupted, causing leaking of the cytosolic content and microbial cell death [94, 95]. This has been reported by chitosan and chitosan derivative microspheres [90, 92].

13.6.2 Formation of a Dense Polymer Film on the Cell Surface

High–molecular weight chitosan may coat porins on the outer membrane of GNB to form a thick polymer film, which prevents the exchange of nutrients and causes microbial cell death [95]. The identification of such a profile was made possible by the thicker appearance of the cell walls, which showed chitosan deposition on the cell surface. When *Salmonella typhimurium* and *E. coli* are exposed to chitosan, the flocculation effect may be detected on their outer membranes using a scanning electron microscope [96].

13.6.3 Interaction with Microbial Genetic Material

Low–molecular weight chitosan and its hydrolysis byproducts can affect DNA/RNA and protein synthesis by entering the cell wall [97]. Xing *et al.* [98] showed that oleoyl-chitosan NPs bind to DNA and RNA while assessing the effect of chitosan NPs on the electrophoretic mobility of nucleic acids. The results suggested that there may be a positive relationship between the interactions between bacterial genomes and the concentration of oleoyl-chitosan NPs. Additionally, *E. coli* DNA and RNA movement was completely inhibited when oleoyl-chitosan NP concentrations approached 1,000 mg/L. Pathogens are inhibited when positively charged amino groups in oleoyl-chitosan NPs interact with negatively charged phosphate groups in DNA and RNA. In a test of -galactosidase expression, Galván Márquez *et al.* [99] discovered that chitosan inhibits protein production [99]. Chitosan oligomers' ability to inhibit DNA transcription is what gives them their antibacterial impact on E. coli cells, which is supported by the finding of chitosan oligomers inside the cell [100].

13.6.4 Chelation of Nutrients by Chitosan

Metal ions present on the surface of bacteria can be chelated by the amino groups of chitosan when the chelation effect outweighs the electrostatic force, which takes place when the mixture's pH is higher than the chitosan's pKa [10, 101, 102]. Divalent cations have the ability to stabilize bacterial cell membranes [103]. When divalent metal ions bind to the teichoic acids in the cell wall of GPB, it can increase the stability of the polymer structure and improve the cell wall's integrity [104, 105]. Divalent cations' binding to the teichoic acids in the cell wall can lessen the differences in osmotic pressure between the two sides of the microbial cell. LPS is a polyanionic chemical produced by several negatively charged phosphate groups in GNB. It is located in the outer membrane's leaflet. The divalent metal cations can maintain the integrity of the bacterial membrane by minimizing the repellent interactions between aggregation of negatively charged phosphate groups [103].

According to Kong *et al.* [9], when the pH of the medium is higher than the pKa value of chitosan or chitosan derivatives, unprotonated amino groups of chitosan can transfer their lone pair of electrons to the metal ions of phosphate groups in the LPS or cell wall teichoic acids (WTAs) on the cell membrane surface to form a metal complex. Positively charged amino groups in chitosan can compete with divalent cations for phosphate groups in the LPS or WTAs on the cell membrane surface [100]. The microbial cell membrane may rupture as a result of the instability of the cell surface potential and the mutual attraction of negatively charged phosphate groups brought on by such a chelation process [106]. Bivalent cations were also shown to decrease the action of chitosan in the following order: $Ba^{2+} > Ca^{2+} > Sr^{2+} > Mg^{2+}$. The findings show that chitosan causes "pores" to develop in the cell membrane and that Ca^{2+} linked to the cell surface releases before chitosan-induced cytosolic content leakage [107].

13.7 Mechanism of Antifungal Activity of Chitosan-Based Nanosystems

Yeasts, mildews, rusts, molds, smuts, and mushrooms are all examples of fungi. One of the most common types of living things on Earth, fungi can be found in plants, water, soil, and animals. The majority of the fungus found in a healthy human stomach was Candida species. Occasionally, environmental factors (such molds) may have an impact on gut ecology.

According to numerous studies, chitosan can bind to the phosphorylated mannosyl side in fungus, causing the plasma membrane to rupture and intracellular components to flow out [98]. Chitosan has been discovered to have an impact on protein production and DNA/RNA expression in fungi [108]. According to studies by [109], fluorescently tagged oligochitosan localizes mostly in the cytoplasm of *P. capsici* and does not engage in binding. Additionally, it has been suggested that oligochitosan may affect the mobility of DNA and RNA during electrophoresis. This raises the possibility that tiny antifungal oligochitosan molecules may be able to slip through the cell membrane and wall of fungi and attach to intracellular DNA or RNA. In contrast to these results, Park *et al.* [110] discovered that oligo-chitosan (1 kD) was less effective against nine fungal species than chitosan with larger molecular weights. In addition, low–molecular weight chitosan and oligochitosan can enter fungal cells and affect mitochondrial activity [111].

The minimum lethal concentrations of chitosan against fungus are closely correlated with the molecular weight and degree of acetylation of chitosan, pH value, and the targeted fungi [112, 113]. Younes *et al.* [112] found a negative correlation between molar weight and fungicidal efficacy and a positive correlation between molar weight and the degree of acetylation. It has been found that the plasma membrane functions as a barrier against chitosan in fungi that are resistant to it but not sensitive to it. The cell membrane of chitosan-sensitive fungus contains more polyunsaturated fatty acids than chitosan-resistant fungi do, which results in increased negative charges, permeabilization, and membrane fluidity [114]. Chitosan is used as an antifungal agent because it prevents the growth of mycelium and the germination of fungal spores, as well as by rupturing cell membranes, destroying cell walls, and inhibiting the production of ribosomes [115].

13.8 Antibiofilm Properties of Chitosan and Chitosan Derivatives

A biofilm is a group of microbial cells embedded in a matrix composed of extracellular polymeric substances (EPS), proteins, and extracellular DNA produced by microbial cells [116]. Antibiotics and microbicides may be able to be blocked by biofilm. The biofilms polymeric matrix prevents antibiotic substances from reaching the target bacterial cell surface. Currently, researchers are focusing on halting the pathogenicity by preventing the production of biofilms and eradicating mature biofilm using a variety of reactive chemicals [117]. But several of the most popular antibiofilm medications now also have cytotoxic characteristics [118, 119]. Therefore, chitosan, an antibiofilm chemical of natural origin that is biocompatible, biodegradable, harmless, non-allergenic, economical, and ecologically friendly, may be a feasible solution.

The primary cause of chitosan's antibiofilm capabilities is the polycationic nature of the amino groups of N-acetylglucosamine monomers. Chitosan's positive charges can hinder the development of microbial biofilms by electrostatically interacting with their negatively charged EPS, extracellular DNA, and proteins components [120, 121]. Chitosan now has access to the biofilm's channel, where it can then kill microorganisms using the aforementioned action modes. Additionally, conjugating chitosan with antimicrobial peptides can produce cationic peptide-polysaccharide to improve electrostatic interactions between the copolymer and microbial cell membrane components and prevent the development of GPB and GNB [122, 123].

13.9 Applications of Chitosan-Based Nanosystems in Food Preservation

Fruit and vegetable freshness is a vital element in determining which product a consumer likes to purchase from the market. Markets struggle to keep the fruit and vegetable fresh and provide customers with better-quality goods. The fruit and vegetable are microbially susceptible and biodegradable. For the food industry, issues with natural ripening and the fruit and vegetable degradation process—primarily through enzymatic reaction—are major concerns. Due to their quick ripening and softening, fruits and vegetables have a limited capacity for storage, handling, and transportation because they are susceptible to rot and perish [124]. Edible coating, a unique food preservation approach, lowers the rates of respiration, oxidative reaction, gas exchange, moisture and solute migration, and other activities by forming physical barriers on the surface of the fruit and vegetable [125]. Biopolymer coating materials are developed to transport active ingredients including antibrowning agents, colorants, flavorings, nutrients, spices, and antimicrobial compounds in order to lengthen product shelf life and reduce the risk of pathogen growth on food surfaces.

Chitosan-processed fresh or chopped fruits, berries, and vegetables have a longer shelf life and are less prone to grow molds and bacteria. Cheese, hummus, and noodle dough do not get germs growing in them when you add chitosan [126, 127]. Chitosan-based film coatings effectively extend the shelf life of tomatoes [128], cucumbers [129], carrots [130, 131], apples [132], pomegranates [133], bell peppers [134], bananas [135], plums [136], strawberries [137], papaya [138], pears, fish [139], meat [140], mango [141], and minced meat [142, 143].

Chitosan-based nanocomposites are excellent replacements for conventional coatings and films to lessen their negative impact on human and environmental health. Additionally, natural chitosan packaging and coatings combined with useful chemicals reduce fruit and vegetable postharvest losses while maintaining their nutritional and sensory qualities. Foods wrapped in chitosan films or coatings can retain their value and have an extended shelf life because they are efficient barriers against the transfer of oxygen and carbon dioxide. It has been demonstrated that adding chitosan to bioactive substances such natural essential oils and extracts results in active films or coatings that are efficient at preserving meat and fish products [144].

Although the usage of chitosan as an antibacterial has received substantial scrutiny, there has not been much attention paid to how chitosan NPs affect food-borne pathogens. According to Centre for Disease Control and Prevention (CDC) estimates, there may be 128,000 hospitalizations, 3,000 fatalities, and 48,000,000 cases of foodborne illness each year in the United States. Additionally, according to estimates from the World Health Organization (WHO), each year, 600,000,000 occurrences of foodborne diseases and 420,000 fatalities can be caused by foodborne infections, with 30% of the fatalities occurring in children under the age of five. These factors have increased the attention of researchers in creating novel antimicrobial drugs to fight foodborne diseases in recent years. Chitosan-based NPs have frequently been used alone and in conjunction with other substances to inhibit food-borne pathogens. The evaluation of the antibacterial activity of low and high–molecular weight chitosan NPs (CSNPs) produced using sodium sulfate or TPP cross-linkers with or without sonication at various energy levels was successful in the eradication of *E. coli* O15:H7. It has been claimed that CSNPs produced by ionic gelation and used as an edible covering on grapes block several FBPs. Under simulated conditions, a vegetable wash made with CSNPs and 1% citric acid reduced the amount of bacteria on lettuce more than commercial formulations. Mohammadi *et al.* [145] reported that they investigated different molecular weight chitosan formulations and discovered that CSNPs had a considerably better antibacterial effect than microparticles against *E. coli*. CSNPs blended with plant essential oils and other naturally occurring antimicrobials like propolis and nisin were also mentioned in the literature. The feline calicivirus, MS2 phage, and murine norovirus have all been demonstrated to be resistant to the antiviral activities of bulk chitosan-based compounds, both with and without plant extracts. Additionally, the effectiveness of chitosan-based products in treating human norovirus, the most typical viral foodborne infection, has been studied. The usefulness of nanochitosan in preventing foodborne viruses has not been studied, though. This shows a possible route for future research into the use of nanochitosan to fight these viruses. The study found that combination products had a stronger effect than CSNPs alone, indicating that chitosan may be used to reduce foodborne pathogens more effectively when coupled with a number of antimicrobial agents.

The food industry has a problem from microbial contamination. Foodborne bacteria can accelerate food deterioration and cause considerable financial losses in addition to causing a number of fatal human illnesses. This invasion and spread of microbes can be stopped and reduced with the use of food packaging (FP). Different biopolymers with acceptable environmental profiles, biodegradability, and biocompatibility have been studied in order to create substitutes for the conventional petroleum-derived materials used for FP. Because the intrinsic properties of chitosan may increase the packaging's antibacterial efficiency, the development of chitosan-based solutions has been the focus of much study. Film packaging materials have been developed due to chitosan's capacity to form films when combined with other natural polysaccharides including starch, pectin, and hydroxypropyl methylcellulose (HPMC). When Möller *et al.* investigated the antibacterial action of a chitosan-HPMC film against Listeria monocytogenes, they offered an example by demonstrating that the bacterial growth was entirely stopped. When the amino groups of chitosan were cross-linked with citric acid, the antibacterial activity of chitosan-HPMC films was drastically decreased, demonstrating the significance of the protonated amino groups of chitosan for the antimicrobial action [146]. Numerous research teams have used chitosan nanofibers as packaging materials because of its various benefits, including biocompatibility, a high surface area, and strong functional and antibacterial qualities. As a result of their capacity to load various active chemicals, including those with limited solubility and stability, nanocarrier systems have come to get more and more attention in the FP industry. Chitosan NPs have been used extensively for this purpose and might potentially work in conjunction with the loaded active component due to their well-known antibacterial capabilities. Feyzioglu and Tornuk conducted a proof-of-concept research to demonstrate the potential use of chitosan NPs loaded with the summer savory essential oil as FP by testing their antibacterial efficacy against three distinct foodborne pathogens (*L. monocytogenes*, *E. coli*, and *S. aureus*). Both chitosan NPs and those filled with summer savory essential oil showed antibacterial action, but the loaded NPs had a stronger impact [27, 147].

13.10 Toxicology, Safety, and Regulatory Aspects of Nanosystem

As the commercialization of nanotechnology has advanced, worries about NP migration and safety impacts have increased. One of the key issues with NPs is their utilization in nanocomposite materials for biomedical and FP applications. Nanofoods offer many advantages, but research on how they could affect our health is still in the feasibility and verification phases. Safety concerns are raised when NPs go from packaging to the food matrix. People are worried about the prospect that NPs might transfer from packaging into foods and that this could harm customers' health [148, 149].

Given the beneficial qualities that enhanced FP based on nanomaterials may bring to conventional FP, its availability in the market is expected to rise. However, due to the potential migration of NPs from FP into foodstuff and the potential dangers to consumers' health that could result from ingesting and being exposed to NPs, which are currently largely unknown, its widespread commercialization must overcome serious obstacles [150]. The most concerning aspect of using better FP made from nanomaterials is the level of migration and potential health risk posed by eating food that contains nanoscale substances that have leached from the packaging [151], and there is constant discussion among scholars who have opposing viewpoints. According to Huang *et al.* [152], the term "migration into foodstuff" refers to the mass transfer process by which ingredients that were once contained in the container are released into the packed food or beverage. Utilizing European and American (FDA) standard migration assays, the migration of NPs from polymer composites for FP into food or food simulants has been evaluated, and European Regulation [153] has specified limitations. However, the current state of knowledge has shown that there are a number of difficulties that might make it difficult to determine and evaluate NPs migration studies from FP to food [150]. The migration rate of a system rises as NP size and polymer dynamic viscosity decrease, according to documented viewpoints [154], and the migration of NPs from FP would be modest and sluggish.

Recent research has demonstrated that NPs can migrate from packaging into food products, and, although European Regulation [153] has established limits through specific migration tests, the current policies and regulations are not yet suitable for allowing extensive application of nanotechnology in the food industry and agricultural sector. Customers' understanding and acceptance of employing innovative nanomaterials as well as the procedures used to make policies are significantly impacted by the public's ongoing worries about the possible harm to their health and concerns about lax policy control. Their safety is still not entirely established, and there are no strict policy regulations in place [155].

13.11 Conclusion

The use of antimicrobials in food production and agriculture has a direct or indirect impact on the emergence of bacteria linked to plants and animals that may infiltrate the food chain through consumption of meat, fish, vegetables, or other food sources. This phenomenon is known as AMR. Due to its biodegradability, biocompatibility, antibacterial activity, non-toxicity, and varied chemical and physical characteristics, chitosan offers a great deal of promise for a variety of applications. A number of foods have been packaged using chitosan films to preserve their quality. Chitosan has strong antibacterial properties against both GPB and GNB, fungus, and other pathogenic and spoilage microorganisms. Over the past 10 years, a lot of work has gone into creating and testing antimicrobial films to increase food safety and shelf life. Given the beneficial characteristics of chitosan-based nanosystem enhanced FP, its availability in the market today is expected to rise. However, due to concerns concerning potential NP migration from FP into foodstuffs and potential harmful effects on consumer health, its widespread commercialization must overcome severe obstacles. A number of studies are presently being conducted to prove the product's safety and assuage public concerns.

References

1. Hao, H., Cheng, G., Iqbal, Z., Ai, X., Hussain, H.I, Huang, L., Yuan, Z., Benefits and risks of antimicrobial use in food-producing animals. *Front. Microbiol.*, 5, 288, 2014.
2. Cheng, G., Ning, J., Ahmed, S., *et al.*, Selection and dissemination of antimicrobial resistance in Agri-food production. *Antimicrob. Resist. Infect. Control*, 8, 158, 2019.
3. Lekshmi, M., Ammini, P., Kumar, S., Varela, M.F., The food production environment and the development of antimicrobial resistance in human pathogens of animal origin. *Microorganisms*, 5, 1, 11, 2017.
4. Van Boeckel, T.P., Brower, C., Gilbert, M., Grenfell, B.T., Levin, S.A., Robinson, T.P., Laxminarayan, R., Global trends in antimicrobial use in food animals. *Proc. Natl. Acad. Sci.*, 112, 18, 5649–5654, 2015.
5. O'neill, J., *Antimicrobial resistance: tackling a crisis for the health and wealth of nations*, Review on Antimicrobial Resistance, 2014, 2018. URL https://books.google.co.uk/books.
6. Andersson, D.I., and Hughes, D., Microbiological effects of sublethal levels of antibiotics. *Nat. Rev. Microbiol.*, 12, 7, 465–478, 2014.

7. Wales, A.D. and Davies, R.H., Co-selection of resistance to antibiotics, biocides and heavy metals, and its relevance to foodborne pathogens. *Antibiotics*, 4, 4, 567–604, 2015.
8. Penesyan, A., Gillings, M., Paulsen, I.T., Antibiotic discovery: combatting bacterial resistance in cells and in biofilm communities. *Molecules*, 20, 4, 5286–5298, 2015.
9. Kong, M., Chen, X.G., Xing, K., Park, H.J., Antimicrobial properties of chitosan and mode of action: a state of the art review. *Int. J. Food Microbiol.*, 144, 51–63, 2010.
10. Rabea, E.I., Badawy, M.E.T., Stevens, C.V., Smagghe, G., Steurbaut, W., Chitosan as antimicrobial agent: Applications and mode of action. *Biomacromolecules*, 4, 1457–1465, 2003.
11. Larsson, D.G.J. and Flach, C.F., Antibiotic resistance in the environment. *Nat. Rev. Microbiol.*, 20, 257–269, 2022.
12. Saxena, S.K., Nyodu, R., Kumar, S., Maurya, V.K., Current Advances in Nanotechnology and Medicine, in: *(eds) NanoBioMedicine*, S. Saxena, and S. Khurana, (Eds.), Springer, Singapore, 2020, https://doi.org/10.1007/978-981-32-9898-9_1.
13. Oh, J.W., Chun, S.C., Chandrasekaran, M., Preparation and *in vitro* characterization of chitosan nanoparticles and their broad-spectrum antifungal action compared to antibacterial activities against phytopathogens of tomato. *Agronomy*, 9, 1, 21, 2019, doi: DOI: 10.3390/agronomy9010021.
14. El-Naggar, N.E.A., Saber, W.I., Zweil, A.M., Bashir, S.I., An innovative green synthesis approach of chitosan nanoparticles and their inhibitory activity against phytopathogenic Botrytis cinerea on strawberry leaves. *Sci. Rep.*, 12, 1, 1–20, 2022.
15. Malerba, M. and Cerana, R., Chitosan effects on plant systems. International. *J. Mol. Sci.*, 17, 7, 996, 2016.
16. Manigandan, V., Karthik, R., Saravanan, R., Marine carbohydrate-based therapeutics for Alzheimer's disease: mini review. *J. Neuro. Neurosci.*, (Special issue), 6, 1–6, 2015.
17. Karthik, R., Manigandan, V., Saravanan, R., Rajesh, R.P., Chandrika, B., Structural characterization and *in vitro* biomedical activities of sulfated chitosan from *Sepia pharaonis*. *Int. J. Biol. Macromol.*, 84, 319–328, 2016.
18. Yilmaz A, H., Antibacterial Activity of Chitosan-Based Systems. In Functional Chitosan. *Drug Deliv. Biomed. Appl.*, 457–489, 2020.
19. Negm, A., Hefni, H.H., Abd-Elaal, A.A., Badr, E.A., Kana, M.T.A., Advancement on modification of chitosan biopolymer and its potential applications. *Int. J. Biol. Macromol.*, 152, 681–702, 2020.
20. Dheyab, M.A., Aziz, A.A., Jameel, M.S., Abu Noqta, O., Mehrdel, B., Synthesis and coating methods of biocompatible iron oxide/gold nanoparticle and nanocomposite for biomedical applications. *Chin. J. Phys.*, 64, 305–325, 2020.

21. Joseph, S.M., Krishnamoorthy, S., Paranthaman, R., Moses, J., Anandharamakrishnan, C., A review on source-specific chemistry, functionality, and applications of chitin and chitosan. *Carbohydr. Polymers Technol. Appl.*, 2, 100036, 2021. https://doi.org/10.1016/j.carpta.2021.100036.
22. Crognale, S., Russo, C., Petruccioli, M., D'Annibale, A., Chitosan Production by Fungi: Current State of Knowledge, Future Opportunities and Constraints. *Fermentation*, 8, 76, 2022, doi: 10.3390/fermentation8020076.
23. Abidin, N.Z., Kormin, F., Abidin, N.Z., Anuar, N.M., Abu Bakar, M., The Potential of Insects as Alternative Sources of Chitin: An Overview on the Chemical Method of Extraction from Various Sources. *Int. J. Mol. Sci.*, 21, 20204978, doi: 10.3390/ijms21144978.
24. Yeul, V.S. and Rayalu, S.S., Unprecedented Chitin and Chitosan: A Chemical Overview. *J. Polym. Environ.*, 21, 606–614, 2012, doi: 10.1007/s10924-012-0458-x.
25. Dutta, P.K., Dutta, J., Tripathi, V.S., Chitin and chitosan: Chemistry, properties and applications. *J. Sci. Ind. Res.*, 63, 20–31, 2004.
26. Knežević-Jugović, Z., Petronijević, Z., Šmelcerović, A., Chitin and chitosan from microorganisms, in: *Chitin, Chitosan, Oligosaccharides and Their Derivatives: Biological Activities and Applications*, pp. 25–34, CRC Press, Boca Raton, 2020.
27. Perinelli, D.R., Fagioli, L., Campana, R., Lam, J.K.W., Baffone, W., Palmieri, G.F., Casettari, L., Bonacucina, G., Chitosan-based nanosystems and their exploited antimicrobial activity. *Eur. J. Pharm. Sci.*, 117, 8–20, 2018.
28. Elsabee, M.Z. and Abdou, E.S., Chitosan-based edible films and coatings: a review. *Mater. Sci. Eng., C*, 33, 1819–1841, 2013.
29. Alishahi, A. and Aïder, M., Applications of chitosan in the seafood industry and aquaculture: a review. *Food Bioprocess Technol.*, 5, 3, 817–830, 2012.
30. Divya, K., Vijayan, S., George, T.K., Jisha, M.S., Antimicrobial properties of chitosan nanoparticles: Mode of action and factors affecting activity. *Fibers Polym.*, 18, 221–230, 2017.
31. Yang, T.C., Chou, C.C., Li, C.F., Antibacterial activity of N-alkylated disaccharide chitosan derivatives. *Int. J. Food Microbiol.*, 97, 237–245, 2005.
32. Xie, Y.J., Liu, X.F., Chen, Q., Synthesis and characterization of water-soluble chitosan derivative and its antibacterial activity. *Carbohydr. Polym.*, 69, 142–147, 2007.
33. El Ghaouth, A., Arul, J., Grenier, J., Asselin, A., Antimicrobial effect of chitosan nanoparticles on Streptococcus. *Phytopathology*, 82, 398–402, 1992.
34. Goy, R.C., Britto, D., Assis, O.B.G., A review of the antimicrobial activity of chitosan. *Polímeros Ciência e Tecnologia*, 19, 241–247, 2009.
35. Wang, M., Sun, R., Li, C., Wang, Q., Zhang, B., MicroRNA expression profiles during cotton (Gossypium hirsutum L) fiber early development. *Sci. Rep.*, 7, 44454, 2017.
36. Chirkov, S.N., The antiviral activity of chitosan. *Appl. Biochem. Microbiol.*, 38, 1–8, 2002.

37. Davydova, V.N., Naqorskaia, V.P., Gorbach, V.I., Kalitnik, A.A., Reunov, A.V., Solov'eva, T.F., Ermak, I.M., Chitosan antiviral activity: Dependence on structure and depolymerization method. *Prikladnaia Biokhimiia i Mikrobiologiia*, 47, 113–118, 2011.
38. Iriti, M. and Varoni, E.M., Chitosan-induced antiviral activity and innate immunity in plants. *Environ. Sci. Pollut. Res.*, 22, 4, 2935–2944, 2015.
39. Iriti, M. and Varoni, E.M., Chitosan-elicited plant innate immunity: focus on antiviral activity, in: *Research Progress in Oligosaccharins*, pp. 65–81, Springer, New York, NY, 2016.
40. Ai, H., Wang, F., Xia, Y., Chen, X., Lei, C., Antioxidant, antifungal, and antiviral activities of chitosan from the larvae of the housefly, *Musca domestica* L. *Food Chem.*, 132, 1, 493–498, 2012.
41. Kulikov, S.N., Chirkov, S.N., Il'ina, A.V., Lopatin, S.A., Varlamov, V.P., Effect of the molecular weight of chitosan on its antiviral activity in plants. *Appl. Biochem. Microbiol.*, 42, 2, 200–203, 2006.
42. Matica, A., Menghiu, G., Ostafe, V., Antifungal properties of chitosans. *New Front. Chem.*, 26, 1, 55–63, 2107.
43. Kassem, A., Ayoub, G.M., Malaeb, L., Antibacterial activity of chitosan nano-composites and carbon nanotubes: A review. *Sci. Total Environ.*, 668, 566–576, 2019.
44. Rozman, N.A.S., Tong, W.Y., Leong, C.R., Tan, W.N., Hasanolbasori, M.A., Abdullah, S.Z., Potential antimicrobial applications of chitosan nanoparticles (ChNP), 2019.
45. Davis, S.S. and Illum, L., Sustained release chitosan microspheres prepared by novel spray drying methods. *J. Microencapsulation*, 16, 3, 343–355, 1999.
46. Calvo, P., Remuñan-López, C., Vila-Jato, J.L., Alonso, M.J., Chitosan and chitosan/ethylene oxide-propylene oxide block copolymer nanoparticles as novel carriers for proteins and vaccines. *Pharm. Res.*, 14, 1431–1436, 1997.
47. Tokumitsu, H., Ichikawa, H., Fukumori, Y., Chitosan-gadopentetic acid complex nanoparticles for gadolinium neutron-capture therapy of cancer: preparation by novel emulsion-droplet coalescence technique and characterization. *Pharm. Res.*, 16, 1830–1835, 1999.
48. Maitra, A., Ghosh, P.K., De, T.K., Sahoo, S.K., U.S. Patent No. 5,874,111, U.S. Patent and Trademark Office, Washington, DC, 1999.
49. Brunel, F., Véron, L., David, L., Domard, A., Delair, T., A novel synthesis of chitosan nanoparticles in reverse emulsion. *Langmuir*, 24, 20, 11370–11377, 2008.
50. Chen, H., Seiber, J.N., Hotze, M., ACS select on nanotechnology in food and agriculture: A perspective on implications and applications. *J. Agric. Food. Chem.*, 62, 6, 1209–1212, 2014.
51. Qian, J., Pan, C., Liang, C., Antimicrobial activity of Fe-loaded chitosan nanoparticles. *Eng. Life Sci.*, 17, 6, 629–634, 2017.

52. Badawy, M.E., Lotfy, T.M., Shawir, S., Preparation and antibacterial activity of chitosan-silver nanoparticles for application in preservation of minced meat. *Bull. Natl. Res. Cent.*, 43, 1, 1–14, 2019.
53. Katas, H., Lim, C.S., Azlan, A.Y.H.N., Buang, F., Busra, M.F.M., Antibacterial activity of biosynthesized gold nanoparticles using biomolecules from *Lignosus rhinocerotis* and chitosan. *Saudi Pharm. J.*, 27, 2, 283–292, 2019.
54. Tamara, F.R., Lin, C., Mi, F.L., Ho, Y.C., Antibacterial effects of chitosan/cationic peptide nanoparticles. *Nanomaterials, 8*, 2, 88, 2018.
55. Huang, L., Cheng, X., Liu, C., Xing, K., Zhang, J., Sun, G., Li, X., Chen, X., Preparation, characterization, and antibacterial activity of oleic acid-grafted chitosan oligosaccharide nanoparticles. *Front. Biol. China*, 4, 321–327, 2009.
56. Sotelo-Boyás, M., Correa-Pacheco, Z., Bautista-Baños, S., Corona-Rangel, M., Physicochemical characterization of chitosan nanoparticles and nanocapsules incorporated with lime essential oil and their antibacterial activity against food-borne pathogens. *LWT*, 77, 15–20, 2017.
57. Ali, M.J., Sharafaldin, A.M., Majid, P., Mahdi, F.R., Kazem, A., Hajar, R., Zuhair, M.H., Mahdi, K.M., Reza, M., Curcumin-loaded chitosan tripolyphosphate nanoparticles as a safe, natural, and effective antibiotic inhibits the infection of *Staphylococcus aureus* and *Pseudomonas aeruginosa in vivo. Iran. J. Biotechnol.*, 12, 1–8, 2014.
58. Sekhon, B.S., Nanotechnology in agri-food production: an overview. *Nanotechnol. Sci. Appl.*, 7, 31–53, 2014.
59. Frewer, L.J., Norde, W., Fischer, A.R.H., Kampers, F.W.H. (Eds.), *Nanotechnology in the Agri-Food Sector: Implications for the Future*, Wiley-VCH, Weinheim, Germany, 2011.
60. Ditta, A., How helpful is nanotechnology in agriculture? Advances in Natural Sciences. *Nanosci. Nanotechnol.*, 3, 3, 033002, 2012.
61. Biswal, S.K., Nayak, A.K., Parida, U.K., Nayak, P.L., Applications of nanotechnology in agriculture and food sciences. *Int. J. Sci. Innov. Discoveries*, 2, 1, 21–36, 2012.
62. Sonkaria, S., Ahn, S.H., Khare, V., Nanotechnology and its impact on food and nutrition: a review. *Recent Pat. Food Nutr. Agric.*, 4, 1, 8–18, 2012.
63. Kuzma, J. and VerHage, P., *Nanotechnology in Agriculture and Food Production: Anticipated Applications*, The Project on Emerging Nanotechnologies, Washington, DC, 2006.
64. Cortes-Lobos, R., *Can Agri-Food Nanotechnology Contribute to Achieve the Millennium Development Goals in Developing Countries?*, Universidad Nacional de General Sarmiento, Los Polvorines, Argentina, 2011.
65. Chen, Q., Jiang, H., Ye, H., Li, J., Huang, J., Preparation, antibacterial, and antioxidant activities of silver/chitosan composites. *J. Carbohydr. Chem.*, 33, 6, 298–312, 2014.
66. Hong, J., Peralta-Videa, J.R., Gardea-Torresdey, J., Nanomaterials in agricultural production: benefits and possible threats?, in: *Sustainable Nanotechnology and the Environment: Advances and Achievements*, N.

Shamim and V.K. Sharma (Eds.), pp. 73–90, American Chemical Society, Washington, DC, 2013.

67. Zargar, V., Asghari, M., Dashti, A., A review on chitin and chitosan polymers: Structure, chemistry, solubility, derivatives, and applications. *ChemBioEng Rev.*, 2, 3, 204–226, 2015.
68. Sathiyabama, M. and Parthasarathy, R., Biological preparation of chitosan nanoparticles and its *in vitro* antifungal efficacy against some phytopathogenic fungi. *Carbohydr. Polym.*, 151, 321–325, 2016.
69. Denkbas¸, E.B., Kilicay, E., Birlikseven, C., Oztürk, E., Magnetic chitosan microspheres: Preparation and characterization. *React. Funct. Polym.*, 50, 3, 225–232, 2002.
70. Yin, H., Du, Y., Dong, Z., Chitin oligosaccharide and chitosan oligosaccharide: Two similar but different plant elicitors. *Front. Plant Sci.*, 7, 522, 2016.
71. Wen, Y., Ma, J., Chen, J., Shen, C., Li, H., Liu, W., Carbonaceous sulfurcontaining chitosan–Fe (III): A novel adsorbent for efficient removal of copper (II) from water. *Chem. Eng. J.*, 259, 372–380, 2015.
72. Elwakeel, K.Z. and Guibal, E., Arsenic (V) sorption using chitosan/Cu (OH) 2 and chitosan/CuO composite sorbents. *Carbohydr. Polym.*, 134, 190–204, 2015.
73. Chen, C., Lee, S.H., Cho, M., Kim, J., Lee, Y., Cross-linked chitosan as an efficient binder for Si anode of Li-ion batteries. *ACS Appl. Mater. Interfaces*, 8, 4, 2658–2665, 2016.
74. Ou, C., Chen, S., Liu, Y., Shao, J., Li, S., Fu, T., Bi, X., Study on the thermal degradation kinetics and pyrolysis characteristics of chitosan-Zn complex. *J. Anal. Appl. Pyrolysis*, 122, 268–276, 2016.
75. Wang, J., Yang, B., Zhang, S., Cao, J., Jiang, W., Effect of thymol on antifungal ability of chitosan coating against Penicillium expansum in Yali pear. *Emir. J. Food Agric.*, 28, 10, 725–731, 2016.
76. Biao, L., Tan, S., Wang, Y., Guo, X., Fu, Y., Xu, F., Liu, Z., Synthesis, characterization and antibacterial study on the chitosan-functionalized Ag nanoparticles. *Mater. Sci. Engineering C*, 76, 73–80, 2017.
77. Mitra, S., Gaur, U., Ghosh, P., Maitra, A., Tumour targeted delivery of encapsulated dextran–doxorubicin conjugate using chitosan nanoparticles as carrier. *J. Control. Release*, 74, 1-3, 317–323, 2001.
78. Agnihotri, S.A., Mallikarjuna, N.N., Aminabhavi, T.M., Recent advances on chitosan-based micro-and nanoparticles in drug delivery. *J. Control. Release*, 100, 1, 5–28, 2004.
79. He, P., Davis, S.S., Illum, L., Chitosan microspheres prepared by spray drying. *Int. J. Pharm.*, 187, 1, 53–65, 1999.
80. Rossman, A.Y., The impact of invasive fungi on agricultural ecosystems in the United States, in: *Ecological impacts of non-native invertebrates and Fungi on terrestrial ecosystems*, pp. 97–107, Springer, Dordrecht, 2008.

81. da Cruz Cabral, L., Pinto, V.F., Patriarca, A., Application of plant derived compounds to control fungal spoilage and mycotoxin production in foods. *Int. J. Food Microbiol.*, 166, 1, 1–14, 2013.
82. Nesic, K., Ivanovic, S., Nesic, V., Fusarial toxins: Secondary metabolites of Fusarium fungi, in: *Reviews of Environmental Contamination and Toxicology*, vol. 228, pp. 101–120, Springer, Cham, 2014.
83. Kaur, P., Duhan, J.S., Thakur, R., Comparative pot studies of chitosan and chitosan-metal nanocomposites as nano-agrochemicals against fusarium wilt of chickpea (Cicer arietinum L.). *Biocatal. Agric. Biotechnol.*, 14, 466–471, 2018.
84. Saharan, V. and Pal, A., *Chitosan based nanomaterials in plant growth and protection*, Springer, New Delhi, India, 2016.
85. Raafat, D., Von Bargen, K., Haas, A., Sahl, H.-G., Insights into the mode of action of chitosan as an antibacterial compound. *Appl. Environ. Microbiol.*, 74, 12, 3764–3773, 2008.
86. Katiyar, D., Hemantaranjan, A., Singh, B., Chitosan as a promising natural compound to enhance potential physiological responses in plant: A review. *Indian J. Plant Physiol.*, 20, 1, 1–9, 2015.
87. Dennis, G., Harrison, W., Agnes, K., Erastus, G., Effect of biological control antagonists adsorbed on chitosan immobilized silica nanocomposite on *Ralstonia solanacearum* and growth of tomato seedlings. *Adv. Res.*, 6, 3, 1–23, 2016.
88. Pospieszny, H. and Atabekov, J.G., Effect of chitosan on the hypersensitive reaction of bean to alfalfa mosaic virus. *Plant Sci.*, 62, 1, 29–31, 1989.
89. Matica, M.A., Aachmann, F.L., Tøndervik, A., Sletta, H., Ostafe, V., Chitosan as a wound dressing starting material: Antimicrobial properties and mode of action. *Int. J. Mol. Sci.*, 20, 20195889.
90. Sahariah, P. and Másson, M., Antimicrobial Chitosan and Chitosan Derivatives: A Review of the Structure-Activity Relationship. *Biomacromolecules*, 18, 3846–3868, 2017.
91. Kong, M., Chen, X.G., Liu, C.S., Liu, C.G., Meng, X.H., Yule, J., Antibacterial mechanism of chitosan microspheres in a solid dispersing system against *E. coli*. *Colloids Surf. B Biointerfaces*, 65, 197–202, 2008.
92. Je, J.Y. and Kim, S.K., Chitosan derivatives killed bacteria by disrupting the outer and inner membrane. *J. Agric. Food Chem.*, 54, 6629–6633, 2006.
93. Liu, D., Li, Z., Zhu, Y., Li, Z., Kumar, R., Recycled chitosan nanofibril as an effective Cu (II), Pb (II) and Cd (II) ionic chelating agent: Adsorption and desorption performance. *Carbohydr. Polym.*, 111, 469–476, 2014.
94. Li, X.-F., Feng, X.-Q., Yang, S., Fu, G.-Q., Wang, T.-P., Su, Z.-X., Chitosan kills Escherichia coli through damage to be of cell membrane mechanism. *Carbohydr. Polym.*, 79, 493–499, 2010.
95. Khalid, S., Piggot, T.J., Samsudin, F., Atomistic and Coarse Grain Simulations of the Cell Envelope of Gram-Negative Bacteria: What Have We Learned? Acc. *Chem. Res.*, 52, 180–188, 2019.

96. Helander, I.M., Nurmiaho-Lassila, E.L., Ahvenainen, R., Rhoades, J., Roller, S., Chitosan disrupts the barrier properties of the outer membrane of Gram-negative bacteria. *Int. J. Food Biol.*, 71, 235–244, 2001.
97. Ke, C.L., Deng, F.S., Chuang, C.Y., Lin, C.H., Antimicrobial Actions and Applications of Chitosan. *Polymers*, 13, 904, 2021. https://doi.org/10.3390/polym13060904.
98. Xing, K., Chen, X.G., Liu, C.S., Cha, D.S., Park, H.J., Oleoyl-chitosan nanoparticles inhibits Escherichia coli and Staphylococcus aureus by damaging the cell membrane and putative binding to extracellular or intracellular targets. *Int. J. Food Microbiol.*, 132, 127–133, 2009.
99. Galván Márquez, I., Akuaku, J., Cruz, I., Cheetham, J., Golshani, A., Smith, M.L., Disruption of protein synthesis as antifungal mode of action by chitosan. *Int. J. Food Microbiol.*, 164, 108–112, 2013.
100. Fei Liu, X., Lin Guan, Y., Zhi Yang, D., Li, Z., De Yao, K., Antibacterial action of chitosan and carboxymethylated chitosan. *J. Appl. Polym. Sci.*, 79, 1324–1335, 2001.
101. Hosseinnejad, M. and Jafari, S.M., Evaluation of different factors affecting antimicrobial properties of chitosan. *Int. J. Biol. Macromol.*, 85, 467–475, 2016.
102. Chien, R.C., Yen, M.T., Mau, J.L., Antimicrobial and antitumor activities of chitosan from shiitake stipes, compared to commercial chitosan from crab shells. *Carbohydr. Polym.*, 138, 259–264, 2016.
103. Clifton, L.A., Skoda, M.W., Le Brun, A.P., Ciesielski, F., Kuzmenko, I., Holt, S.A., Lakey, J.H., Effect of divalent cation removal on the structure of Gram-negative bacterial outer membrane models. *Langmuir*, 31, 404–412, 2015.
104. Wickham, J.R., Halye, J.L., Kashtanov, S., Khandogin, J., Rice, C.V., Revisiting magnesium chelation by teichoic acid with phosphorus solid-state NMR and theoretical calculations. *J. Phys. Chem.*, 113, 2177–2183, 2009.
105. Kern, T., Giffard, M., Hediger, S., Amoroso, A., Giustini, C., Bui, N.K., Joris, B., Bougault, C., Vollmer, W., Simorre, J.P., Dynamics characterization of fully hydrated bacterial cell walls by solid-state NMR: Evidence for cooperative binding of metal ions. *J. Am. Chem. Soc.*, 31, 10911–10919, 2010. https://doi.org/10.1021/ja104533w.
106. Feng, P., Luo, Y., Ke, C., Qiu, H., Wang, W., Zhu, Y., Hou, R., Xu, L., Wu, S., Chitosan-Based Functional Materials for Skin Wound Repair: Mechanisms and Applications. *Front. Bioeng. Biotechnol.*, 9, 650598, 2021.
107. Young, D.H. and Kauss, H., Release of Calcium from Suspension-Cultured Glycine max Cells by Chitosan, Other Polycations, and Polyamines in Relation to Effects on Membrane Permeability. *Plant Physiol.*, 73, 698–702, 1983.
108. Abd-Elsalam, K.A., Vasil'kov, A.Y., Said-Galiev, E.E., Rubina, M.S., Khokhlov, A.R., Naumkin, A.V., Shtykova, E.V., Alghuthaymi, M.A., Bimetallic blends and chitosan nanocomposites: Novel antifungal agents against cotton seedling damping-off. *Eur. J. Plant Pathol.*, 151, 57–72, 2018.

109. Xu, J., Zhao, X., Wang, X., Zhao, Z., Du, Y., Oligochitosan inhibits Phytophthora capsici by penetrating the cell membrane and putative binding to intracellular targets. *Pestic. Biochem. Physiol.*, 88, 167–175, 2007. https://doi.org/10.1016/j.pestbp.2006.10.010.
110. Park, Y., Kim, M.H., Park, S.C., Cheong, H., Jang, M.K., Nah, J.W., Hahm, K.S., Investigation of the antifungal activity and mechanism of action of LMWS-chitosan. *J. Microbiol. Biotechnol.*, 18, 1729–1734, 2008.
111. Robles-Martínez, L., Guerra-Sánchez, M.G., Hernández-Lauzardo, A.N., Pardo, J.P., Velázquez-del Valle, M.G., Effects of chitosan and oligochitosan on development and mitochondrial function of Rhizopus stolonifer. *J. Basic Microbiol.*, 54, S42–S49, 2014.
112. Younes, I., Sellimi, S., Rinaudo, M., Jellouli, K., Nasri, M., Influence of acetylation degree and molecular weight of homogeneous chitosans on antibacterial and antifungal activities. *Int. J. Food Microbiol.*, 185, 57–63, 2014.
113. Atai, Z., Atai, M., Amini, J., *In vivo* study of antifungal effects of low-molecular-weight chitosan against *Candida albicans*. *J. Oral Sci.*, 59, 425–430, 2017.
114. Palma-Guerrero, J., Lopez-Jimenez, J., Pérez-Berná, A., Huang, I.C., Jansson, H.B., Salinas, J., Villalaín, J., Read, N., Lopez-Llorca, L.V., Membrane fluidity determines sensitivity of filamentous fungi to chitosan. *Mol. Microbiol.*, 75, 1021–1032, 2010.
115. Meng, D., Garba, B., Ren, Y., Yao, M., Xia, X., Li, M., Wang, Y., Antifungal activity of chitosan against Aspergillus ochraceus and its possible mechanisms of action. *Int. J. Biol. Macromol.*, 158, 1063–1070, 2020.
116. Bhattacharya, S.P., Bhattacharya, A., Sen, A., A comprehensive and comparative study on the action of pentacyclic triterpenoids on *Vibrio cholerae* biofilms. *Microb. Pathogen.*, 149, 104493, 2020.
117. Khan, F., Manivasagan, P., Lee, J.-W., Pham, D.T.N., Oh, J., Kim, Y.-M., Fucoidan-stabilized gold nanoparticle-mediated biofilm inhibition, attenuation of virulence and motility properties in *Pseudomonas aeruginosa* PAO1. *Mar. Drugs*, 17, 208, 2019.
118. Aleksic, I., Petkovic, M., Jovanovic, M., Milivojevic, D., Vasiljevic, B., Nikodinovic-Runic, J., Senerovic, L., Anti-biofilm properties of bacterial di-rhamnolipids and their semi-synthetic amide derivatives. *Front. Microbiol.*, 8, 20172454.
119. Brahma, U., Kothari, R., Sharma, P., Bhandari, V., Antimicrobial and anti-biofilm activity of hexadentated macrocyclic complex of copper (II) derived from thiosemicarbazide against *Staphylococcus aureus*. *Sci. Rep.*, 8, 1–8, 2018.
120. Jamil, B., Habib, H., Abbasi, S.A., Ihsan, A., Nasir, H., Imran, M., Development of Cefotaxime Impregnated Chitosan as Nano-antibiotics: De Novo Strategy to Combat Biofilm Forming Multi-drug Resistant Pathogens. *Front. Microbiol.*, 7, 330, 2016. https://doi.org/10.3389/fmicb.2016.00330.

121. Jiang, F., Deng, Y., Yeh, C.K., Sun, Y., Quaternized chitosans bind onto preexisting biofilms and eradicate pre-attached microorganisms. *J. Mater. Chem.*, 2, 8518–8527, 2014. https://doi.org/10.1039/C4TB01131G.
122. Su, Y., Tian, L., Yu, M., Gao, Q., Wang, D., Xi, Y., Yang, P., Lei, B., Ma, P.X., Li, P., Cationic peptidopolysaccharides synthesized by 'click' chemistry with enhanced broad-spectrum antimicrobial activities. *Polym. Chem.*, 8, 3788–3800, 2017. https://doi.org/10.1039/C7PY00528H.
123. Li, P., Zhou, C., Rayatpisheh, S., Ye, K., Poon, Y.F., Hammond, P.T., Duan, H., Chan-Park, M.B., Cationic peptidopolysaccharides show excellent broad-spectrum antimicrobial activities and high selectivity. *Adv. Mater.*, 24, 4130–4137, 2012. https://doi.org/10.1002/adma.201104186.
124. Hu, Z.J., Tang, C.X., He, Z.B., *et al.*, 1-methylcyclopropene (MCP)-containing cellulose paper packaging for fresh fruit and vegetable preservation: a review. *BioResources*, 12, 1, 2234–2248, 2017.
125. Arnon, H., Zaitsev, Y., Porat, R., *et al.*, Effects of carboxymethyl cellulose and chitosan bilayer edible coating on postharvest quality of citrus fruit. *Postharvest Biol. Technol.*, 87, 21–26, 2014, doi: DOI: 10.1016/j.postharvbio.2013.08.007.
126. Friedman, M. and Juneja, V.K., Review of antimicrobial and antioxidative activities of chitosans in food. *J. Food Prot.*, 73, 9, 1737–1761, 2010.
127. Romanazzi, G., Feliziani, E., Baños, S.B., Sivakumar, D., Shelf life extension of fresh fruit and vegetables by chitosan treatment. *Crit. Rev. Food Sci. Nutr.*, 57, 3, 579–601, 2015.
128. Leandro, D.S.P., Bitencourt, T.A., Saltoratto, A.L., Seleghim, M.H., Assis, O.B., Antifungal activity of chitosan and its quaternized derivative in gel form and as an edible coating on cut cherry tomatoes. *J. Agric. Sci.*, 63, 3, 271–285, 2018.
129. Olawuyi, I.F. and Lee, W., Influence of chitosan coating and packaging materials on the quality characteristics of fresh-cut cucumber. *Korean J. Food Preserv.*, 26, 4, 371–380, 2019.
130. Leceta, I., Molinaro, S., Guerrero, P., Kerry, J.P., de la Caba, K., Quality attributes of MAP-packaged ready-to-eat baby carrots by using chitosan-based coatings. *Postharvest Biol. Technol.*, 100, 142–150, 2015.
131. Villafañe, F., Edible coatings for carrots. *Food Rev. Int.*, 33, 1, 84–103, 2016.
132. Li, H., Wang, Y., Liu, F., Yang, Y., Wu, Z., Cai, H., Zhang, Q., Wang, Y., Li, P., Effects of chitosan on control of postharvest blue mold decay of apple fruit and the possible mechanisms involved. *Sci. Hortic.*, 186, 77–83, 2015.
133. Varasteh, F., Arzani, K., Barzegar, M., Zamani, Z., Pomegranate (Punica granatum L.) Fruit Storability Improvement Using Pre-storage Chitosan Coating Technique. *J. Agric. Sci. Technol.*, 19, 2, 389–400, 2017.
134. Ali, A., Noh, N.M., Mustafa, M.A., Antimicrobial activity of chitosan enriched with lemongrass oil against anthracnose of bell pepper. *Food Packag. Shelf Life*, 3, 56–61, 2015.

135. Zahoorullah, S.M., Dakshayani, L., Rani, A.S., Venkateswerlu, G., Effect of Chitosan Coating on the Post-Harvest Quality of Banana during Storage. *Asian J. Biotechnol. Bioresour. Technol.*, 1, 1, 1–10, 2017.
136. Kumar, P., Sethi, S., Sharma, R.R., Srivastav, M., Varghese, E., Effect of chitosan coating on postharvest life and quality of plum during storage at low temperature. *Sci. Hortic.*, 226, 104–109, 2017.
137. Badawy, M.E.I., Rabea, E.I., El-Nouby, M.A.M., Ismail, R.I., Taktak, N.E.M., Strawberry Shelf Life, Composition, and Enzymes Activity in Response to Edible Chitosan Coatings. *Int. J. Fruit Sci.*, 17, 2, 117–136, 2016.
138. Bhanushree, L.S., Vasudeva, K.R., Suresha, G.J., Sadananda, G.K., Mohamad Tayeebulla, H., Influence of chitosan on postharvest behavior of papaya (*Carica papaya* L.) Fruits under different storage conditions. *J. Pharmacogn. Phytochem.*, 7, 2, 2010–2014, 2018.
139. Remya, S., Mohan, C.O., Bindu, J., Sivaraman, G.K., Venkateshwarlu, G., Ravishankar, C.N., Effect of chitosan-based active packaging film on the keeping quality of chilled stored barracuda fish. *J. Food Sci. Technol.*, 53, 1, 685–693, 2015.
140. Chang, W., Liu, F., Sharif, H.R., Huang, Z., Goff, H.D., Zhong, F., Preparation of chitosan films by neutralization for improving their preservation effects on chilled meat. *Food Hydrocoll.*, 90, 50–61, 2018.
141. Jongsri, P., Wangsomboondee, T., Rojsitthisak, P., Seraypheap, K., Effect of molecular weights of chitosan coating on postharvest quality and physicochemical characteristics of mango fruit. *LWT*, 73, 28–36, 2016.
142. Chantarasataporn, P., Tepkasikul, P., Kingcha, Y., Yoksan, R., Pichyangkura, R., Visessanguan, W., Chirachanchai, S., Water-based oligochitosan and nanowhisker chitosan as potential food preservatives for shelf life extension of minced pork. *Food Chem.*, 159, 463–470, 2014.
143. Tayel, A.A., Ibrahim, S.I., Al-Saman, M.A., Moussa, S.H., Production of fungal chitosan from date wastes and its application as a biopreservative for minced meat. *Int. J. Biol. Macromol.*, 69, 471–475, 2014.
144. Casariego, A.B.W.S., Souza, B.W.S., Cerqueira, M.A., Teixeira, J.A., Cruz, L., Díaz, R., Vicente, A.A., Chitosan/clay films' properties as affected by biopolymer and clay micro/nanoparticles' concentrations. *Food Hydrocoll.*, 23, 7, 1895–1902, 2009.
145. Mohammadi, A., Hashemi, M., Hosseini, S.M., Effect of chitosan molecular weight as micro and nanoparticles on antibacterial activity against some soft rot pathogenic bacteria. *LWT Food Sci. Technol.*, 71, 347–355, 2016.
146. Möller, H., Grelier, S., Pardon, P., Coma, V., Antimicrobial and physicochemical properties of chitosan–HPMC-based films. *J. Agric. Food. Chem.*, 52, 6585–6591, 2004.
147. Feyzioglu, G.C. and Tornuk, F., Development of chitosan nanoparticles loaded with summer savory essential oil for antimicrobial and antioxidant delivery applications. *LWT Food Sci. Technol.*, 70, 104–110, 2016.

148. Cwiek-Ludwicka, K. and Ludwicki, J.K., Nanomaterials in food contact materials; considerations for risk assessment. *Rocz. Panstw. Zakl. Hig*, 68, 321–329, 2017.
149. Chaudhary, P., Fatima, F., Kumar, A., Relevance of Nanomaterials in Food Packaging and its Advanced Future Prospects. *J. Inorg. Organomet. Polym. Mater.*, 30, 5180–5192, 2020.
150. Enescu, D., Cerqueira, M.A., Fucinos, P., Pastrana, L.M., Recent advances and challenges on applications of nanotechnology in food packaging: A literature review. *Food Chem. Toxicol.*, 134, 110814, 2019.
151. Bumbudsanpharoke, N. and Ko, S., Nano-Food Packaging: An Overview of Market, Migration Research, and Safety Regulations. *J. Food Sci.*, 80, R910–R923, 2015.
152. Huang, J.-Y., Li, X., Zhou, W., Safety assessment of nanocomposite for food packaging application. *Trends Food Sci. Technol.*, 45, 2, 187–199, 2015.
153. Commission Regulation (EC) No. 10/2011 on plastic materials and articles intended to come into contact with food. *Off. J. Eur. Union*, 12, 1–89, 2011.
154. Cushen, M., Kerry, J., Morris, M., Cruz-Romero, M., Cummins, E., Nanotechnologies in the food industry—recent developments, risks, and regulation. *Trends Food Sci. Technol.*, 24, 30–46, 2012.
155. Garduno-Balderas, L.G., Urrutia-Ortega, I.M., Medina-Reyes, E.I., Chirino, Y.I., Difficulties in establishing regulations for engineered nanomaterials and considerations for policymakers: Avoiding an imbalance between benefits and risks. *J. Appl. Toxicol.*, 35, 10, 1073–1085, 2015.

14

Significant of Nanochitosan in the Management of Biotic Stress

Willy Barinem Vidona[1], Charity Willy-Vidona[2*] and Charles Oluwaseun Adetunji[2]

[1]*Department of Anatomy, Edo State University Uzairue, Edo State, Nigeria*
[2]*Department of Microbiology, Edo State University Uzairue, Edo State, Nigeria*

Abstract

In order to maintain global food security through sustainable crop production, regulating biotic stress in agriculture is a major challenge that this study aims to address. Accordingly, nanochitosan holds great promise for improving farming methods and reducing the negative impacts of biotic stress on crop output and overall farming stability. The purpose of the preliminary study is to clarify the function of nanochitosan and how it interacts with intricate biological processes in plants. Because of its smaller size and greater surface area, nanochitosan is more effective and bioavailable than conventional chitosan. The study highlights how versatile nanochitosan is and how effective it is against different kinds of biotic stresses and crops in different agricultural settings. The study explores the properties of nanochitosan, highlighting the fact that it is a nanoparticle with a diameter ranging from 1 nm to 100 nm. The special qualities of nanoparticles, like their size-dependent traits, are examined, and different approaches to the manufacture of nanochitosan, like polyelectrolyte complex and ionotropic gelation, are talked about. The study focuses on the characteristics of chitosan, a naturally occurring polysaccharide that is the precursor to nanochitosan, such as its solubility and structural changes. The work highlights the significance of plants' innate defense mechanisms, such as inducible defenses and the activation of regulatory machinery, in managing biotic stress circumstances. Chitosan, particularly in the form of nanochitosan, shows promise as a biotic stress management tool because it boosts the plant's defenses, produces antioxidant qualities, and stops fungus growth. Finally, by tackling the issues raised by biotic stress, the study offers insightful

**Corresponding author*: vidonanem@gmail.com

Charles Oluwaseun Adetunji, Maulin P. Shah, Yerima Mohammed Bello, Daniel Hefft, Jay Singh, Shyam S. Pandey and Ravindra Pratap Singh (eds.) Nanochitosan Applications for Enhanced Crop Production and Food Security, (331–344) © 2025 Scrivener Publishing LLC

information about how nanochitosan might transform agriculture. Because of its many uses, nanochitosan is a viable option for ecologically friendly and sustainable agricultural methods that will increase global food security.

***Keywords*:** Nanochitosan, biotic stress, sustainable agriculture, plant defense mechanisms, antifungal properties, nanoparticle synthesis

14.1 Introduction

Maintaining food security for farmers throughout the world by means of sustainable crop production is still a top priority, and managing biotic stress sustainably in agriculture is one of the major obstacles standing in the way of this objective. More focus is being placed on developing novel and ecologically sustainable solutions as biotic stress persists in endangering crop output and the general stability of farming systems. Here, the nanoparticle known as nanochitosan shows great promise as a means of transforming farming methods and mitigating the detrimental effects of biotic stress. Comparing nanochitosan to regular chitosan reveals that the former is more effective and has improved bioavailability due to its larger surface area and smaller size [1].

The purpose of this preliminary investigation is to clarify the function of nanochitosan and how it interacts with the complex biological systems in plants. Through the activation of certain enzymes, intricate signaling pathways, and gene expression, this interaction fortifies plants' innate defensive mechanisms and eventually increases their resistance to biotic stresses. Defense gene activation and the ensuing synthesis of antimicrobial chemicals are important components of plant resistance to biotic stress [2].

Because recent research has provided important insights into the application of nanochitosan in a variety of agricultural contexts, it is imperative to understand that its role in alleviating biotic stress pertains to a variety of crops and biotic stressor types. El Hadrami *et al.* [1], for instance, demonstrated the antifungal properties of nanochitosan in the treatment of Botrytis cinerea, a common and dangerous fungal infection. Their research highlighted how effective it is as a workable replacement for chemical fungicides, reducing the environmental impact of agricultural practices.

14.2 Nanochitosan Compound

According to Allemann *et al.* [3], nanoparticles are compact colloidal substances that have a diameter of 1 nm to 100 nm also categorized as nanocapsules and nanospheres. In contrast to their bulk substitutes, the unique properties of nanoparticles are primarily attributed to their diminutive size, which is almost on the atomic scale [4]. Whichever size they are, these characteristics remain largely unchanged. Due to studies demonstrating that, when size decreases, the proportion of surface atoms to the total substance increases, it is known that characteristics of materials alter at the nanoscale [5]. Nanoparticles can have a manufactured or natural polymeric matrix. Because they are non-toxic, biodegradable, and biocompatible, natural polymers are typically selected. Given to their ability to endure unfavorable processing conditions, synthesized nanoparticles have become comparatively more significant [6]. Size reduction techniques are used to create nanoparticles like procedures that work from the top down, such as sonication, homogenization process, and milling, or from the bottom up, such as reactive precipitation and solvent displacement [7].

Chitosan is a naturally occurring polysaccharide that is mostly made up of a certain amount of N-acetylglucosamine units and alternating linked glucosamine units of β. Occasionally does chitosan exist in nature as a changed biological polymers produced by homogeneous deacetylation of chitin [8] particularly the water soluble kind seen in mollusk and crustacean shells, fungal cell walls, and insect coats [9]. As stated [23], nanochitosan is a nitrogenous, rigid, white, and inelastic polysaccharide. The morphology of chitosan is affected by the deacetylation procedure conducted in either a homogeneous or heterogeneous environment [10]. The solubility of chitosan is one of the many properties that are determined by this structural variation. As reported [11], chitosan produced through a heterogeneous process is not soluble in water, but it might turn soluble in an acidic environment. Three functional groups are present in chitosan: an amino group and hydroxyl groups at positions C2, C3, and C6, respectively [12].

The features of both chitosan and nanoparticles, including contact and interface implications, small size, and quantum size effects, are present in nanochitosan particles [13].

Another environmentally beneficial way to combat the pollution that the seafood processing industry causes is with chitosan. As a result of its antibacterial, biodegradable, and nontoxic qualities, chitosan has so many applications; hence, it is used in the food, biomedical, agricultural, paper, genetic engineering, water pollution control, water treatment systems, and photography industries, among other fields [14].

14.3 Methods of Preparation

Due to their elevated level of free energy, chitosan nanoparticles have a natural tendency to coalesce; this must be considered in the process employed to create the nanoparticles. The methods utilized to create this particles include ionotropic gelation, complicated coacervation/reverse micellar approach, emulsion and microemulsion techniques, and polyelectrolyte complex of self-assembling of hydrophobically modified chitosan, which are the ways used to produce chitosan nanoparticles [15].

Ionotropic gelation: This process, which was invented [16], electrostatic interaction between the amine group of chitosan and a negatively charged polyanion, like tripolyphosphate.
Polyelectrolyte complex: It is produced when plasmid DNA and cationic charged polymer self-assemble, which occurs when the hydrophilicity of the cationic polymer decreases as a result of charge neutralization with the DNA. Mechanical stirring at room temperature could lead to in the polymerization of chitosan after it has been dissolved in a solution of acetic acid when DNA solution is added to form spontaneously [17].
Microemulsion approach: Glutaraldehyde was added to the hexane/surfactant combination while it was continuously stirred at room temperature. N-hexane and acetic solution were used to dissolve a surfactant and chitosan [18].
Reverse micellar approach: The main advantage of this technology is that it produces micro-nanoparticles in a constrained range of sizes and does not require the use of harmful organic solvents or cross linking agents [19].
Solvent diffusion method for emulsification: El-Shabouri *et al.* [20] initially reported on this technique, after high-pressure homogenization and mechanical stirring. To create an emulsion, an organic phase was added to a chitosan solution that contained a stabilizing agent, notably poloxamer, after which the emulsion is greatly diluted with water. When an organic solvent diffuses into water, polymer precipitation takes place, creating nanoparticles.

14.4 Significance of Nanochitosan

Considering its combination of antimicrobial and eliciting properties, chitosan is an adaptable compound with broad-spectrum antibacterial features that are effective against fungi, bacteria, and viruses. It is also used to impede the transmission of diseases and to equally promote plant defense mechanisms [21]. Additionally, chitosan stimulates the production of secondary metabolites, which fortifies plants' defensive mechanisms against infection [22]. Chitosan can be used to inhibit a wide range of microbes, including Erwinia carotovora [23], Xanthomonas, Pseudomonas syringae [24], and Agrobacterium tumefaciens. These species can be broadly classified as Gram-positive and Gram-negative bacteria, as well as sensitive and resistant fungus [25]. However, it seems that chitosan works better against fungi than it does against bacteria.

By triggering a hypersensitive reaction that reduces the viability of phage bacterial cells, neutralizes the infectious potential of phage particles, and restricts the growth of virulent phages, chitosan also prevents viral infections in plants. With its antibacterial and antifungal qualities, the molecular size, concentration, and rate of deacetylation of chitosan all affect its antiviral activity.

At several stages of pathogen development, such as the sporulation process, mycelial development, sprouting, and spore viability, chitosan nanoparticles have been shown to exhibit a broad-spectrum fungal effect *in vitro* in a variety of species, including Rhizopus stolonifera, Colleotrichum gleosporoide, and Botrytis cinerea. Fungal actions have been observed in a number of plant pathogen systems, such as the pear shown to inhibit Physalophora piricola and Alternaria kikuchiana [26], the grapevine and strawberry shown to inhibit Botrytis cinerea [27, 28], and the dragon fruit shown to inhibit Colletotrichum gleosporoides [29]. Additionally, certain nanostructures are thought to be suitable substitutes for managing phytopathogenic fungi because of their potent antifungal qualities. In particular, metal nanoparticles have been thoroughly investigated, tested, and yielded significant results due to their exceptional antifungal properties. Plants have been used in extensive studies to examine the various characteristics of chitosan-based nanoparticles, such as surface coating, antibacterial activity, and inhibition of reactive oxygen species.

While chitosan therapy is highly responsive in beans and peas, nanochitosan has also been observed to have antiviral activity in Fabaceae. Some endemic illnesses that are resistant to chitosan in cabbage are the radish and cauliflower mosaic viruses [30].

Given that they are known to have a significant impact on the reactivity and toxicity of different plants depending on their kind, formulations in various concentrations, and seed priming techniques, chitosan preparations are employed selectively by combining them with appropriate amounts for different plants.

It has been extremely significant to utilize environmentally friendly nanofertilizer formulae as an alternative to the expensive synthetic fertilizers that cause pollution, endanger human health, and worsen environmental problems. Consequently, nano-based smart pesticide applications are the most welcome innovations in today's highly specialized farming. These carriers effectively transfer a range of physiologically active nutrients, including vitamins, plant growth regulators, and micronutrients, at low concentrations [31]. Tests using a range of inorganic mediators, such as nanotubes, metallic nanoparticles, nanoemulsions, and nanobeads, have shown the efficacy of biosynthesized chitosan and nanochitosan particles in effective plant absorption, nutrient delivery, and controlled release mechanisms. Because nanochitosan are naturally polycatalytic, they have the special capacity to deliver NPK nanofertilizers to the plants in a regulated manner [32].

14.5 Biotic Stress Mechanism

Damage caused to plants by other living things, including insects, pathogens, and herbivores, is known as biotic stress. As a result of a shortage of nutrients brought on by biotic stress, plants' growth becomes stunted and eventually die. Biotic stress may additionally be the cause of after-harvest damage to fruits and cereals. The crop's environment's climate plays a major role in causing this stress. To combat this, the plant has an inherent defense system that it can activate at the first signal of hazard. The plant's genetic makeup incorporates this protection mechanism.

A variety of diseases, parasites, and pests can cause pathogenic infestation plants and leading to biotic stress. Fungal parasites can be classified as biotrophic that feeds on living host cells or necrotrophic that kills host cell by toxin release. Doughari *et al.* and Sobiczewski *et al.* [33, 34] postulate that they have the ability to cause vascular wilts and leaf spots in plants. Nematode infestation is mostly responsible for diseases transmitted by soil that results in lack of nutrients, poor development, and wilting because they feed on plant parts [35–37]. Likewise, viruses can cause harm both locally and systemically, which can lead to chlorosis [38].

Conversely, insects damage plants by either depositing eggs on them or by direct feeding on them and may additionally serve as vectors of several germs [39]. To withstand these kinds of concern, plants have evolved a complex immune system [40, 41] as an initial line of defense to ward off infections through development of structures such as trichomes, wax, and cuticles. Plants can also produce chemical defenses to fend off invading infections. Furthermore, plants use two layers of pathogen detection to initiate resistance against biological invaders [42].

Plant diseases are caused by fungi, which are plant pathogens; hence, these fungal infections are the cause of the majority of vegetation diseased conditions. By destroying its healthy cells, it stresses plants as the fungi spread to other plants via contaminated seeds, soil, crop waste, other crops, or weeds. In addition to farming implements and machinery, fungi can also be carried by wind or water. Due to the buildup of toxins within the plant, fungi have a detrimental effect on plant health. It reduces overall agricultural productivity and, on occasion, even kills plants. Fungal infections include rusts, downy mildew, Fusarium rots, and white blisters. Plant pathogenic viruses can seriously harm crops by infecting them through lesions on the plants and it is challenging to get rid of them without damaging the host plant. Therefore, it is safer to prevent viral infections than allow it cause great harm. Plants can contract viruses or its infestattion from one another through farming practices such as grafting, trimming, or propagation of afflicted plants, or from other organisms. Plant infections are caused by pathogenic microorganisms in many ways of which rain, wind, birds, and even insects can serve as agents to disperse these germs. Fungal and bacterial diseases are comparable and thus common bacterial plant diseases include wilts, leaf spots, blights, cankers, and soft rots in fruits and roots.

14.6 Management of Biotic Stress Conditions

When a plant detects biotic stress factors, a series of molecular and cell-level events are initiated [43, 44], which eventually sets off an interplay of structural, functional, and biochemical processes [40, 45]. In order to combat biotic attack and maintain growth, better yield, and ultimate survival, plants have evolved advanced sensory mechanisms [46]. They have also developed defensive systems to ward off attacks from a wide range of pathogenic microbes including viruses, fungi, nematodes, and plant-eating insects [47].

In order to manage biotic stress, defensive mechanisms that are inducible must be activated or exerted at where poisonous secondary metabolic products accumulate. This happens through the activation of regulatory machinery, which ultimately results in the interaction of several signals to produce a suitable response.

According to studies [48, 49], chitosan has demonstrated its effectiveness as an antioxidant by scavenging reactive oxygen species and chelating metal ions by supplying a hydrogen atom or one electron. Chitosan's amino and hydroxyl functional molecules bind to metal ions to cause a variety of reactions, including chelation, ion exchange, and adsorption [50]. Nanochitosan displayed antioxidant properties as demonstrated in a study [51]. In a separate study, it was noted that chitosan displayed the potentials to scavenging hydroxyl radicals and chelate ion, owing to its semicrystalline structure and robust hydrogen bonds, which prevent its dissociation from metal ions [52].

Chitosan has demonstrated antifungal efficacy against plant pathogens [53, 54] and has ability as a biocontrol agent [55]. It works as a potent bio-elicitor to tackle host-pathogen interactions, which offer a number of fascinating technical options that lead to PR protein-mediated systemic resistance-induced mutations in plants. This was supported by Chun *et al.* [56], who noted that size reduction raises the surface area–to–volume ratio indirectly and enhances the contact area for microbial penetration into host cells. According to evaluations, nanochitosan particles are a strong inducer of defense and antioxidant enzymes that entail a complicated response of gene set activation encoding various proteins. The activation of defense genes leads to the accumulation of various enzymes and metabolites, which includes the synthesis and buildup of pathogenesis-related proteins in response to invasive pathogen conditions. Research by Alscher *et al.* [57] demonstrates that nanochitosan particles enhance the expression of defense-related genes, thereby boosting the plant's defense mechanism. The accumulation of defense-related enzymes and stress-induced metabolites is essential for activating plant defense genes. Additionally, in stressful situations, it is vital for plants to produce and store pathogenesis-related proteins to defend against invasive pathogens.

The activation of defense genes leads to the accumulation of various enzymes and metabolites, which includes the synthesis and buildup of pathogenesis-related proteins in response to invasive pathogen conditions. In controlled experiments involving potted plants, the application of Cu-chitosan nanoparticles to the leaves of maize plants triggered the activation of plant defense mechanisms through enzymatic activity. Likewise, in challenging conditions, plants need to synthesize and store pathogenesis-related proteins in order to protect themselves from invasive pathogens. The superoxide dismutase (SOD) enzyme plays a crucial role in safeguarding plant cells against singlet oxygen radicals while also converting them into peroxide which is noted [58]. Saharan *et al.* [59] further demonstrated that the enzyme catalase exhibits a higher capacity for scavenging, as it breaks down peroxide into water and oxygen, thereby enhancing stress tolerance, thereby demonstrating a positive correlation between increased CAT expression and improved stress resistance.

Thus, the increased production of SOD and CAT as a result of treatment with CS or CNP may provide plants with the essential defense against oxidative stress brought on by pathogenic incursion. Increased SOD and CAT activity after NP treatments probably contribute significantly to the plant's ability to withstand oxidative stress during a pathogen attack by preserving equilibrium, halting degeneration, and scavenging ROS. Cu-chitosan nanoparticles were obserced to be the most effective of the different nanoformulations; at a 0.1% concentration, they demonstrated an amazing 89.5%, 63.1%, and 60.0% growth suppression of Alternaria alternata, Macrophomina phaseolina, and Rhizoctonia solani, respectively. These nanoparticles also demonstrated the highest spore germination inhibition rate of 87.4% for A. alternata at the same dose [60]. These nanoparticles, which were present at a concentration of 0.12%, showed significant inhibitory effects on mycelia growth and spore germination when examined in a controlled *in vitro* environment. In particular, they reduced the growth of mycelia by 70.5% and 73.5% and the germination of spores by 61.5% and 83.0% for A. solani and F. oxysporum, respectively. Furthermore, the application of Cu-chitosan nanoparticles at the same dose produced impressive disease control efficacy in pot trials using tomato plants. Pal and Saharan [61] demonstrated 87.7% efficacy in early blight and 61.1% efficacy in controlling Fusarium wilt. Additionally, Li *et al.* [62] investigated the application of Cu-chitosan nanoparticles to strengthen maize defensive mechanisms against Curvularia leaf spot disease.

Numerous studies have shown that chitosan oligosaccharides, which are produced from chitosan polymers, offer protection against biotic stressors [63–66]. This chitosan polymer has been shown [67] to inhibit pathogenic infiltration in plants by upregulating the expression of many genes involved in the plant's defense mechanisms. According to one theory, the poly-d-glucosamine units in chitosan attach to receptors that resemble those involved in pathogenic invasion. This causes the plant to activate feedback mechanisms, which include increased phytoalexin biosynthesis. Phytoalexins confer resistance against biotic stresses by serving as protective agents for the plant's innate immunity. Ganan *et al.* [68] investigated the potential of chitosan oligosaccharides as an antibiotic against a range of gram-positive (+ve) and gram-negative (−ve) bacteria, including E. coli, Aggregatibacter actinomycetemcomitans, Vibrio vulnificus, and P. aeruginosa, as well as S. aureus, M. luteus, B. subtilis, B. cereus, and Lactiplantibacillus plantarum. Respectively, it was also discovered that chitosan oligosaccharides had comparable cytotoxic effects on a number of significant fungi, such as many important fungi, including Saccharomyces cerevisiae, Rhodotorula glutinis, Rhodotorula mucilaginosa, B. cinerea, Candida albicans, and Candida krusei.

References

1. El Hadrami, A., Adam, L.R., El Hadrami, I., Daayf, F., Chitosan in plant protection. *Mar. Drugs*, 8, 4, 968–987, 2010.
2. Liu, D., Anderson, C.T., Williams, C.E., Antimicrobial properties of chitosan and mode of action: A state of the art review. *Int. J. Food Microbiol.*, 291, 245–252, 2019.
3. Allemann, E., Gurny, R., Doelker, E., Drug-loaded nanoparticles – preparation methods and drug targeting issues. *Eur. J. Pharm. Biopharm.*, 39, 173–191, 1993.
4. Ravishankar Rai, V. and Jamuna Bai, A., Nanoparticles and their potential application as antimicrobials, science against microbial pathogens: communicating current research and technological advances, in: *Formatex, Microbiology Series*, vol. 3, Méndez-Vilas, A. (Ed.), pp. 197–209, 2011.
5. Gupta, A.K., Naregalkar, R.R., Vaidya, V.D., Gupta, M., Recent advances on surface engineering of magnetic iron oxide nanoparticles and their biomedical applications. *Nanomed. (Lond)*, 2, 1, 23–29, 2007.
6. Whitesides, G.M., The 'right' size in nanobiotechnology. *Nat. Biotechnol.*, 21, 10, 1161–1165, 2003.
7. Vauthier, C.D., Chauvierre, C., Brigger, I., Couvreur, P., Drug delivery to resistant tumors: the potential of poly(alkyl cyanoacrylate) nanoparticles. *J. Control. Release*, 93, 151–160, 2003.
8. Kurita, K., Koyama, Y., Nishimura, S., Kamiya, M., Facile preparation of water soluble from chitosan. *Chem. Lett.*, 18, 9, 1597–1598, 1989.
9. Muzzarelli, R.A.A., *Chitin*, Pergamon Press, Oxford, 1977.
10. Roberts, G.A.F., *Chitin Chemistry*, Macmillan Press, London, 1992.
11. Sannan, T., Kurita, K., Iwakura, Y., Studies on Chitin, 1. Solubility of Chitin in Organic Solvents. *Makromol. Chem.*, 177, 3589–3600, 1976.
12. Divya, K. and Jisha, M.S., Chitosan nanoparticles preparation and applications. *Environ. Chem. Lett.*, 16, 1, 101–112, 2017.
13. Ingle, A., Gade, A., Pierrat, S., Sonnichsen, C., Rai, M., Mycosynthesis of silver nanoparticles using the fungus fusarium acuminatum and its activity against some human pathogenic bacteria. *Curr. Nanosci.*, 4, 2, 141–144, 2008.
14. Cheba, B.A., Chitin and chitosan: marine biopolymers with unique properties and versatile applications. *Glob. J. Biotechnol. Biochem.*, 6, 3, 149–153, 2011.

15. Peniche, H. and Peniche, C., *Chitosan nanoparticles: a contribution to nanomedicine*, vol. 60, pp. 883–889, Polymer International, Hoboken, New Jersey, 2011.
16. Calvo, P., Remunanae-Lopez, C., Vila-Jato, J.L., Alonso, M.J., Novel hydrophilic chitosan polyethylene oxide nanoparticles as protein carriers. *J. Appl. Polym. Sci.*, 63, 1, 125–132, 1997.
17. Erbacher, P., Zou, S., Bettinger, T., Stefan, A.M., Remy, J.S., Chitosan-based vector/DNA complexes for gene delivery: biophysical characteristics and transfection ability. *Pharm. Res.*, 15, 9, 1332–1339, 1998.
18. De, T.K., Ghosh, P.K., Maitra, A., Sahoo, S.K., Process for the preparation of highly monodispersed polymeric hydrophilic nanoparticles. *Google Patents*, 1999.
19. Brunel, F., Varon, L., David, L., Domard, A., Delair, T., A novel synthesis of chitosan nanoparticles in reverse emulsion. *Langmuir*, 24, 20, 11370–11377, 2008.
20. El-Shabouri, M.H., Positively charged nanoparticles for improving the oral bioavailability of cyclosporin-A. *Int. J. Pharm.*, 249, 1, 101–108, 2002.
21. Xing, K., Xing, Y., Liu, Y.F., Zhang, Y., Shen, X.Q., Li, X.Y., Fungicidal effect of chitosan via inducing membrane disturbance against Ceratocystis fimbriata. *Carbohydr. Polym.*, 192, 95–103, 2018.
22. Rkhaila, A., Chtouki, T., Erguig, H., el Haloui, N., Ounine, K., Chemical Properties of Biopolymers (Chitin/Chitosan) and Their Synergic Effects with Endophytic Bacillus Species: Unlimited Applications in Agriculture. *Molecules*, 26, 1117, 2021.
23. Badawy, M.E., II and Rabea, E., II, A biopolymer chitosan and its derivatives as promising antimicrobial agents against plant pathogens and their applications in crop protection. *Int. J. Carbohydr. Chem.*, 2011, 1–29, 2011.
24. Mansilla, A., Albertengo, L., Rodríguez, M., Debbaudt, A., Zúñiga, A., Casalongué, C., Evidence on antimicrobial properties and mode of action of a chitosan obtained from crustacean exoskeletons on Pseudomonas syringae pv. tomato DC3000. *Appl. Microbiol. Biotechnol.*, 97, 6957–6966, 2013.
25. Torres, J., Reyes, J., Castellanos, T., Angulo, C., Quiñones, E., Hernandez, L., A biopolymer with antimicrobial properties and plant resistance inducer against phytopathogens: Chitosan. *Not. Bot. Horti. Agrobot.-Cluj-Napoca*, 49, 12231, 2021.
26. Meng, D., Garba, B., Ren, Y., Yao, M., Xia, X., Li, M., Wang, Y., Antifungal activity of chitosan against Aspergillus ochraceus and its possible mechanisms of action. *Int. J. Biol. Macromol.*, 158, 1063–1070, 2020.
27. Reglinski, T., Elmer, P., Taylor, J., Wood, P.N., Hoyte, S., Inhibition of Botrytis cinerea growth and suppression of botrytis bunch rot in grapes using chitosan. *Plant Pathol.*, 59, 882–890, 2010.
28. Feliziani, E., Landi, L., Romanazzi, G., Preharvest treatments with chitosan and other alternatives to conventional fungicides to control postharvest decay of strawberry. *Carbohydr. Polym.*, 132, 111–117, 2015.

29. Zahid, N., Maqbool, M., Siddiqui, Y., Manickam, S., Ali, A., Regulation of inducible enzymes and suppression of anthracnose using submicron chitosan dispersions. *Sci. Hortic.*, 193, 381–388, 2015.
30. Qin, C., Li, H., Xiao, Q., Liu, Y., Zhu, J., Du, Y., Water-solubility of chitosan and its antimicrobial activity. *Carbohydr. Polym.*, 63, 367–374, 2006.
31. Santo Pereira, A.E., Silva, P.M., Oliveira, J.L., Oliveira, H.C., Fraceto, L.F., Chitosan nanoparticles as carrier systems for the plant growth hormone gibberellic acid. *Colloids Surf. B Biointerfaces*, 150, 141–152, 2017.
32. Sathiyabama, M. and Parthasarathy, R., Biological preparation of chitosan nanoparticles and its *in vitro* antifungal efficacy against some phytopathogenic fungi. *Carbohydr. Polym.*, 151, 321–325, 2017.
33. Doughari, J., An overview of plant immunity. *J. Plant Pathol. Microbiol.*, 6, 104172, 2015.
34. Sobiczewski, P., Iakimova, E., Mikiciński, A., Wêgrzynowicz-Lesiak, E., Dyki, B., Necrotrophic behaviour of Erwinia amylovora in apple and tobacco leaf tissue. *Plant Pathol.*, 66, 842–855, 2017.
35. Osman, H.A., Ameen, H.H., Mohamed, M., Elkelany, U.S., Efficacy of integrated microorganisms in controlling root-knot nematode Meloidogyne javanica infecting peanut plants under field conditions. *Bull. Natl. Res. Cent.*, 44, 1–10, 2020.
36. Bernard, G.C., Egnin, M., Bonsi, C., The impact of plant-parasitic nematodes on agriculture and methods of control, in: *Nematology-Concepts, Diagnosis and Control*, M.M. Shah and M. Mahamood (Eds.), IntechOpen, London, 2017.
37. Lambert, K. and Bekal, S., Introduction to plant-parasitic nematodes. *Plant Health Instruct.*, 10, 1094–1218, 2002.
38. Pallas, V. and García, J.A., How do plant viruses induce disease? Interactions and interference with host components. *J. Gen. Virol.*, 92, 2691–2705, 2011.
39. Schumann, G.L. and D'Arcy, C.J., *Essential Plant Pathology*, APS Press, London, 2006.
40. Saijo, Y. and Loo, E.P., II, Plant immunity in signal integration between biotic and abiotic stress responses. *New Phytol.*, 225, 87–104, 2020.
41. Taiz, L. and Zeiger, E., Secondary metabolites and plant defense. *Plant Physiol.*, 4, 315–344, 2006.
42. Dangl, J.L. and McDowell, J.M., Two modes of pathogen recognition by plants. *Proc. Natl. Acad. Sci. U.S.A.*, 103, 8575–8576, 2006.
43. Lamers, J., Van Der Meer, T., Testerink, C., How plants sense and respond to stressful environments. *Plant Physiol.*, 182, 1624–1635, 2020.
44. Rejeb, I.B., Pastor, V., Mauch-Mani, B., Plant responses to simultaneous biotic and abiotic stress: molecular mechanisms. *Plants*, 3, 458–475, 2014.
45. Nejat, N. and Mantri, N., Plant immune system: crosstalk between responses to biotic and abiotic stresses the missing link in understanding plant defence. *Curr. Issues Mol. Biol.*, 23, 1–16, 2017.

46. Rizhsky, L., Liang, H., Shuman, J., Shulaev, V., Davletova, S., Mittler, R., When defense pathways collide. The response of Arabidopsis to a combination of drought and heat stress. *Plant Physiol.*, 134, 1683–1696, 2004.
47. Hammond-Kosack, K. and Jones, J.D.G., Responses to plant pathogens. *Biochem. Mol. Biol. Plants*, 1, 1102–1156, 2000.
48. Rajalakshmi, A., Krithiga, N., Jayachitra, A., Antioxidant activity of the chitosan extracted from shrimp exoskeleton. *Middle East J. Sci. Res.*, 16, 10, 1446–1451, 2013.
49. Lin, S.B., Chen, S.H., Peng, K.C., Preparation of antibacterial chito-oligosaccharide by altering the degree of deacetylation of β-chitosan in a Trichoderma harzianum chitinase-hydrolysing process. *J. Sci. Food Agric.*, 89, 2, 238–244, 2009.
50. Onsosyen, E. and Skaugrud, O., Metal recovery using chitosan. *J. Chem. Technol. Biotechnol.*, 49, 4, 395–404, 1990.
51. Huang, Y.C. and Li, R.Y., Preparation and characterization of antioxidant nanoparticles composed of chitosan and fucoidan for antibiotics delivery. *Mar. Drugs*, 12, 8, 4379–439, 2014.
52. Xie, W., Xu, P., Liu, Q., Antioxidant activity of water-soluble chitosan derivatives. *Bioorg. Med. Chem. Lett.*, 11, 13, 1699–1701, 2001.
53. Kumaraswamy, R.V., Kumari, S., Choudhary, R.C., Pal, A., Raliya, R., Biswas, P., Saharan, V., Engineered chitosan-based nanomaterials: bioactivities, mechanisms and perspectives in plant protection and growth. *Int. J. Biol. Macromol.*, 113, 494–506, 2018.
54. Luan, F., Wei, L., Zhang, J., Mi, Y., Dong, F., Li, Q., Antioxidant activity and antifungal activity of chitosan derivatives with propane sulfonate groups. *Polymers*, 10, 395, 2018.
55. Soliman, M.H. and El Mohamedy, R.S.R., Induction of defense-related physiological and antioxidant enzyme response against powdery mildew disease in okra (Abelmoschus esculentus L.) plant by using chitosan and potassium salts. *Mycobiology*, 45, 409–420, 2017.
56. Chun, S.C. and Chandrasekaran, M., Chitosan and chitosan nanoparticles induced expression of pathogenesis-related proteins genes enhances biotic stress tolerance in tomato. *Int. J. Biol. Macromol.*, 125, 948–954, 2019.
57. Alscher, R.G., Erturk, N., Heath, L.S., Role of superoxide dismutases (SODs) in controlling oxidative stress in plants. *J. Exp. Bot.*, 53, 1331–1341, 2002.
58. Sharma, P., Jha, A.B., Dubey, R.S., Reactive oxygen species, M., oxidative damage, and antioxidative defense mechanism in plants under stressful conditions. *J. Bot.*, 2012, 1–26, 2012.
59. Saharan, V., *et al.*, Synthesis of chitosan based nanoparticles and their *in vitro* evaluation against phytopathogenic fungi. *Int. J. Biol. Macromol.*, 62, 677–683, 2013.
60. Saharan, V., *et al.*, Synthesis and *in-vitro* antifungal efficacy of Cu-chitosan nanoparticles against pathogenic fungi of tomato. *Int. J. Biol. Macromol.*, 75, 346–353, 2015.

61. Pal, A. and Saharan, V., Chitosan Based Nano-shield to Combat Biotic Stress in Plants. *Acta Sci. Agric.*, 2.2, 21–22, 2018.
62. Li, K., Xing, R., Liu, S., Li, P., Chitin and chitosan fragments responsible for plant elicitor and growth stimulator. *J. Agric. Food. Chem.*, 68, 12203–12211, 2020.
63. Lan, W., Wang, W., Yu, Z., Qin, Y., Luan, J., Li, X., Enhanced germination of barley (Hordeum vulgare L.) using chitooligosaccharide as an elicitor in seed priming to improve malt quality. *Biotechnol. Lett*, 38, 1935–1940, 2016.
64. Wang, M., Chen, Y., Zhang, R., Wang, W., Zhao, X., Du, Y., Effects of chitosan oligosaccharides on the yield components and production quality of different wheat cultivars (Triticum aestivum L.) in Northwest China. *Field Crops Res.*, 172, 11–20, 2015.
65. Mukarram, M., Naeem, M., Aftab, T., Khan, M.M.A., Chitin, chitosan, and chitooligosaccharides: recent advances and future perspectives, in: *Radiation-processed polysaccharides: Emerging Roles in Agriculture*, M. Naemm, T. Aftab, M.M.A. Khan, (Eds.), pp. 339–353, Academic Press, Elsevier Inc, Cambridge, Massachusetts, 2022.
66. Kim, S.K. and Rajapakse, N., Enzymatic production and biological activities of chitosan oligosaccharides (COS): A review. *Carbohydr. Polym.*, 62, 4, 357–368, 2005.
67. Benchamas, G., Huang, G., Huang, S., Huang, H., Preparation and biological activities of chitosan oligosaccharides. *Trends Food Sci. Technol.*, 107, 38–44, 2021.
68. Ganan, M., Lorentzen, S.B., Agger, J.W., Heyward, C.A., Bakke, O., Knutsen, S.H., *et al.*, Antifungal activity of well-defined chito-oligosaccharide preparations against medically relevant yeasts. *PLoS One*, 14, 1, e0210208, 2019.

15

Relevance of Nanochitosan in Food Sector and Food Packaging Sectors: Current Trends

John Tsado Mathew[1]*, Charles Oluwaseun Adetunji[2], Abel Inobeme[3], Abdulfatai Aideye Otori[4], Musah Monday[1], Yakubu Azeh[1], Elijah Yanda Shaba[5], Muhammad Aishetu Ibrahim[1], Amos Mamman[1] and Oluwafemi Adebayo Oyewole[6]

[1]*Department of Chemistry, Ibrahim Badamasi Babangida University Lapai, Niger State, Nigeria*
[2]*Applied Microbiology, Biotechnology and Nanotechnology Laboratory, Department of Microbiology, Edo University Uzairue, Auchi, Edo State, Nigeria*
[3]*Department of Chemistry, Edo University Uzairue, Auchi, Edo State, Nigeria*
[4]*Department of Chemical Engineering, Federal Polytechnic, Bida, Niger State, Nigeria*
[5]*Department of Chemistry, Federal University of Technology Minna, Nigeria*
[6]*Department of Microbiology, Federal University of Technology, Minna, Nigeria*

Abstract
Nanotechnology involves manipulating the size and shape of the materials at the nanometer scale in order to create, produce, and use structures. Additionally, nanoparticles have already been used in numerous aspects of human existence. Nanocomposites already have produced a number of advances that may find use in the food packaging industry. The shelf-life of many different types of food is anticipated to be significantly increased by the application of nanocomposite compositions. Less material is required to achieve the same or superior barrier qualities, which can result in lighter packages. These days, chitosan nanocomposites, which combine chitosan with inorganic or organic nanoparticles in the nanoscale,

**Corresponding author*: johntsadom@gmail.com; jmathew@ibbu.edu.ng

Charles Oluwaseun Adetunji, Maulin P. Shah, Yerima Mohammed Bello, Daniel Hefft, Jay Singh, Shyam S. Pandey and Ravindra Pratap Singh (eds.) Nanochitosan Applications for Enhanced Crop Production and Food Security, (345–360) © 2025 Scrivener Publishing LLC

are receiving a lot of attention. Following processing, food commodities are typically packaged and moved hygienically to protect and preserve them from any unpleasant changes, especially in the level of amount before it is provided to the consumer. Additionally, food packaging has been noted as one of the most recent advancements in material technology and science, which also represents a more significant aspect that may affect customer demand. A high level of food product with the following important qualities, such as an extended shelf life, natural quality, little processing, assured safety, and minimal processing, is typically requested by consumers today, according to observations. Therefore, there is a need to look for a cutting-edge and environmentally friendly technology that may greatly develop the field of food packaging, particularly through the science of nanomaterials and nanotechnology. Numerous aspects of food science, including processing, storage, packaging, functionality, and food safety, have undergone significant change as a result of breakthroughs in nanotechnology. Numerous nanomaterials, including inorganic metals, metal oxides, and their nanocomposites, along with nanoorganic material combined with bioactive compounds, have been used in the food business. This chapter aims to give comprehensive information on recent developments in the use of nanochitosan in the food packaging industry. Furthermore, detailed information on the application of nanochitosan in food packaging is also highlighted.

Keywords: Food packaging, nanochitosan, nanomaterials, nanocomposites

15.1 Introduction

Due to its remarkable biocompatibility and bioactivities, the natural polymer chitosan (CS), which is nontoxic and biodegradable, is widely used in the food and biomedical industries. Based on the polycationic nature of CS, superior gelling is possible when anionic components are present. CS nanoparticles (CSNPs) of different sizes (84–600 nm) have been produced using the ionic gelation method with the help of mechanical stirring or ultrasound, and their encapsulation efficiency ranges from 23% to 97%. The consistency during storage, prolonged release, and increased bioactivities like antioxidant and antibacterial activities compared to the free form were all improved by the CSNPs loaded with active chemicals. The main uses of various CSNPs as encapsulating agents, food additives, and edible coatings will also be covered in this review. Last but not least, the effectiveness of CSNPs for medicinal applications is discussed, including lowering glucose or fat levels, promoting wound healing, and treating cancer [33].

Over the past couple of years, there has been a strong push to reduce packaging waste by substituting eco-friendly materials derived from renewable sources for plastic food packaging. The most prevalent biopolymer is chitosan, which is composed of 2-amino-2-deoxy-D-glucose (glucosamine) units connected through 1,4-linkage. Because of its outstanding qualities such as excellent high biodegradability, low toxicity, and antibacterial activity, its application has been boosted in food sectors. Its protecting wall could prevent ripening and minimize water loss as well as extend the shelf life of food products. Based on its functions, such as color stabilization, emulsification, antioxidant activity, and dietary fiber-like property assisting water retention and fat entrapment, its role in foods can be seen in broad categories, providing health benefits. By fusing nanoparticles, organic substances as well as metal oxides, like lauric acid, acetic acid, propionic acid, and cinnamaldehyde into the chitosan matrix, the advancement of nanostructured materials has improved the beneficial properties of chitosan and led to the development of chitosan-based components, which have impressive uses in the food packaging industries. This chapter covers background material on nanochitosan and its structure, several ways for producing nanochitosan nanoparticles, their characterization and use, chitosan-based bio-nanocomposites, and their use in food packaging [2, 3, 7].

The creation of chitosan micro- and nanoparticles and its derivatives cleared the way for the more efficient and cost-effective use of these biomolecules, as well as for the expansion of their uses beyond what was first anticipated. Chitosan nanoparticles, like all other forms of nanoparticles made from other materials, have unique chemical, physical, as well as morphological properties that ultimately decide their applicability. The bioactivities but also behavioral traits of chitosan nanoparticles in various applications and systems are largely attributable to the techniques used to prepare them. The main methods of preparation of chitosan nanoparticles are emulsion-droplet coalescence, emulsion cross-linking, ionotropic gelation, coacervation/precipitation, template polymerization, molecular self-assembly, as well as reverse micelles (Figure 15.1). Regarding the characteristics of the nanoparticles, each of these processes has advantages as well as disadvantages. Moreover, careful chitosan nanoparticle preparation might offer a greater affinity for negatively charged biological membranes and site-specific targeting *in vivo*, enabling their use as incorporating materials for enzymes, DNA, as well as drugs; their use in controlled release systems; and their use as coatings of wound dressings to speed healing.

Chitosan nanoparticle–based films are utilized by the food industry to prevent food from becoming contaminated with microorganisms, extend shelf life, and improve mechanical stability of food packing materials.

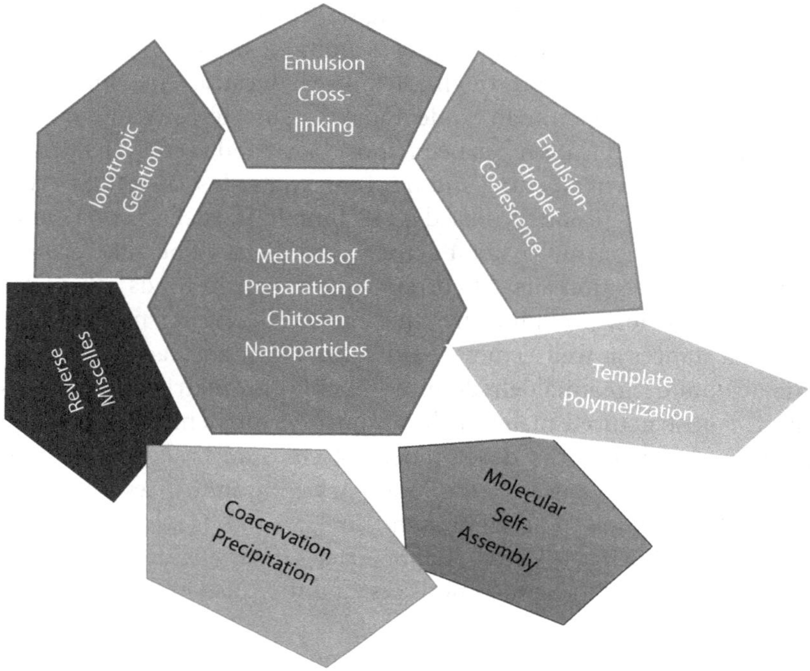

Figure 15.1 The main methods of preparation of chitosan nanoparticles.

Despite the fact that chitosan nanoparticles seem to be safe in some applications, further research is needed to better understand the hazards associated with their uses in food and medicine [12, 27].

Assuring the safety of packaged food items as they are delivered to customers or other end users has long been a top priority for the food business. Over the years, there have been a number of epidemics linked to the intake of spoiled food or infections found in food. The food industry has received a major boost today in attempting to address not only food toxicity and, moreover, increasing the quality, shelf-life, and acceptability of packaged foods thanks to the application of nanobiomaterials for food packaging that can observe the toxicants, freshness, as well as microbial contamination rate. In recent years, the food industry has made significant investments in the use of packaging materials that are biodegradable and biocompatible with sensors to replace the original non-degradable packaging materials, such as plastics comprising polyvinylchloride and polyethene, which have previously been proven to have toxic effects. The detection of heavy metals in food, nutritional levels, alcohol presence, toxins, and microorganisms in food are now all possible with biological nanoparticle-based sensors.

This device is affordable, is very sensitive, and offers a convenient on-site or online assessment of the various packaged goods' safety [4, 5].

15.2 Chitosan Nanoparticles and Food Packaging

Chitosan nanoparticles (CNPs), which are mostly formed from seafood by-products, are characterized as biocompatible, nontoxic, and functionalized nanostructures and are biodegradable. In the reinforcement of several biodegradable composites for biomedical applications and food packaging, as green fillers, CNPs demonstrated the necessary capacity. The differences in the mechanical, physical, structural, thermal, morphological, as well as antibacterial properties of nanobiocomposites as a consequence of the addition of CNPs are described after evaluating various production methods and CNP characterization methodologies. Additionally discussed is the impact of hybrid CNPs and bioactive-loaded CNPs with metal nanoparticles, montmorillonite, and graphene in nanocomposites. Additionally, the safety features of CNPs-loaded materials are emphasized to assess their application in biomedical and food packaging systems. In conclusion, CNPs are promising nanomaterials to enhance different structural, operational, and antibacterial properties of biocomposites for numerous applications in food packaging, biomedical uses, and delivery systems, despite a few limitations [16].

In the field of food packaging, polymer nanocomposite has received more focus than other contemporary nanotechnologies. It has better packaging qualities like barrier qualities, moisture permeability, solvent resistance, oxygen permeability, anti-microbial, mechanical toughness, as well as thermal stability. Biopolymer-based packaging films have now been promoted as alternatives to the commonly employed plastic packaging material made from petroleum. Traditional polymeric materials made of petroleum could cause serious environmental and health problems if they are discarded. A different type of packaging material is required. Given that they are biocompatible, biodegradable, non-toxic, affordable, renewable, as well as edible, polysaccharide-based biopolymers are receiving a lot of interest in the packaging business [14].

Nano-sized materials change their magnetic, optical, electrical, and some other properties and, for this purpose, are extensively used in food packaging. Nanoparticles (NPs), nanoclays (NCs), nanocomposites (NCs), nanosensors (NSs), nanostructures (NSTs), and nanoemulsions (NEs) are some of the significant nanomaterials that have been employed in food preservation and packaging. Through active and intelligent packaging, edible coatings, and the creation of several versatile nanosystems, nanomaterials can provide answers for food packaging and preservation. As a result, nanoparticles can be seen as crucial instruments and effective choices for regulating, limiting, and enhancing food quality and safety parameters, all of which are widely desired in the field of food technology. Novel nanoparticles even enable real-time food quality monitoring, offering a useful choice for applications involving food preservation. The migratory phenomenon and the emergence of harmful consequences on the human body exposed to nanomaterials in food packaging, especially, edible nanopackaging, offer major toxicological risks [9].

Since the last decade, there has been a concerted push to reduce packaging waste by substituting eco-friendly materials derived from renewable sources for plastic food packaging. The most prevalent biopolymer is chitosan, which is composed of 2-amino-2-deoxy-D-glucose (glucosamine) units connected through 1,4-linkage. Its application has increased in the food industry as a result of its outstanding qualities, which include strong antibacterial activity, high biodegradability, and low toxicity. Its protective wall can delay ripening, stop water loss, and lengthen the shelf life of food products. Depending on its characteristics, including emulsification, color stabilization, dietary fiber-like property assisting water retention, as well as fat entrapment and antioxidant activity, its role in foods can be seen in broad categories, providing health benefits. By fusing nanoparticles, organic substances, and metal oxides such as lauric acid, acetic acid, propionic acid, and cinnamaldehyde into the chitosan matrix, the invention of nanotechnology has improved the beneficial applications of chitosan and led to the development of chitosan-based materials, which have remarkable uses in the food packaging industries. This chapter covers background material on chitosan and its structure, several ways for producing chitosan nanoparticles, their characterization and use, chitosan-based bio-nanocomposites, and their use in food packaging [17].

Due to its excellent antioxidant, antibacterial, and enzyme inhibitory action as well as its degradation rate, chitosan is a multifunctional biopolymer that is frequently employed in the medical and food industries. As a novel food preservation substance, chitosan's biological activity has steadily gained scientific attention. In order to examine the possible applications potential of active chitosan-based food packaging, this paper covers recent research on the bioactive mechanism of chitosan and provides ways for changing and using chitosan for preserving food as well as various preservation methodologies. In order to give a theoretical foundation and a scientific point of reference for future research, concerns and perspectives on the role of chitosan in improving the freshness of food products are provided [20].

Due to its adaptable qualities, chitosan and chitosan-derived nanoparticles and nanocomposites are particularly promising and have found use in a variety of fields, including biomedicine, food packaging, personal care, and food. Due primarily to their biodegradability and anti-microbial qualities, several studies indicate that chitosan nanoparticles and their derivatives are among the best film coating and barrier materials for protecting the quality of foods and for providing skin care products. The current review provides information on the advancements made in the use of chitosan nanoparticles, particularly in food packaging and food, personal care items, and the biomedical industry. Future research will focus on the industrialization of chitosan nanoparticles, as well as the safety and quality of these chitosan-derived products for human use [28, 36].

Chitosan and other biologically active molecules have a lot of potential in the food industry, especially given the need to preserve food products and the growing concern over conventional packaging materials' harmful environmental effects. Chitosan is a biopolymer that is produced from the leftovers of industrial fishing and is derived from chitin. Due to its unique chemical structure, which consists of a linear polycation chain with a reactive hydroxyl, a high charge density, amino groups, and significant hydrogen bonding, this substance offers a wide range of applications. These unique physical and chemical characteristics, as well as their availability and fast biocompatibility with biodegradability, human tissues, and antibacterial and antifungal activity, have great potential for application in food safety and preservation. Chitosan nanoparticle synthesis using eco-friendly technology and its application as a novel food additive are shown and addressed. These chitosan nanoparticles were examined for their potential as an inventive and all-natural antibacterial agent while taking into account variations in the polymer's molecular weight and degree of deacetylation [35].

15.3 Potential Application of Nanochitosan in Food Packaging

One important area of sophisticated technology that the food, pharmaceutical, and agricultural industries may contribute to is nanotechnology, which they can also exploit to their advantage. Nanomaterials may play a both a quantitative and a qualitative leadership role in the production of high-quality, healthier, safer, and semi-perishable functional foods. Compared to conventional food processing techniques, nanotechnologies yield food that is of superior quality and has a longer shelf life. They also stay clear of infections. The current state and future uses of cutting-edge nanomaterials in the food sector are covered in this in-depth review of nanotechnologies for the creation of functional foods, taking into account processing, security, storage, and packaging. Through modifications to the particle size, surface charge, and possible cluster formation of food nanomaterials, nanotechnology applications enhance the meal's flavor, bioavailability, consistency as well as texture. Additionally, the discussion includes the synergistic action of nanomaterials in food protection, nanodelivery-mediated nutraceuticals, the use of nanosensors in smart food packaging to evaluate the quality of stored foods, as well as the typical approaches used to determine the effects of nanomaterials in biological systems (Figure 15.2) [24].

Before reaching the final customer, food products are packaged and transported in a hygienic manner to prevent and maintain them from whatever undesirable changes in quality. Food packaging is always changing to keep up with advances in technology, material science, and customer demand [18]. Food with natural quality, ensured safety, little processing, prolonged shelf life, and a ready-to-eat concept is currently in demand from modern customers in competitive economies. Innovative packaging

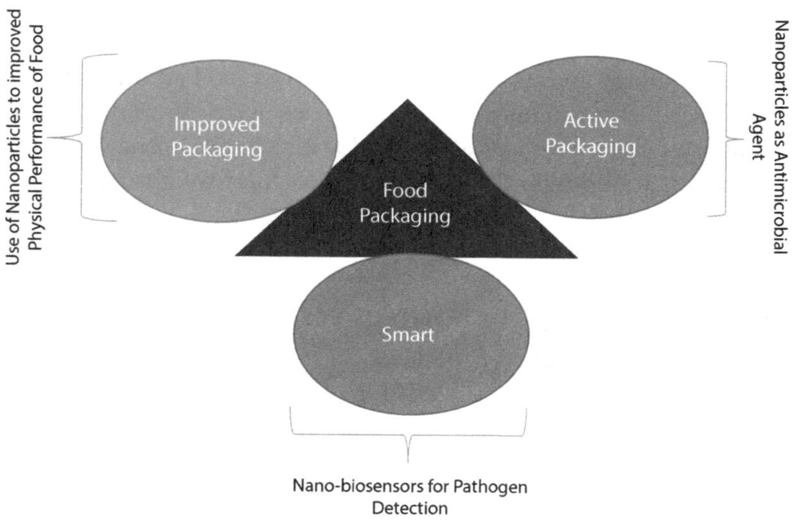

Figure 15.2 Nanotechnology for food packaging in food industry.

techniques ensure effective distribution and transportation preservation while also facilitating consumer communication. The study of nanomaterials, or nanotechnology, has been crucial to the technological developments in food packaging in the 21st century. Nanotechnology enables the creation and modification of materials at the nanoscale that have scientific and economic relevance. The application of nanotechnology in the food packaging business has mostly solved worries about food quality, safety, and stability. Furthermore, food product packaging using nanotechnology tells consumers about the true quality of the goods within [25]. Nanotechnology has also been investigated for the controlled release of antimicrobials and preservatives, extending the shelf life of the product inside the package. The encouraging reports for using nanotechnology in food packaging have made this a separate priority research subject. Consumers have responded favorably to food packaging composed of nanoparticles because it provides superior barrier and mechanical properties as well as food preservation. On the other hand, recent improvements and advancements in this field have given rise to a number of ethical, environmental, and safety problems. The inclusion of nanoparticles in food packaging is subject to policies and regulations that are being examined [22, 32].

This review focuses on improvements in various nanomaterials used for food packaging, with special attention paid to the negative effects of plastic packaging on humans and the environment. Nanomaterials are also used to create biocompatible packaging that offers an alternative to the majority of commonly used non-degradable polymer materials. The worldwide community would benefit greatly from the commercialization of such environmentally friendly food packaging. Additionally, the nation would find it simpler to combat food processing, quick access, and storage waste, lowering the matrix of poverty index. The adequate antibacterial activity shelf life and lowering the risk of microplastic ingestion speak for themselves. Additionally, the development of such technology will open up the possibility of using diverse nanomaterials as food packaging, making it both environmentally beneficial and pleasing to consumers [31].

By delivering protection and extending shelf life, food packaging is essential in providing consumers with high-quality, safe food products. Food packaging is a broad field based on engineering, food science, chemistry, and microbiology. These fields have all made significant contributions to maintaining the physicochemical characteristics of food and its raw materials, including flavor, color, texture as well as moisture content, in addition to ensuring freedom from microbial deterioration and oxidation. In addition to serving as standard food packaging, antimicrobial food packaging technologies are made to stop microbial development on food surfaces, improving food quality and stability [1]. In order to increase the shelf life of packaged food products, nanomaterials with special antibacterial and physiochemical properties are extensively researched for use in food packaging as preservatives and antimicrobials. Food packaging uses a variety of nanomaterials, such as alginate, gelatin, chitosan-based polymeric nanoparticles, cellulose, NE, lipid nanoparticles, nanosponges, nanofibers, and nanoliposomes, as well as nanocomposites containing nanoparticles such as copper, silver, titanium dioxide, gold, zinc oxide, and magnesium oxide (Figure 15.3). Systems for encapsulating antimicrobial nanomaterials are created to be more effective against microorganisms. Various research teams have recently looked into smart food packaging systems that can detect rotting and pathogenic microbes. The current paper covers recent developments in the field of food packaging technology related to various nanomaterials, with prospective uses as antioxidant and antimicrobial materials integrated with technology granting intelligent functions and processes in food packaging [34].

Environmental problems arise because most materials used to package food these days are not biodegradable. Environmentally friendly food packaging materials have been made using a range of biopolymers. Furthermore, biopolymer-based composites have only been employed in a few specific situations because of their frequently poor mechanical and barrier properties, which might be improved by adding reinforcing elements (fillers). In reinforced materials, poor matrix–filler interactions are

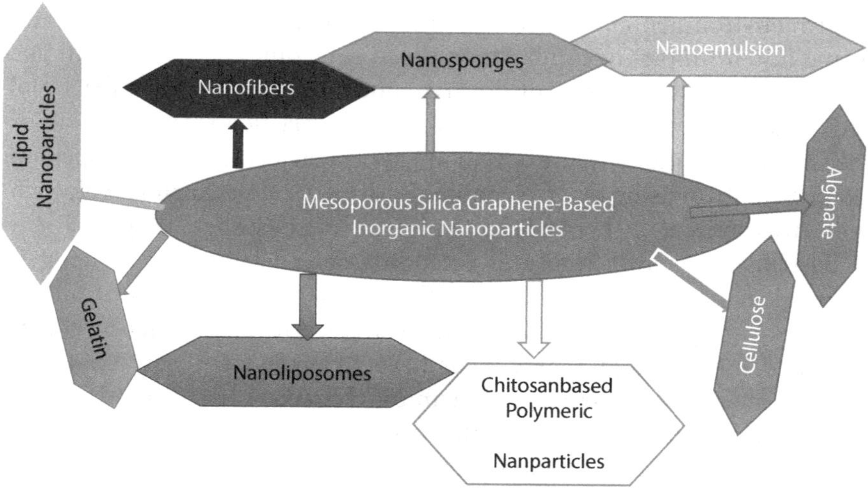

Figure 15.3 Mesoporous silica graphene–based inorganic nanoparticles.

typical; however, they generally improve with decreasing filler dimensions. Fillers with at least one nanoscale dimension, or nanoparticles, are utilized to create nanocomposites. Due to the proportionately greater surface area of nanoparticles compared to their microscale counterparts, interactions between the filler and matrix are encouraged, and the final material performs better. In addition to acting as nanoreinforcements, a polymer can use nanoparticles for biosensing, immobilizing enzymes, antibacterial action, and other purposes. Overview of the primary types of nanoparticles investigated for application in food packaging systems, along with their effects as well as applications [23].

Food quality and texture can be improved with the help of nanotechnology, which can also be utilized to create packaging that keeps food fresh. The major purpose of packing is to preserve a product's safety and quality throughout transportation and storage, as well as to increase its shelf life by fending off harmful agents such germs, chemical pollutants, air, moisture, and light [21]. The purpose of this paper is to highlight the successes of nanotechnology in the field of food packaging by providing an overview of the polymers that are frequently used during food packaging along with methods for enhancing their physical characteristics, such as thermal stability, mechanical strength, and gas barrier. It became evident from studying recent publication research that nanomaterials such nanopolymers are vying to take the place of traditional materials in food packaging. The existence of pollutants, microtoxins, and microbes in food can be demonstrated using nanosensors [13, 29].

Interest in ecologically friendly packaging materials has grown as a result of growing awareness of the issues plastic packaging poses to the environment. One of the numerous polysaccharides, chitosan, has been extensively studied and used to make food packaging that degrades naturally. To add fresh capabilities and improve the performance of packaging materials, many researchers have combined chitosan with other biomaterials using various techniques. This chapter offers a brief overview of recent developments in this developing field and focuses on the creation of chitosan-based bionanocomposite materials for food packaging [8, 30].

In the food packaging sector, there is a lot of interest in the use of nanotechnology to create nanoparticles. It pledges to produce food packaging with enhanced features that help food products last longer on the shelf. The most prevalent nanoparticles used in food packaging are discussed in this overview, along with the considerable changes that they make to the properties of packaging materials and the range of commercially accessible nano-based packaging options. Nanoparticles are used in the development of active packaging, improved packaging, and intelligent packaging to ensure food quality and traceability across the supply chain. Because of their numerous qualities, including UV impermeability, oxygen scavenging ability, and antibacterial activity, nanoparticles are helpful in the creation of nanocomposites. Researching how nanoparticles migrate through and interact with the polymer matrix is essential when developing packaging materials, as high surface area–to–volume ratios might occasionally be the cause of nanoparticle toxicity [6, 10, 19, 26].

Because it contains reactive amino and hydroxyl functional groups in its structure, chitosan is an appealing natural biopolymer made from renewable resources. Chitosan can be used in a variety of industrial applications, including magnetic-field assisted drug delivery, enzyme or cell immobilization or enzyme, and many others because of its excellent biocompatibility. The preparation of immobilized enzyme carriers has attracted significant research attention in the last 10 years in the field of nanotechnology. This investigation examines the properties, uses, and chitosan nanoparticles and potential of chitosan as adequate support systems for enzyme immobilization. The results showed that chitosan nanoparticles performed better in immobilizing enzymes and retaining their activity than chitosan macro- and microparticles. When used to encapsulate enzymes, chitosan nanoparticles significantly increase the stability of the immobilized enzymes and make it simple for them to be removed from the reaction mixture at the conclusion of the biochemical process [11, 15].

15.4 Nanochitosan in Food Packaging Sector

Due to its excellent antibacterial, antioxidant, and enzyme inhibitory action as well as its degradability, chitosan is a multifunctional biopolymer that is frequently employed in the food and medical industries (Table 15.1). As a novel food preservation substance, chitosan's biological activity has steadily gained scientific attention. In order to assess the potential application value of active chitosan-based food packaging, this paper summarizes recent research on the bioactive mechanism of chitosan and presents techniques for modifying and using chitosan for food preservation as well as alternative preservation methodologies. Concerns and viewpoints on the function of chitosan in boosting the freshness of food products are offered in order to provide a theoretical framework and a scientific point of reference for further investigation [20].

Table 15.1 The use of nanochitosan in the food packaging sector.

Application	Benefits	Challenges	Examples of use	References
Barrier properties	Improved gas and moisture barrier	Cost of production	Coating for paper-based packaging to enhance shelf life	[15]
Antimicrobial properties	Inhibits microbial growth and extends food freshness	Stability of antimicrobial activity	Incorporation in films for meat and seafood packaging	[6]
Biodegradability	Environmentally friendly and reduces plastic waste	Standardization of biodegradation rates	Composite films for single-use packaging	[3]
Mechanical strength	Enhanced strength and flexibility	Uniform dispersion in packaging materials	Reinforcement in plastic packaging for fragile products	[10]

15.5 Conclusions and Future Perspectives

The manufacture of high-quality food requires sophisticated packaging, processing, and long-term storage techniques. Nanotechnology has a significant impact on the food business by order to improve food quality by enhancing texture and flavor. Nanochitosan has been widely employed in food packaging during the past few decades. In the food industry, packaging is crucial for protecting the product. Food packaging serves a variety of purposes and is linked to food waste. While the use of nanoparticles greatly enhances the properties of nanochitosan packaging, things are not quite as they seem. When different types of nanoparticles find their way into food and are consumed over time, it can be harmful to a person's health. However, the way that food is consumed in the future indicates that nanotechnology is the most advanced and effective way to package food.

References

1. Abimbola, O.F., Adetunji, C.O., Dare, B.J., Mathew, J.T., Inobeme, A., Ghazanfar, S., Olaniyan, O.T., Ajiboye, M.D., Dauda, W., Packaging of Animal-Based Food Products: The Need for Intelligent Freshness Sensors, in: *Sensing and Artificial Intelligence Solutions for Food Manufacturing*, pp. 89–102, CRC Press, Boca Raton, Florida, USA, 2023.
2. Adetunji, C.O., Anani, O.A., Olaniyan, O.T., Inobeme, A., Bodunrinde, R.E., Mathew, J.T., Adetunji, J.B., Hefft, D., II, Prospect of Utilizing Novel Nanomaterials for Effective Food Security, Safety, Packaging, and Preservation, in: *The Science of Nanomaterials: Basics and Applications*, Apple Academic Press, Inc. Co-published with CRC Press (Taylor & Francis), Oakville, Ontario, Canada, 2022.
3. Adetunji, C.O., Ogundolie, F.A., Ajiboye, M.D., Mathew, J.T., Inobeme, A., Olotu, T., Olaniyan, O., Ijabadeniyi, O.A., Ajayi, O.O., Dauda, W.P., *et al.*, Bio- and Nanosensors in the Food Industry, in: *Bio- and Nano-sensing Technologies for Food Processing and Packaging*, pp. 22–36, The Royal Society of Chemistry, London, United Kingdom, 2022.
4. Adetunji, C.O., Ogundolie, F.A., Olaniyan, O., Mathew, J.T., Inobeme, A., Olotu, T., Ghazanfar, S., Ijabadeniyi, O.A., Ajiboye, M.D., Ajayi, O.O., *et al.*, Nanobiomaterials for Food Packaging Sensor Applications, in: *Bio- and Nano-sensing Technologies for Food Processing and Packaging*, pp. 167–180, The Royal Society of Chemistry, London, United Kingdom, 2022.
5. Adetunji, C.O., Ogundolie, F.A., Ajiboye, M.D., Mathew, J.T., Inobeme, A., Dauda, W.P., Ghazanfar, S., Olotu, T., Olaniyan, O., Ijabadeniyi, O.A., *et al.*, Nano-engineered Sensors for Food Processing, in: *Bio- and Nano-sensing Technologies for Food Processing and Packaging*, pp. 151–166, The Royal Society of Chemistry, London, United Kingdom, 2022.
6. Adetunji, C.O., Inobeme, A., Olaniyan, O.T., Anani, O.A., Mathew, J.T., Bodunrinde, R.E., Adetunji, J.B., Hefft, D., II, Nanomaterials as Effective Tools for Detection and Microbial Diagnosis of Foodborne Pathogens, in: *The Science of Nanomaterials: Basics and Applications*, Apple Academic Press, Inc. Co-published with CRC Press (Taylor & Francis), Oakville, Ontario, Canada, 2022.
7. Ahmed, S., Bio-based Materials for Food Packaging (Green and Sustainable Advanced Packaging Materials), in: *Chitosan-Based Nanocomposites in Food Packaging*, pp. 269–285, 2018.
8. Agarwal, A., Shaida, B., Rastogi, M., Singh, S.B., Food Packaging Materials with Special Reference to Biopolymers-Properties and Applications. *Chem. Afr.*, 23, 1–28, 2022.
9. Agriopoulou, S., Stamatelopoulou, E., Skiada, V., Tsarouhas, P., Varzakas, T., Emerging Nanomaterial Applications for Food Packaging and Preservation: Safety Issues and Risk Assessment. *The 1st International Electronic Conference on Food Science and Functional Foods*, vol. 707, 2020.

10. Alweera, A., Nazia, K., Samra, F., Anjum, Z., Kaiser, Y., Application of nanotechnology in food packaging: Pros and Cons. *J. Agric. Food Res.*, 7, 100270, 2020.
11. Aranaz, I., Alcántara, A.R., Civera, M.C., Arias, C., Elorza, B., Heras Caballero, A., Acosta, N., Chitosan: An Overview of Its Properties and Applications. *Polymers*, *13*, 19, 3256, 2021.
12. Bashir, S.M., Ahmed, R.G., Patrício, A., Haq, Z., Sheikh, A.A., Shah, M.Z.H., Singh, H., Khan, A.A., Imtiyaz, S., Ahmad, S.B., Chitosan Nanoparticles: A Versatile Platform for Biomedical Applications. *Mat.*, 15, 6521, 2022.
13. Bratovcic, A., Odobašić, A., Ćatić, S., Šestan, I., Application of polymer nanocomposite materials in food packaging. *Croatian J. Food Sci. Tech.*, 7, 86–94, 2015.
14. Ediyilyam, S., George, B., Shankar, S.S., Dennis, T.T., Wacławek, S., Černík, M., Padil, V.V.T., Chitosan/Gelatin/Silver Nanoparticles Composites Films for Biodegradable Food Packaging Applications. *Pol.*, 13, 1680, 2021.
15. Fatullayeva, S., Tagiyev, D., Zeynalov, N., Recent advances of chitosan-based polymers in biomedical applications and environmental protection. *J. Polym. Res.*, 29, 259, 2022.
16. Garavand, F., Cacciotti, I., Vahedikia, N., Salara, A.R., Tarhan, A., Akbari-Alavijeh, S., Shaddel, R., Rashidinejad, A., Seid, M., A comprehensive review on the nanocomposites loaded with chitosan nanoparticles for food packaging. *Crit. Rev. Food Sci. Nutr.*, 32, 1–34, 2020.
17. Hooda, R., Batra, B., Kalra, V., Singh, J.R., Sharma, M., Chitosan-Based Nanocomposites in Food Packaging: Green and Sustainable Advanced Packaging Materials, in: *Chitosan-Based Nanocomposites in Food Packaging*, pp. 269–285, 2018.
18. Inobeme, A., Adetunji, C.O., Mathew, J.T., Ajai, A., II, Mann, A., Inobeme, J., Oyedolapo, B., Adekoya, M.A., Onyeaku, S., Diversity, Biodegradation and Bioremediation of Polycyclic Aromatic Hydrocarbons, in: *Microbial Technologies in Industrial Wastewater Treatment*, pp. 31–50, Springer Nature Singapore, Singapore, 2023.
19. Lamri, M., Bhattacharya, T., Boukid, F., Chentir, I., Dib, A.L., Das, D., Djenane, D., Gagaoua, M., Nanotechnology as a Processing and Packaging Tool to Improve Meat Quality and Safety. *Foods*, 10, 2633, 2021.
20. Liu, T., Li, J., Tang, Q., Qiu, P., Gou, D., Zhao, J., Chitosan-Based Materials: An Overview of Potential Applications in Food Packaging. *Foods*, 11, 1490, 2022.
21. Mathew, J.T., Dauda, B.E.N., Mann, A., Ndamitso, M.M., Etsuyankpa, M.B., Nasirudeen, M.B., Nutrient And Anti-Nutrient Assessment of Fermented and Unfermented Seed of *Haematostaphis barteri* from Niger State, Nigeria. *Int'l. J. Appl. Biol. Res.*, 9, 55–66, 2018.
22. Modi, B., Hari, T., Sobika, B., Ashma, A., Sangita, P., Prakash, D.K., Dhan, B.G.C., Niranjan, P., Current Trends of Food Analysis, Safety, and Packaging. *Int'l. J. Food Sci.*, 20, 1–15, 2021.
23. Ncube, L.K., Ude, A.U., Ogunmuyiwa, E.N., Zulkifli, R., Beas, I.N., Environmental Impact of Food Packaging Materials: A Review of

Contemporary Development from Conventional Plastics to Polylactic Acid Based Materials. *Mat.*, *13*, 4994, 2020.

24. Nile, S.H., Baskar, V., Selvaraj, D., Nile, A., Xiao, J., Kai, G., Nanotechnologies in Food Science: Applications, Recent Trends, and Future Perspectives. *Nano Micro Lett.*, 12, 45, 2020.
25. Ogundolie, F.A., Adetunji, C.O., Mathew, J.T., Inobeme, A., Ghazanfar, S., Olaniyan, O.T., Ajiboye, M.D., Dauda, W.P., Adetunji, J.B., Akinbo, O., Intelligent freshness indicators in packaging of plant-based food products, in: *Evaporative Coolers for the Postharvest Management of Fruits and Vegetables*, pp. 323–331, Academic Press, Cambridge, Massachusetts, USA, 2023.
26. Pereda, M., Marcovich, N.E., Ansorena, M.R., Nanotechnology in Food Packaging Applications: Barrier Materials, Antimicrobial Agents, Sensors, and Safety Assessment, in: *Handbook of Ecomaterials*, Springer, Cham, 2019.
27. Priyanka, T., Sharma, A., Nayak, V., Verma, R., Singh, J., Chapter 4 - Properties and types of chitosan-based nanomaterials, in: *Nanomaterial-Plant Interactions, Role of Chitosan and Chitosan-Based Nanomaterials in Plant Sciences*, S. Kumar and S.V. Madihally (Eds.), pp. 85–117, Academic Press, Cambridge, Massachusetts, USA, 2022.
28. Radhakrishnan, Y., Gopal, G., Lakshmanan, C.C., Nandakumar, K.S., Chitosan Nanoparticles for Generating Novel Systems for Better Applications: A Review. *J. Mol. Gen. Med.*, 005, 1–10, 2015.
29. Rahul, B., Mahabub, A., Animesh, S., Md Ismail, H., Md. Moinul, H., Mominul, H., Application of nanotechnology in food: processing, preservation, packaging and safety assessment. *Heliyon*, 8, e11795, 2022.
30. Rahul, R.G., Hema, G., Anupam, C., Wazed, A., Chitosan-based bionanocomposites for food packaging applications, Woodhead Publishing Series in Composites Science and Engineering, in: *Bionanocomposites for Food Packaging Applications*, pp. 181–200, Woodhead Publishing, Cambridge, United Kingdom, 2022.
31. Sangeetha, V., Supriya, V., Deepa, P.V., Nano Packaging and its Potential Application: A Review. *Biomed. J. Sci. Tech. Res.*, 41, 33061–33063, 2022.
32. Sharma, C., Dhiman, R., Rokana, N., Panwar, H., Nanotechnology: An Untapped Resource for Food Packaging. *Front. Microbiol.*, 8, 1735, 2017.
33. Singh, A., Mittal, A., Benjakul, S., Chitosan nanoparticles: preparation, food applications and health benefits. *Sci. Asian*, 47, 1–10, 2021.
34. Suvarna, V., Nair, A., Mallya, R., Khan, T., Omri, A., Antimicrobial Nanomaterials for Food Packaging. *Antibiotics*, 11, 729, 2022.
35. Teixeira-Costa, B.E. and Andrade, C.T., Chitosan as a Valuable Biomolecule from Seafood Industry Waste in the Design of Green Food Packaging. *Biomol.*, 11, 1599, 2021.
36. Yanat, M. and Karin, S., Preparation methods and applications of chitosan nanoparticles; with an outlook toward reinforcement of biodegradable packaging. *React. Funct. Polym.*, 161, 104849, 2021.

16

Application of Nanochitosan for the Biodegradation of Agricultural Wastes

John Tsado Mathew[1]*, Charles Oluwaseun Adetunji[2], Abel Inobeme[3], Musah Monday[1], Yakubu Azeh[1], Etsuyankpa Muhammad Bini[4], Abdulfatai Aideye Otori[5], Isah Mohammed Chado[6], Elijah Yanda Shaba[7], Jibrin Noah Akoji[8], Amos Mamman[1] and Oluwafemi Adebayo Oyewole[9]

[1]*Department of Chemistry, Ibrahim Badamasi Babangida University, Lapai, Niger State, Nigeria*
[2]*Applied Microbiology, Biotechnology and Nanotechnology Laboratory, Department of Microbiology, Edo University Uzairue, Auchi, Edo State, Nigeria*
[3]*Department of Chemistry, Edo University, Uzairue, Auchi, Edo State, Nigeria*
[4]*Department of Chemistry, Federal University of Lafia, Nasarawa State, Nigeria*
[5]*Department of Chemical Engineering, Federal Polytechnic, Bida, Niger State, Nigeria*
[6]*Department of Biology, Ibrahim Badamasi Babangida University, Lapai, Niger State, Nigeria*
[7]*Department of Chemistry, Federal University of Technology, Minna, Nigeria*
[8]*Department of Chemistry, Baze University, Abuja, Nigeria*
[9]*Department of Microbiology, Federal University of Technology, Minna, Nigeria*

Abstract

Agricultural waste management poses significant challenges due to its environmental impact and potential health hazards. Traditional waste disposal methods such as landfilling and incineration are neither sustainable nor cost-effective. As a promising alternative, nanotechnology offers a range of solutions for the efficient degradation of agricultural waste. This abstract highlights the application of nanochitosan, a nanoscale derivative of chitosan, for the biodegradation of agricultural wastes. Nanochitosan, derived from chitosan, a natural biopolymer obtained from chitin, exhibits unique properties such as high surface area,

**Corresponding author*: johntsadom@gmail.com; jmathew@ibbu.edu.ng

Charles Oluwaseun Adetunji, Maulin P. Shah, Yerima Mohammed Bello, Daniel Hefft, Jay Singh, Shyam S. Pandey and Ravindra Pratap Singh (eds.) Nanochitosan Applications for Enhanced Crop Production and Food Security, (361–382) © 2025 Scrivener Publishing LLC

porosity, and biocompatibility, making it an ideal candidate for waste management applications. By leveraging these characteristics, nanochitosan-based systems have shown remarkable potential in addressing the challenges associated with agricultural waste. Nanochitosan, derived from chitosan, a natural biopolymer obtained from chitin, exhibits unique properties such as high surface area, porosity, and biocompatibility, making it an ideal candidate for waste management applications. By leveraging these characteristics, nanochitosan-based systems have shown a remarkable potential in addressing the challenges associated with agricultural waste. The application of nanochitosan for the biodegradation of agricultural wastes offers numerous advantages, including increased degradation efficiency, reduced environmental impact, and improved waste management practices. The use of nanochitosan-based systems can lead to the production of value-added products such as biofuels, biopolymers, and organic fertilizers, contributing to the development of a circular economy. However, nanochitosan demonstrates a great potential as an effective and sustainable solution for the biodegradation of agricultural wastes. Further research and development in this field are essential to optimize nanochitosan-based systems, address challenges, and explore new applications. The integration of nanotechnology into waste management strategies has the potential to revolutionize the agricultural sector, promoting a greener and more sustainable future.

Keywords: Nanochitosan, agricultural waste, biodegradation, nanotechnology, waste management

16.1 Introduction

In recent years, there has been a growing concern about the environmental impact of agricultural waste generated from various farming practices. These wastes, such as crop residues, animal manure, and food processing by-products, pose significant challenges in terms of their disposal and management. Traditional methods of waste treatment and disposal, such as landfilling and incineration, are not only costly but also contribute to environmental pollution. As a result, there is a pressing need for sustainable and eco-friendly solutions to tackle agricultural waste [61, 62].

Nanochitosan exhibits significant potential in the biodegradation of agricultural wastes. Due to its nanoscale size, it offers an expanded surface area, facilitating enhanced microbial interactions and enzymatic activity during waste degradation. The positive charge of nanochitosan promotes the adsorption of negatively charged pollutants, aiding in the removal of contaminants from agricultural residues. Its biocompatibility supports a favorable environment for microbial growth, optimizing the degradation process. The use of nanochitosan presents a sustainable and effective solution for agricultural waste management, contributing to environmental conservation and promoting a circular economy by converting waste into valuable resources through enhanced biodegradation [44, 50].

The use of nanochitosan in the biodegradation of agricultural wastes offers several advantages. Firstly, nanochitosan possesses a large surface area and high porosity, which enhance its adsorption and binding capabilities with organic waste materials. This property enables efficient encapsulation and immobilization of agricultural waste, preventing their dispersion and leaching into the surrounding environment. Secondly, nanochitosan acts as a biostimulant and enhances the activity of microbial populations responsible for the biodegradation process. It provides a favorable microenvironment for the growth and proliferation of beneficial microorganisms, such as bacteria and fungi, which play a crucial role in breaking down complex organic compounds present in agricultural waste [46, 57].

Moreover, nanochitosan can modify the physicochemical properties of agricultural waste, improving their bioavailability and accessibility to microbial degradation. By altering the surface characteristics and structure of waste materials, nanochitosan facilitates enzymatic reactions and accelerates the biodegradation process. Furthermore, nanochitosan-based systems can be tailored to release specific nutrients or bioactive compounds, promoting the growth of beneficial microorganisms and optimizing the biodegradation process. This controlled release mechanism ensures the sustained availability of essential elements required for microbial metabolic activities, thereby enhancing the overall efficiency of waste biodegradation [33, 70].

Therefore, the application of nanochitosan for the biodegradation of agricultural wastes holds a great potential in addressing the environmental challenges associated with waste management in the agricultural sector. The unique properties of nanochitosan enable effective encapsulation, immobilization, and modification of waste materials, facilitating their degradation by beneficial microorganisms. By utilizing nanochitosan-based systems, we can promote the sustainable and eco-friendly management of agricultural wastes, contributing to a cleaner and healthier environment [18, 64]. This chapter discusses the biodegradation of agricultural waste using nanochitosans (Figure 16.1). It presents the mechanisms and environment impact of this process and future perspectives in this regard.

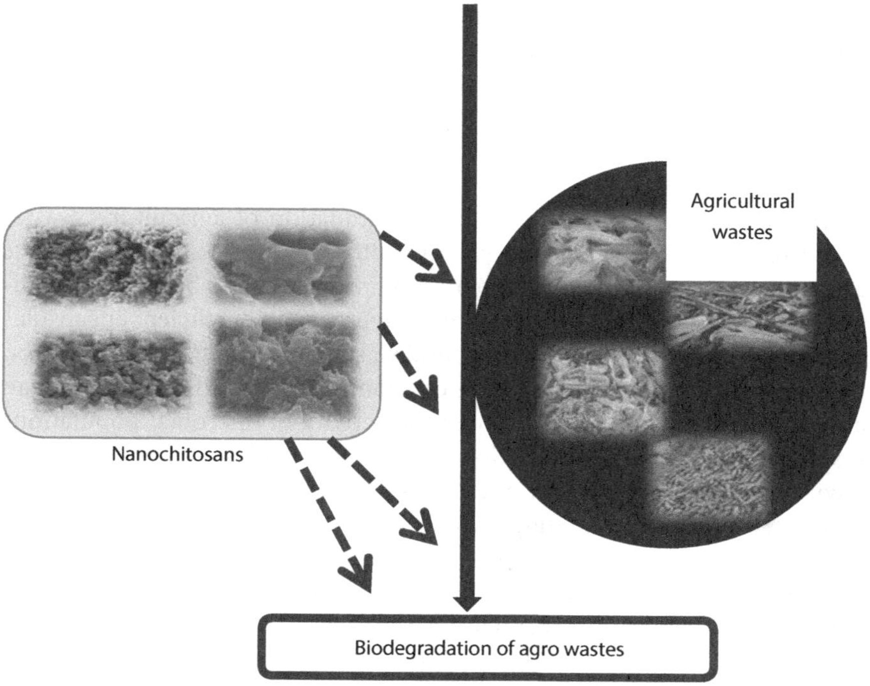

Figure 16.1 Schematic representation of chapter.

16.2 Role of Nanochitosan in Waste Management Nanochitosan

Nanochitosan plays a significant role in waste management due to its unique properties and versatile applications. Here are some key roles of nanochitosan in waste management: Adsorption and Removal of Contaminants: Nanochitosan possesses a high surface area and porosity, allowing it to effectively adsorb and bind various contaminants present in waste streams. It can capture heavy metals, organic pollutants, and dyes, thereby reducing their concentration in the waste and preventing their release into the environment [25, 37].

Enhanced Biodegradation: Nanochitosan acts as a biostimulant and enhances the biodegradation process. When incorporated into waste materials, nanochitosan provides a favorable microenvironment for the growth and activity of microorganisms responsible for waste degradation. It can enhance enzymatic reactions, increase microbial accessibility to

waste compounds, and improve the overall efficiency of biodegradation processes [15, 72].
Waste Encapsulation and Stabilization: Nanochitosan can encapsulate or stabilize hazardous waste materials, preventing their leaching and dispersion. It can form a protective coating around waste particles, immobilizing contaminants and reducing their mobility. This encapsulation process improves waste containment and minimizes the risk of environmental contamination [39, 48].
Controlled Release of Nutrients: Nanochitosan-based systems can be designed to release nutrients or bioactive compounds gradually. This controlled release mechanism ensures the sustained availability of essential elements for microbial activities during waste biodegradation. It optimizes nutrient utilization, promotes microbial growth, and enhances the overall efficiency of waste management processes [20, 71].
Waste Valorization: Nanochitosan can facilitate the valorization of waste materials by converting them into value-added products. It can be used as a catalyst or support material in various waste conversion processes, such as biomass pyrolysis, fermentation, or enzymatic reactions. This enables the production of biofuels, biochemicals, or other valuable products from waste materials [28, 49].
Biodegradable Packaging: Nanochitosan can be incorporated into biodegradable packaging materials, improving their mechanical properties, barrier properties, and biodegradability. This application reduces the environmental impact of packaging waste and promotes the use of sustainable alternatives. However, nanochitosan plays a versatile role in waste management by adsorbing contaminants, enhancing biodegradation processes, encapsulating hazardous waste, facilitating controlled nutrient release, aiding in water treatment, promoting waste valorization, and improving biodegradable packaging materials. Its unique properties make it a valuable tool in the development of sustainable and eco-friendly waste management strategies [17, 52].

16.3 Unique Properties of Nanochitosan

Nanochitosan possesses several unique properties that make it an excellent candidate for the biodegradation of agricultural wastes. These properties contribute to its effectiveness in enhancing waste degradation processes. Nanochitosans, therefore, have numerous applications due to their unique properties (Figure 16.2). Here are some of the key unique properties of nanochitosan:

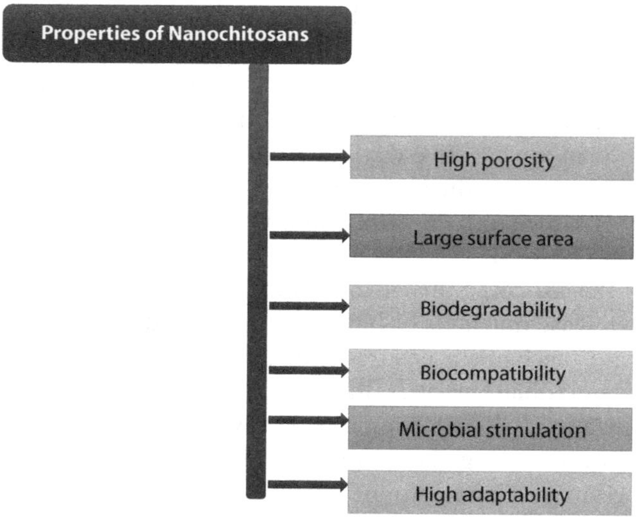

Figure 16.2 Unique properties of nanochitosans.

Biodegradability: Nanochitosan is a biodegradable material derived from chitosan, which is itself a biopolymer obtained from the shells of crustaceans. This property ensures that nanochitosan itself can be broken down by microbial activity over time, minimizing its environmental impact [65].
Large Surface Area and High Porosity: Nanochitosan exhibits a large surface area and high porosity, providing a significant contact area for interaction with agricultural waste materials. This allows for efficient adsorption, encapsulation, and binding of waste compounds, improving their accessibility to microbial degradation [21].
Biocompatibility and Non-Toxicity: Nanochitosan is biocompatible and non-toxic, making it safe for use in various environmental applications. Its use in agricultural waste biodegradation does not pose harm to the environment, ecosystems, or human health [19].
Microbial Stimulation: Nanochitosan acts as a biostimulant, promoting the growth and activity of beneficial microorganisms involved in waste degradation. It provides a favorable microenvironment by offering physical support, nutrients, and a surface for microbial attachment. This stimulation enhances microbial metabolic activities, accelerating the biodegradation process [47, 76].
Surface Modification and Enzyme Activation: Nanochitosan can modify the surface characteristics of agricultural waste materials, making them more amenable to enzymatic reactions. This modification can increase the

accessibility of waste compounds to microbial enzymes, facilitating their breakdown into simpler, more easily degradable substances [58].
Controlled Release Mechanism: Nanochitosan-based systems can be designed to release nutrients or bioactive compounds in a controlled manner. This controlled release mechanism ensures the sustained availability of essential elements required for microbial growth and metabolic activities during waste biodegradation [60].
Environmental Adaptability: Nanochitosan exhibits good environmental adaptability and stability under a wide range of conditions. It can withstand variations in temperature, pH, and moisture content, making it suitable for diverse agricultural waste management scenarios. By leveraging these unique properties, nanochitosan can effectively enhance the biodegradation of agricultural wastes. Its biodegradability, large surface area, biocompatibility, and ability to stimulate microbial activity contribute to its efficiency in encapsulating waste, modifying waste surfaces, and promoting the controlled release of nutrients. As a result, nanochitosan offers a great potential in improving the sustainability and effectiveness of agricultural waste management strategies [29].

16.4 Application of Nanochitosan in Agricultural Waste Biodegradation

Nanochitosan finds application in agricultural waste biodegradation by enhancing microbial activity, promoting enzymatic degradation, and improving adsorption of pollutants. Its nanoscale properties contribute to increased surface area, fostering efficient degradation. Nanochitosan's biocompatibility supports a sustainable approach to agricultural waste management, offering a promising solution for environmental sustainability (Table 16.1). There are some major applications of nanochitosan in agricultural waste biodegradation:

Waste Encapsulation and Stabilization: Nanochitosan can be used to encapsulate and stabilize agricultural waste materials, such as crop residues or animal manure. The encapsulation process helps to immobilize the waste, preventing the release of harmful substances and controlling the rate of decomposition. This application is particularly useful in managing and utilizing agricultural waste as a nutrient-rich soil amendment [51, 74].
Biostimulation of Microbial Activity: Nanochitosan acts as a biostimulant by promoting the growth and activity of beneficial microorganisms

Table 16.1 Some possible applications of nanochitosan in agricultural waste biodegradation.

Agricultural waste	Nanochitosan application	Benefits	References
Crop Residues	Nanochitosan-coated composting material	Enhanced composting process and improved nutrient release	[42, 67]
Animal Manure	Nanochitosan-based additives for anaerobic digestion	Increased biogas production, reduced odor, and improved nutrient availability	[35, 45]
Agrochemical Waste	Nanochitosan-mediated bioremediation	Enhanced degradation of pesticides and reduced environmental contamination	[26, 69]
Food Processing Waste	Nanochitosan encapsulation for enzyme immobilization	Improved enzyme stability and enhanced waste degradation efficiency	[1, 26, 36]

involved in waste degradation. It provides a conducive microenvironment by enhancing microbial attachment, improving nutrient availability, and facilitating enzymatic reactions. This biostimulation enhances the efficiency of waste biodegradation processes [2, 3, 24].

Nutrient Release and Soil Fertility Enhancement: Nanochitosan-based systems can be designed to gradually release nutrients from agricultural waste materials. This controlled release mechanism ensures the sustained availability of essential nutrients, promoting microbial growth and metabolic activities during waste degradation. The released nutrients contribute to improved soil fertility, supporting sustainable agriculture practices [60].

Odor Control: Agricultural waste often produces unpleasant odors during decomposition. Nanochitosan can help mitigate odor issues by binding and neutralizing odor-causing compounds, reducing the emission of foul odors from waste materials. This odor control property enhances the environmental and social acceptability of agricultural waste management practices [14].

Enhanced Composting: Nanochitosan can be incorporated into composting processes to enhance the efficiency and quality of compost production. It accelerates the decomposition of organic matter, promotes microbial activity, and improves the overall stability and maturity of the compost. Nanochitosan-treated compost exhibits enhanced nutrient content, increased water retention capacity, and improved soil amendment properties [41].

Soil Erosion Control: Nanochitosan can be utilized as a soil stabilizer to control soil erosion caused by agricultural activities. When applied to erodible soils or mixed with organic materials, nanochitosan forms a protective film or gel-like matrix that helps retain moisture, reduce surface runoff, and prevent soil erosion. This application contributes to soil conservation and sustainable land management [32].

Water Treatment: Nanochitosan can be used in water treatment processes for the removal of pollutants and organic compounds derived from agricultural waste runoff. It can effectively adsorb heavy metals, pesticides, and other contaminants, reducing their concentration in water bodies and protecting aquatic ecosystems [4, 5, 55].

These applications demonstrate the potential of nanochitosan in enhancing the biodegradation of agricultural wastes. By leveraging its unique properties, nanochitosan offers opportunities to address waste management challenges, improve soil fertility, reduce environmental impact, and promote sustainable agricultural practices.

16.5 Mechanisms of Nanochitosan in Biodegradation

Nanochitosan facilitates biodegradation through various mechanisms. Its nanoscale size enhances surface area, promoting enzyme accessibility and microbial interactions and accelerating the breakdown of organic compounds in agricultural wastes. The positive charge of nanochitosan attracts negatively charged pollutants, aiding in adsorption and subsequent microbial degradation. Additionally, nanochitosan's biocompatibility supports microbial growth, fostering a conducive environment for enzymatic activity. Overall, nanochitosan acts synergistically, enhancing biodegradation processes by improving substrate accessibility, adsorption capacity, and microbial support, making it a promising candidate for sustainable agricultural waste management [75]. Most of the main mechanisms are as follows:

Adsorption and Enzymatic Degradation: Nanochitosan has a high surface area and a positively charged surface, which enables it to adsorb various pollutants, such as heavy metals, dyes, and organic compounds. The adsorption process involves electrostatic interactions, hydrogen bonding, and van der Waals forces. Once adsorbed, nanochitosan can facilitate enzymatic degradation of the adsorbed pollutants by providing a favorable environment for the action of enzymes [21].

Reactive Oxygen Species (ROS) Generation: Nanochitosan can generate reactive oxygen species, such as hydrogen peroxide (H2O2), hydroxyl radicals (•OH), and superoxide radicals (•O2−), through the Fenton-like reaction. These ROS are highly reactive and can induce oxidative stress in microorganisms or pollutants, leading to their degradation. The ROS generation capability of nanochitosan enhances its effectiveness in biodegradation processes. Antimicrobial Activity: Nanochitosan exhibits inherent antimicrobial properties, attributed to its cationic nature and the ability to disrupt the cell membranes of microorganisms. This antimicrobial activity can be utilized in biodegradation applications to control microbial populations and prevent the growth of harmful bacteria or fungi, thereby facilitating the degradation process [27].

Nanostructured Physical Barrier: Nanochitosan can form a nanostructured physical barrier or coating on the surface of materials, such as soil or wastewater treatment membranes. This barrier can help prevent the adhesion or fouling of pollutants, bacteria, or other contaminants, thus enhancing the efficiency of biodegradation processes [44, 54].

Stimulation of Microbial Activity: Nanochitosan can act as a substrate or a carbon source for microbial growth. By providing a nutrient-rich environment, nanochitosan can stimulate the activity and proliferation of beneficial microorganisms, such as bacteria or fungi that possess biodegradative capabilities. These microorganisms can utilize nanochitosan as a food source and degrade organic pollutants more efficiently. It is important to note that the specific mechanisms of nanochitosan in biodegradation can vary depending on the target pollutant or the environmental conditions. The interactions between nanochitosan and the pollutants, as well as the surrounding microorganisms, play a crucial role in determining the overall effectiveness of nanochitosan-based biodegradation processes [16, 56].

16.6 Nanochitosan-Based Biodegradation Techniques

Nanochitosan-based biodegradation techniques utilize the unique properties of nanochitosan to enhance the degradation of pollutants or facilitate environmental remediation processes. Here are some commonly employed techniques:

Nanochitosan Adsorption: Nanochitosan can be used as an adsorbent to remove pollutants from water or soil. It can effectively adsorb heavy metals, dyes, organic compounds, and other contaminants due to its high surface area and positive charge. The adsorption process involves mixing nanochitosan with the contaminated medium, allowing the pollutants to adsorb onto the nanochitosan particles. The nanochitosan can then be separated from the medium, taking the adsorbed pollutants with it [21].
Nanochitosan-Enhanced Enzymatic Degradation: Nanochitosan can enhance enzymatic degradation of pollutants by creating a favorable environment for enzyme activity. It can adsorb and concentrate enzymes or provide protection against harsh conditions that may otherwise inactivate the enzymes. Nanochitosan can also improve enzyme stability, prolonging their activity and enhancing the overall degradation efficiency [23].
Nanochitosan-Facilitated Microbial Degradation: Nanochitosan can stimulate microbial activity and provide a conducive environment for microorganisms to degrade pollutants. It can act as a substrate or carbon source for microbial growth, promoting the proliferation of microorganisms with biodegradative capabilities. Nanochitosan can also protect the microorganisms from adverse conditions, such as pH fluctuations or toxic substances, and enhance their viability and efficiency in pollutant degradation [31].
Nanochitosan-Coated Remediation Materials: Nanochitosan can be used to coat various remediation materials, such as membranes, filters, or porous materials. The nanochitosan coating acts as a physical barrier, preventing the adhesion or fouling of pollutants on the surface of the materials. This coating also facilitates the adsorption or degradation of pollutants by providing a reactive surface and enhancing the surface area available for interactions [30].
Nanochitosan-Mediated Advanced Oxidation Processes (AOPs): Advanced oxidation processes involve the generation of highly reactive species, such as hydroxyl radicals (•OH), to degrade pollutants. Nanochitosan can be used as a catalyst or a support material in AOPs to enhance the generation of reactive species. It can facilitate the production of reactive oxygen species through the Fenton-like reaction, which leads to the degradation of pollutants [12]. These techniques can be applied in various environmental remediation scenarios, including wastewater treatment, soil remediation, air purification, and the cleanup of contaminated sites. The choice of

technique depends on the specific pollutant, the target environment, and the desired outcome of the biodegradation process.

16.7 Environmental Benefits of Nanochitosan-Based Biodegradation

Nanochitosan-based biodegradation offers several environmental benefits, making it a promising approach for pollution control and environmental remediation [13, 40]. Some of the main environmental benefits associated with the use of nanochitosan are as follows:

Effective Pollutant Removal: Nanochitosan has a high surface area and a positive charge, which enables it to efficiently adsorb a wide range of pollutants. It can effectively remove heavy metals, dyes, organic compounds, and other contaminants from water or soil systems. By facilitating the removal of pollutants, nanochitosan-based biodegradation techniques help to improve the quality of water and soil, reducing the harmful impacts on ecosystems and human health [6–11, 30].

Reduced Toxicity and Ecotoxicity: The presence of pollutants in the environment can pose risks to both aquatic and terrestrial organisms. Nanochitosan-based biodegradation techniques can help reduce the toxicity and ecotoxicity of pollutants by effectively removing them from the environment. By minimizing the exposure of organisms to toxic substances, these techniques contribute to the protection and preservation of biodiversity [22, 38].

Enhanced Biodegradation Efficiency: Nanochitosan can enhance the efficiency of biodegradation processes by providing a favorable environment for microbial growth and enzymatic activity. It can stimulate the activity and proliferation of microorganisms with biodegradative capabilities, facilitating the degradation of organic pollutants. The improved efficiency of biodegradation reduces the persistence of pollutants in the environment and promotes the restoration of ecosystems [69].

Renewable and Biodegradable Material: Chitosan, the precursor of nanochitosan, is derived from natural sources such as crustacean shells or fungal cell walls. It is a renewable resource and can be produced sustainably. Moreover, chitosan and nanochitosan are biodegradable materials, which means that they can be broken down by natural processes over time. This property reduces the accumulation of non-biodegradable materials in the environment, contributing to the sustainability of pollution control strategies [59].

Potential for Green and Sustainable Technologies: Nanochitosan-based biodegradation techniques align with the principles of green and sustainable technologies. The use of nanochitosan can reduce the reliance on traditional, chemical-intensive methods for pollutant removal. It offers an eco-friendly and economically viable alternative, minimizing the use of harsh chemicals and energy-intensive processes. This promotes the transition toward more sustainable practices in environmental remediation [13]. Versatility and Adaptability: Nanochitosan can be modified and tailored to specific applications, making it a versatile material for biodegradation purposes. Its properties can be optimized to target specific pollutants or environmental conditions. Additionally, nanochitosan can be incorporated into various forms, such as films, beads, or coatings, allowing for its utilization in diverse environmental settings. Overall, nanochitosan-based biodegradation techniques provide effective and environmentally friendly approaches to tackle pollution and remediate contaminated environments. These techniques contribute to the preservation of ecosystems, the protection of human health, and the promotion of sustainable practices in pollution control and environmental management [43, 53].

16.8 Challenges and Future Perspectives of Nanochitosan Biodegradation of Agricultural Wastes

Nanochitosan-based biodegradation of agricultural wastes holds promise for addressing the environmental challenges associated with the disposal of agricultural residues. However, there are also several challenges that need to be addressed and future perspectives that can further enhance the application of nanochitosan in this field.

16.8.1 Challenges of Nanochitosan Biodegradation of Agricultural Wastes

Cost-Effectiveness: The cost-effectiveness of nanochitosan biodegradation of agricultural wastes faces several challenges. First, the production of nanochitosan involves complex processes, requiring sophisticated equipment and skilled personnel, contributing to high production costs. Additionally, the synthesis of nanochitosan often relies on chitosan, which itself may be expensive. Moreover, the scalability of nanochitosan production for large-scale agricultural waste treatment remains a hurdle. The need for precise control over particle size and distribution adds to the expenses. Developing efficient, cost-friendly production methods and optimizing the use of nanochitosan in agricultural waste management are crucial for overcoming these economic challenges [64].

Scale-Up and Feasibility: Scaling up nanochitosan-based biodegradation processes to treat agricultural waste on a larger scale can be challenging. The feasibility of implementing these techniques in real-world agricultural settings, such as farms or processing facilities, needs to be carefully evaluated. Factors such as process integration, compatibility with existing waste management systems, and infrastructure requirements should be considered [73].

Optimization of Degradation Parameters: The efficiency of nanochitosan-based biodegradation can be influenced by various parameters, including pH, temperature, contact time, and nanochitosan dosage. Optimizing these parameters for different types of agricultural wastes is crucial to maximize the degradation efficiency. However, identifying the optimal conditions and establishing standardized protocols for various agricultural waste types can be complex and require extensive research [34].

Residue Management: The management of residues generated during nanochitosan-based biodegradation processes is an important consideration. The treated agricultural residues may still contain residual nanochitosan particles or by-products, which need to be properly handled to prevent any potential adverse effects on the environment or human health. Developing appropriate disposal or recovery methods for these residues is essential [66].

16.8.2 Future Perspectives of Nanochitosan Biodegradation of Agricultural Wastes

Nanochitosan Modification and Functionalization: Further research on modifying and functionalizing nanochitosan can expand its capabilities in agricultural waste biodegradation. Tailoring the surface properties, charge, or structure of nanochitosan particles can enhance their adsorption capacity, stability, or selectivity toward specific agricultural pollutants. This can lead to improved efficiency and specificity in the biodegradation process [50].

Integration With Other Treatment Technologies: Combining nanochitosan-based biodegradation with other treatment technologies, such as composting, anaerobic digestion, or biochar production, can offer synergistic benefits. Integrated approaches can enhance the overall degradation efficiency, promote resource recovery, and generate value-added products from agricultural wastes [63].

Development of Nanochitosan-Based Delivery Systems: Nanochitosan can serve as a carrier for bioactive substances, such as enzymes or microorganisms that facilitate the degradation of agricultural wastes. Developing nanochitosan-based delivery systems can help protect and deliver these bioactive agents to the target waste, enhancing their effectiveness and stability during the biodegradation process [18].

Environmental Risk Assessment: Conducting comprehensive environmental risk assessments is crucial to ensure the safe and sustainable application of nanochitosan-based biodegradation techniques. Assessing the potential impacts of nanochitosan particles, their degradation by-products, and any released contaminants on ecosystems and human health is essential for responsible deployment [68].

Technology Transfer and Adoption: Bridging the gap between research and practical application is essential for the widespread adoption of nanochitosan-based biodegradation technologies for agricultural wastes. Collaborations between researchers, industry, and policymakers can facilitate technology transfer, support pilot-scale demonstrations, and promote the development of guidelines and regulations for safe and effective implementation [34].

16.9 Conclusion

The nanochitosan-based biodegradation has shown a great potential as an eco-friendly and sustainable approach to managing agricultural wastes. Nanochitosan particles have unique physicochemical properties that enable them to efficiently adsorb, degrade, and detoxify various agricultural pollutants, such as pesticides, heavy metals, and organic compounds. Furthermore, nanochitosan-based biodegradation can generate value-added products, such as biofertilizers or biochar, from the treated agricultural residues, contributing to circular economy principles. Despite the promising results, several challenges need to be addressed to enable the widespread application of nanochitosan-based biodegradation in agricultural waste management. These include optimizing degradation parameters, ensuring cost-effectiveness, managing residues, and scaling up the technology. Moreover, exploring future perspectives, such as nanochitosan modification, integration with other treatment technologies, and the development of delivery systems, can further enhance the capabilities and efficiency of nanochitosan-based biodegradation.

References

1. Abimbola, O.F., Adetunji, C.O., Dare, B.J., Mathew, J.T., Inobeme, A., Ghazanfar, S., Olaniyan, O.T., Ajiboye, M.D., Dauda, W., Packaging of Animal-Based Food Products: The Need for Intelligent Freshness Sensors, in: *Sensing and Artificial Intelligence Solutions for Food Manufacturing*, pp. 89–102, CRC Press, Boca Raton, Florida, USA, 2023.
2. Adetunji, C.O., Anani, O.A., Olaniyan, O.T., Inobeme, A., Bodunrinde, R.E., Mathew, J.T., Adetunji, J.B., Hefft, D., II, Prospect of Utilizing Novel Nanomaterials for Effective Food Security, Safety, Packaging, and Preservation, in: *The Science of Nanomaterials: Basics and Applications*, Apple Academic Press, Inc. Co-published with CRC Press (Taylor & Francis), Oakville, Ontario, Canada, 2022a.
3. Adetunji, C.O., Ogundolie, F.A., Mathew, J.T., Inobeme, A., Olotu, T., Olaniyan, O.T., Ijabadeniyi, O.A., Ajiboye, M.D., Ajayi, O.O., Dauda, W., *et al.*, Graphene-based nanomaterials for targeted drug delivery and tissue engineering, in: *Novel Platforms for Drug Delivery Applications*, pp. 277–288, Elsevier Ltd, Oakville, Ontario, Canada, 2022b.
4. Adetunji, C.O., Ogundolie, F.A., Olaniyan, O., Mathew, J.T., Inobeme, A., Olotu, T., Ghazanfar, S., Ijabadeniyi, O.A., Ajiboye, M.D., Ajayi, O.O., *et al.*, Nanobiomaterials for Food Packaging Sensor Applications, in: *Bio- and Nano-sensing Technologies for Food Processing and Packaging*, pp. 167–180, The Royal Society of Chemistry, London, United Kingdom, 2022c.
5. Adetunji, C.O., Ogundolie, F.A., Ajiboye, M.D., Mathew, J.T., Inobeme, A., Olotu, T., Olaniyan, O., Ijabadeniyi, O.A., Ajayi, O.O., Dauda, W.P., *et al.*, Bio- and Nanosensors in the Food Industry, in: *Bio- and Nano-sensing Technologies for Food Processing and Packaging*, pp. 22–36, The Royal Society of Chemistry, London, United Kingdom, 2022d.
6. Adetunji, C.O., Olaniyan, O.T., Olugbenga, S.M., Hefft, D., II, Shariati, M.A., Rebezov, M., Ajayi, O.O., Sanavar, A., Bodunrinde, R.E., Adetunji, J.B., *et al.*, Medicinal Properties and Pharmacological Effectiveness of Aspalathuslinearis (Burm. f.) R. Dahlgren, in: *Phytochemical Composition and Pharmacy of Medicinal Plants*, vol. 2, T. Pullaiah (Ed.), pp. 419–433, Apple Academic Press, Inc. Co-published with CRC Press (Taylor & Francis), Oakville, Ontario, Canada, 2023a.
7. Adetunji, C.O., Olaniyan, O.T., Shariati, M.A., Hefft, D., II, Olugbenga, S.M., Rebezov, M., Ajayi, O.O., Baibalinova, G., Bodunrinde, R.E., Adetunji, R.B., *et al.*, Medicinal and Pharmacological Attributes of Acalyphahispida Burm. F, in: *Phytochemical Composition and Pharmacy of Medicinal Plants*, vol. 2, T. Pullaiah (Ed.), pp. 435–443, Apple Academic Press, Inc. Co-published with CRC Press (Taylor & Francis), Oakville, Ontario, Canada, 2023b.
8. Adetunji, C.O., Olaniyan, O.T., Shariati, M.A., Hefft, D., II, Olugbenga, S.M., Rebezov, M., Ajayi, O.O., Anichkina, O., Bodunrinde, R.E., Adetunji, R.B., *et al.*, Medicinal Properties of Sennaalata (L.) Roxb. (Syn. Cassia alata L.)

and Its Biological Activities, in: *Phytochemical Composition and Pharmacy of Medicinal Plants*, vol. 2, T. Pullaiah (Ed.), pp. 445–459, Apple Academic Press, Inc. Co-published with CRC Press (Taylor & Francis), Oakville, Ontario, Canada, 2023c.

9. Adetunji, C.O., Olaniyan, O.T., Olugbenga, S.M., Shariati, M.A., Osarenkhoe, O.O., Odoh, U.E., Oyedara, O.O., Rebezov, M., Ajayi, O.O., Babaeva, M., *et al.*, Pharmacology, Photochemical, and Healing Effectiveness of Chromolaenaodorata (L.) R.M. King and H. Robinson, in: *Phytochemical Composition and Pharmacy of Medicinal Plants*, vol. 2, T. Pullaiah (Ed.), pp. 461–482, Apple Academic Press, Inc. Co-published with CRC Press (Taylor & Francis), Oakville, Ontario, Canada, 2023d.
10. Adetunji, C.O., Olaniyan, O.T., Shariati, M.A., Inobeme, A., Mathew, J.T., Olugbenga, S.M., Adetunji, J.B., Hefft, D., II, Rebezov, M., Ajayi, O.O., *et al.*, Phytochemical Constituents and Medicinal Effectiveness of Buchholziacoriacea Engl, in: *Phytochemical Composition and Pharmacy of Medicinal Plants*, vol. 2, T. Pullaiah (Ed.), pp. 483–498, Apple Academic Press, Inc. Co-published with CRC Press (Taylor & Francis), Oakville, Ontario, Canada, 2023e.
11. Adetunji, C.O., Olaniyan, O.T., Shariati, M.A., Bodunrinde, R.E., Ajayi, O.O., Adetunji, J.B., Hefft, D., II, Olugbenga, S.M., Rebezov, M., Khayrullin, M., *et al.*, Pharmacology, Phytochemistry, and Medicinal Usefulness of Basella alba L, in: *Phytochemical Composition and Pharmacy of Medicinal Plants*, vol. 2, T. Pullaiah (Ed.), pp. 499–516, Apple Academic Press, Inc. Co-published with CRC Press (Taylor & Francis), Oakville, Ontario, Canada, 2023f.
12. Ahmad, S., Almehmadi, M., Janjuhah, H.T., Kontakiotis, G., Abdulaziz, O., Saeed, K., Ahmad, H., Allahyani, M., Aljuaid, A., Alsaiari, A.A., *et al.*, The Effect of Mineral Ions Present in Tap Water on Photodegradation of Organic Pollutants: Future Perspectives. *Water*, *15*, 175, 2023.
13. Ajose, D.J., Abolarinwa, T.O., Oluwarinde, B.O., Montso, P.K., Fayemi, O.E., Aremu, A.O., Ateba, C.N., Application of Plant-Derived Nanoparticles (PDNP) in Food-Producing Animals as a Bio-Control Agent against Antimicrobial-Resistant Pathogens. *Biomed.*, *10*, 2426, 2022.
14. Arunachalam, K., Krishnan, G.P., Sethuraman, S., Abraham, S.V.P., II, Krishnan, S.T., Venkateswar, A., Arunkumar, J., Shi, C., MubarakAli, D., Exploring Possible Ways to Enhance the Potential and Use of Natural Products through Nanotechnology in the Battle against Biofilms of Foodborne Bacterial Pathogens. *Pathog.*, *12*, 270, 2023.
15. Atanasova, N., Stoitsova, S., Paunova-Krasteva, T., Kambourova, M., Plastic Degradation by Extremophilic Bacteria. *Int. J. Mol. Sci.*, *22*, 5610, 2021.
16. Atay, Y.H., Antibacterial Activity of Chitosan-Based Systems. *Funct. Chitosan Drug Deliv. Biomed. Appl.*, 2, 457–489, 2020.
17. Baghi, F., Gharsallaoui, A., Dumas, E., Ghnimi, S., Advancements in Biodegradable Active Films for Food Packaging: Effects of Nano/Microcapsule Incorporation. *Foods*, *11*, 760, 2022.

18. Bandara, S., Du, H., Carson, L., Bradford, D., Kommalapati, R., Agricultural and Biomedical Applications of Chitosan-Based Nanomaterials. *Nanomat.*, 10, 1903, 2020.
19. Baranwal, J., Barse, B., Fais, A., Delogu, G.L., Kumar, A., Biopolymer: A Sustainable Material for Food and Medical Applications. *Polym.*, *14*, 983, 2022.
20. Basavegowda, N. and Baek, K.H., Current and future perspectives on the use of nanofertilizers for sustainable agriculture: the case of phosphorus nanofertilizer. *3 Biotech*, *11*, 357, 2021.
21. Benettayeb, A., Seihoub, F.Z., Pal, P., Ghosh, S., Usman, M., Chia, C.H., Usman, M., Sillanpää, M., Chitosan Nanoparticles as Potential Nano-Sorbent for Removal of Toxic Environmental Pollutants. *Nanomat.*, *13*, 447, 2023.
22. Boros, B.V. and Ostafe, V., Evaluation of Ecotoxicology Assessment Methods of Nanomaterials and Their Effects. *Nanomat.*, *10*, 610, 2020.
23. Boroumand, H., Badie, F., Mazaheri, S., Seyedi, Z.S., Nahand, J.S., Nejati, M., Baghi, H.B., Abbasi-Kolli, M., Badehnoosh, B., Ghandali, M., Chitosan-Based Nanoparticles Against Viral Infections. *Front. Cell. Infec. Micro.*, *11*, 643953, 2021.
24. Castiglione, A.M., Mannino, G., Contartese, V., Bertea, C.M., Ertani, A., Microbial Biostimulants as Response to Modern Agriculture Needs: Composition, Role and Application of These Innovative Products. *Plants*, *10*, 1533, 2021.
25. Chandra, D., Molla, M.T.H., Bashar, M.A., Chitosan-based nano-sorbents: synthesis, surface modification, characterisation and application in Cd (II), Co (II), Cu (II) and Pb (II) ions removal from wastewater. *Sci. Rep.*, 13, 6050, 2023.
26. Chaudhary, P., Xu, M., Ahamad, L., Chaudhary, A., Kumar, G., Adeleke, B.S., Verma, K.K., Hu, D.-M., Kumar, P., Application of Synthetic Consortia for Improvement of Soil Fertility, Pollution Remediation, and Agricultural Productivity: A Review. *Agron.*, 13, 643, 2023.
27. Collin, F., Chemical Basis of Reactive Oxygen Species Reactivity and Involvement in Neurodegenerative Diseases. *Int. J. Mol. Sci.*, *20*, 2407, 2019.
28. Das, A., Verma, M., Mishra, V., Food waste to resource recovery: a way of green advocacy. *Environ. Sci. Pollut. Res.*, 10, 17874–17886, 2023.
29. Desai, N., Rana, D., Salave, S., Gupta, R., Patel, P., Karunakaran, B., Sharma, A., Giri, J., Benival, D., Kommineni, N., Chitosan: A Potential Biopolymer in Drug Delivery and Biomedical Applications. *Pharmac.*, *15*, 1313, 2023.
30. Ethaib, S., Al-Qutaifia, S., Al-Ansari, N., Zubaidi, S.L., Function of Nanomaterials in Removing Heavy Metals for Water and Wastewater Remediation: A Review. *Environ.*, *9*, 123, 2022.
31. Fadiji, T., Rashvand, M., Daramola, M.O., Iwarere, S.A., A Review on Antimicrobial Packaging for Extending the Shelf Life of Food. *Proc.*, *11*, 590, 2023.

32. Fatehi, H., Ong, D.E.L., Yu, J., Chang, I., Biopolymers as Green Binders for Soil Improvement in Geotechnical Applications: A Review. *Geosci.*, 11, 291, 2021.
33. Ferreira, P.G., Ferreira, V.F., da Silva, F.C., Freitas, C.S., Pereira, P.R., Paschoalin, V.M.F., Chitosans and Nanochitosans: Recent Advances in Skin Protection, Regeneration, and Repair. *Pharmac.*, *14*, 1307, 2022.
34. Fouda-Mbanga, B.G., Prabakaran, E., Pillay, K., Carbohydrate biopolymers, lignin based adsorbents for removal of heavy metals (Cd^{2+}, Pb^{2+}, Zn^{2+}) from wastewater, regeneration and reuse for spent adsorbents including latent fingerprint detection: A review. *Biotech. Rep.*, *30*, e00609, 2021.
35. Fugol, M., Prask, H., Szlachta, J., Dyjakon, A., Pasławska, M., Szufa, S., Improving the Energetic Efficiency of Biogas Plants Using Enzymatic Additives to Anaerobic Digestion. *Ener.*, *16*, 1845, 2023.
36. Gał ̨azka, A., Jankiewicz, U., Szczepkowski, A., Biochemical Characteristics of Laccases and Their Practical Application in the Removal of Xenobiotics from Water. *Appl. Sci.*, 13, 4394, 2023.
37. Gamage, A., Jayasinghe, N., Thiviya, P., Wasana, M.L.D., Merah, O., Madhujith, T., Koduru, J.R., Recent Application Prospects of Chitosan Based Composites for the Metal Contaminated Wastewater Treatment. *Poly.*, *15*, 1453, 2023.
38. Gambardella, C. and Pinsino, A., Nanomaterial Ecotoxicology in the Terrestrial and Aquatic Environment: A Systematic Review. *Tox.*, *10*, 393, 2022.
39. Gupta, V., Biswas, D., Roy, S., A Comprehensive Review of Biodegradable Polymer-Based Films and Coatings and Their Food Packaging Applications. *Mater.*, *15*, 5899, 2022.
40. Hameed, A.Z., Raj, S.A., Kandasamy, J., Baghdadi, M.A., Shahzad, M.A., Chitosan: A Sustainable Material for Multifarious Applications. *Polym.*, *14*, 2335, 2022.
41. Hernández-García, E., Vargas, M., González-Martínez, C., Chiralt, A., Biodegradable Antimicrobial Films for Food Packaging: Effect of Antimicrobials on Degradation. *Foods*, *10*, 1256, 2021.
42. Imran, A., Sardar, F., Khaliq, Z., Nawaz, M.S., Shehzad, A., Ahmad, M., Yasmin, S., Hakim, S., Mirza, B.S., Mubeen, F., *et al.*, Tailored Bioactive Compost from Agri-Waste Improves the Growth and Yield of Chili Pepper and Tomato. *Front. Bioenginee. Biotech*, *9*, 787764, 2022.
43. Mathew, J.T., Okonkwo, S., Ekwoba, L., Ajai, A., II, Bernard, E., Inobeme, J., Agbugui, M.M., Singh, K.R.B., Chemometric approach in environmental pollution analysis: A critical review. *J. Environm. Manag.*, 309, 1–18, 2022.
44. Inobeme, A., Mathew, J.T., Adetunji, C.O., Ajai, A.I., Inobeme, J., Okonkwo, S., Adekoya, M.A., Bamigboye, M.O., Jacob, J.O., *et al.*, Recent advances in nanotechnology for remediation of heavy metals. *Environm. Monit. Assess.*, 195, 1–24, 2023.

45. Jasińska, A., Grosser, A., Meers, E., Possibilities and Limitations of Anaerobic Co-Digestion of Animal Manure—A Critical Review. *Ener.*, 16, 3885, 2023.
46. Jiménez-Gómez, C.P. and Cecilia, J.A., Chitosan: A Natural Biopolymer with a Wide and Varied Range of Applications. *Mol.*, *25*, 3981, 2020.
47. Kaczmarek-Szczepańska, B., Sionkowska, M.M., Mazur, O., Świątczak, J., Brzezinska, M.S., The role of microorganisms in biodegradation of chitosan/tannic acid materials. *Int. J. Biol. Macrom.*, *184*, 584–592, 2021.
48. Kang, B.R., Park, J.S., Ryu, G.R., Jung, W.-J., Choi, J.-S., Shin, H.-M., Effect of Chitosan Coating for Efficient Encapsulation and Improved Stability under Loading Preparation and Storage Conditions of Bacillus Lipopeptides. *Nanomat.*, *12*, 4189, 2022.
49. Kaur, P., Thakur, M., Tondan, D., Bamrah, G.K., Misra, S., Kumar, P., Pandohee, J., Kulshrestha, S., Nanomaterial conjugated lignocellulosic waste: cost-effective production of sustainable bioenergy using enzymes. *Biotech.*, *11*, 480, 2021.
50. Kedir, W.M., Abdi, G.F., Goro, M.M., Tolesa, L.D., Pharmaceutical and drug delivery applications of chitosan biopolymer and its modified nanocomposite: A review. *Heliyon*, *8*, e10196, 2022.
51. Kumar, R., Kumar, N., Rajput, V.D., Mandzhieva, S., Minkina, T., Saharan, B.S., Kumar, D., Sadh, P.K., Duhan, J.S., Advances in Biopolymeric Nanopesticides: A New Eco-Friendly/Eco-Protective Perspective in Precision Agriculture. *Nanomat.*, *12*, 3964, 2022.
52. Liu, T., Li, J., Tang, Q., Qiu, P., Gou, D., Zhao, J., Chitosan-Based Materials: An Overview of Potential Applications in Food Packaging. *Foods*, *11*, 1490, 2022.
53. Maliki, S., Sharma, G., Kumar, A., Moral-Zamorano, M., Moradi, O., Baselga, J., Stadler, F.J., García-Peñas, A., Chitosan as a Tool for Sustainable Development: A Mini Review. *Polym.*, *14*, 1475, 2022.
54. Mamba, F.B., Mbuli, B.S., Ramontja, J., Recent Advances in Biopolymeric Membranes towards the Removal of Emerging Organic Pollutants from Water. *Memb.*, *11*, 798, 2021.
55. Manchala, S., Gandamalla, A., Rudrarapu, A., Makhlouf, A.S.H., Biodegradable Nanocelluloses for Removal of Hazardous Organic Pollutants from Wastewater, in: *Handbook of Biodegradable Materials*, G.A.M. Ali, (Ed.), Springer, Cham, 2023.
56. Mathew, J.T., Adetunji, C.O., Inobeme, A., Musah, M., Azeh, Y., Otori, A.A., Shaba, E.Y., Mamman, A., Tanko, E., Removal of Heavy Metals Using Bio-remedial Techniques, in: *Modern Approaches in Waste Bioremediation*, M.P. Shah (Ed.), pp. 117–130, Springer Nature Switzerland AG, Cham, Switzerland, 2023.
57. Muhammad, S., Abdul Khalil, H.P.S., Abd Hamid, S., Albadn, Y.M., Suriani, A.B., Kamaruzzaman, S., Mohamed, A., Allaq, A.A., Yahya, E.B., Insights into Agricultural-Waste-Based Nano-Activated Carbon Fabrication and Modifications for Wastewater Treatment Application. *Agric.*, 12, 1737, 2022.

58. Nasrollahzadeh, M., Sajjadi, M., Iravani, S., Varma, R.S., Starch, cellulose, pectin, gum, alginate, chitin and chitosan derived (nano)materials for sustainable water treatment: A review. *Carbohyd. Polym.*, *251*, 116986, 2021.
59. Nawawi, W. M. F. B. W., Jones, M., Murphy, R.J., Lee, K.Y., Kontturi, E., Bismarck, A., Nanomaterials Derived from Fungal Sources-Is It the New Hype? *Biomacromol.*, *21*, 30–55, 2020.
60. Nongbet, A., Mishra, A.K., Mohanta, Y.K., Mahanta, S., Ray, M.K., Khan, M., Baek, K.-H., Chakrabartty, I., Nanofertilizers: A Smart and Sustainable Attribute to Modern Agriculture. *Plants*, *11*, 2587, 2022.
61. Ogundolie, F.A., Adetunji, C.O., Mathew, J.T., Inobeme, A., Ghazanfar, S., Olaniyan, O.T., Ajiboye, M.D., Dauda, W.P., Adetunji, J.B., Akinbo, O., *et al.*, Intelligent freshness indicators in packaging of plant-based food products, in: *Evaporative Coolers for the Postharvest Management of Fruits and Vegetables*, pp. 323–331, Academic Press, Cambridge, Massachusetts, USA, 2023.
62. Oluseun, A., II and Adebukola, A.O., Agricultural Solid Wastes: Causes, Effects, and Effective Management. *IntechOpen*, 23, 1–12, 2021.
63. Osman, A., II, Fawzy, S., Farghali, M., El-Azazy, M., Elgarahy, A.M., Fahim, R.A., Maksoud, M., II, Ajlan, A.A., Yousry, M., Saleem, Y., *et al.*, Biocharfor agronomy, animal farming, anaerobic digestion, composting, water treatment, soil remediation, construction, energy storage, and carbon sequestration: a review. *Environ. Chem. Lett.*, *20*, 2385–2485, 2022.
64. Picos-Corrales, L.A., Morales-Burgos, A.M., Ruelas-Leyva, J.P., Crini, G., García-Armenta, E., Jimenez-Lam, S.A., Ayón-Reyna, L.E., Rocha-Alonzo, F., Calderón-Zamora, L., Osuna-Martínez, U., *et al.*, Chitosan as an Outstanding Polysaccharide Improving Health-Commodities of Humans and Environmental Protection. *Polym.*, *15*, 526, 2023.
65. Piekarska, K., Sikora, M., Owczarek, M., Jóźwik-Pruska, J., Wiśniewska-Wrona, M., Chitin and Chitosan as Polymers of the Future—Obtaining, Modification, Life Cycle Assessment and Main Directions of Application. *Polym.*, *15*, 793, 2023.
66. Raza, M.H., Abid, M., Faisal, M., Yan, T., Akhtar, S., Adnan, K.M.M., Environmental and Health Impacts of Crop Residue Burning: Scope of Sustainable Crop Residue Management Practices. *Int. J. Envir. Res. Pub. Health*, *19*, 4753, 2022.
67. Sayara, T., Basheer-Salimia, R., Hawamde, F., Sánchez, A., Recycling of Organic Wastes through Composting: Process Performance and Compost Application in Agriculture. *Agron.*, *10*, 1838, 2020.
68. Shelar, A., Nile, S.H., Singh, A.V., Rothenstein, D., Bill, J., Xiao, J., Chaskar, M., Kai, G., Patil, R., Recent Advances in Nano-Enabled Seed Treatment Strategies for Sustainable Agriculture: Challenges, Risk Assessment, and Future Perspectives. *Nano Micro Lett.*, *15*, 54, 2023.
69. Singh, Y. and Saxena, M.K., Insights into the recent advances in nano-bioremediation of pesticides from the contaminated soil. *Front. Microb.*, *13*, 982611, 2022.

70. Stasińska-Jakubas, M. and Hawrylak-Nowak, B., Protective, Biostimulating, and Eliciting Effects of Chitosan and Its Derivatives on Crop Plants. *Mol.*, *27*, 2801, 2022.
71. Sun, Y., Zhu, G., Zhao, W., Jiang, Y., Wang, Q., Wang, Q., Rui, Y., Zhang, P., Gao, L., Engineered Nanomaterials for Improving the Nutritional Quality of Agricultural Products: A Review. *Nanomat.*, *12*, 4219, 2022.
72. Tamoor, M., Samak, N.A., Jia, Y., Mushtaq, M.U., Sher, H., Bibi, M., Xing, J., Potential Use of Microbial Enzymes for the Conversion of Plastic Waste Into Value-Added Products: A Viable Solution. *Front. Microb.*, *12*, 777727, 2021.
73. Waqas, M., Hashim, S., Humphries, U.W., Ahmad, S., Noor, R., Shoaib, M., Naseem, A., Hlaing, P.T., Lin, H.A., Composting Processes for Agricultural Waste Management: A Comprehensive Review. *Proc.*, *11*, 731, 2023.
74. Zabot, G.L., Schaefer Rodrigues, F., PolanoOdy, L., ViníciusTres, M., Herrera, E., Palacin, H., Córdova-Ramos, J.S., Best, I., Olivera-Montenegro, L., Encapsulation of Bioactive Compounds for Food and Agricultural Applications. *Polym.*, *14*, 4194, 2022.
75. Zainol-Abidin, N.A., Kormin, F., Zainol-Abidin, N.A., Mohamed Anuar, N.A.F., Abu Bakar, M.F., The Potential of Insects as Alternative Sources of Chitin: An Overview on the Chemical Method of Extraction from Various Sources. *Int. J. Mol. Sci.*, *21*, 4978, 2020.
76. Zhang, M., Zhang, F., Li, C., An, H., Wan, T., Zhang, P., Application of Chitosan and Its Derivative Polymers in Clinical Medicine and Agriculture. *Polym.*, *14*, 958, 2022.

17

Application of Nanochitosan in the Detection of Pesticide Residues and Degradation

Japhet Gaius Yakubu[1]*, Oluwafemi Adebayo Oyewole[1], Ummulkhair Salamah Ilyasu[1], Opeyemi Isaac Ayanda[2] and Charles Oluwaseun Adetunji[3]

[1]*Department of Microbiology, Federal University of Technology, Minna, Nigeria*
[2]*Department of Biological Sciences, Covenant University, Ota, Ogun State, Nigeria*
[3]*Department of Microbiology, Edo State University, Uzairue, Nigeria*

Abstract

Chemicals called pesticides are used to manage pests like weeds, rodents, and insects. They are commonly used in agriculture to safeguard animals and crops from pests and illnesses. The usage of pesticides, however, can potentially have detrimental effects on the environment and public health. Pesticides can harm non-target creatures and contaminate the soil, water, and air. In addition, exposure to pesticides can result in a variety of medical issues in people, such as cancer, neurological conditions, and reproductive issues. Chitosan is a natural polymer formed from crab shells. Nanochitosan is chitosan in a nanoscale form. It has several distinctive qualities that make it a potential material for a variety of applications, including the detection and degradation of pesticides, such as large surface area, biocompatibility, and non-toxicity. There are numerous ways to detect pesticides using nanochitosan. Utilizing sensors made of nanochitosan is one typical strategy. Usually, nanochitosan is immobilized onto a substrate, like a gold electrode, or nanochitosan-based colorimetric assays are used to create these sensors. Because there is currently no environmentally benign method of pesticide remediation, scientists have looked for other ways to prevent pesticides from having a negative impact on the ecosystem. In order to degrade pesticides, nanochitosan

**Corresponding author*: japhetgaiusyakubu12@gmail.com

Charles Oluwaseun Adetunji, Maulin P. Shah, Yerima Mohammed Bello, Daniel Hefft, Jay Singh, Shyam S. Pandey and Ravindra Pratap Singh (eds.) Nanochitosan Applications for Enhanced Crop Production and Food Security, (383–408) © 2025 Scrivener Publishing LLC

has been used in a variety of ways. These include the usage of nanochitosan-based adsorbents, which are typically created by adding a functional group with a strong affinity for pesticides to nanochitosan. The pesticide is subsequently removed from the environment by the functional group after it attaches to it. Because nanochitosan can be impacted by environmental factors including pH and temperature, there are restrictions on its usage for bioremediation. Nanochitosan provides a number of benefits over other materials used for the detection and degradation of pesticides, notwithstanding its limits. As a result, it may be investigated to reduce the effects of pesticides on the environment.

Keywords: Nanochitosan, pesticide, biodegradation, nanoremediation, chitosan

17.1 Introduction

Agriculture, which is constrained by a variety of biotic and abiotic factors, serves as the economic backbone of many developing countries because it generates both food and gross domestic product (GDP) [1]. However, with the enormous demand for food in line with the ever-increasing global population, there has been a rise in additional exploration to produce enough food with sustainable agriculture to satisfy consumers' current nutritional needs, increasing challenges like nutrient deficiencies, pest, soil erosion, water scarcity, and related diseases [2]. Pesticides and fertilizers have been used improperly and excessively to address pest-related issues, which has led to their classification as pollutants [3]. Moreover, the massive use of extremely harmful chemical pesticides, which have a negative impact on ecosystems and human health due to deposition, sorption with leaching tendencies that result in massive accumulation inside soil particles, biological molecules, and metabolic transmutation, is a major problem [3]. Furthermore, pesticide residues on produce have negative health effects [4]. A pesticide is a general term that refers to a large variety of synthetic and natural substances that are frequently used to manage pests. These substances include insecticides, fungicides, herbicides, plant growth regulators, and others. These include chloroacetanilides, organochlorines, organophosphorus, phenoxy derivatives, pyrethrins, pyrethroids, triazines, carbamates, dipyridyl derivatives, glycine derivatives, dithiocarbamates, benzimidazoles, and other unspecified substances [5].

Pathogens, insects, and weeds are responsible for crop losses of 13%, 14%, and 13%, respectively, according to a 2009 report. Herbicides account for 44% of all crop management chemicals used globally, followed by fungicides and bactericides (27%), insecticides (22%), and various other chemicals (7%) [6]. A pesticide should specifically be lethal to targeted pests and not non-targeted pests, although this declaration has brought up questions about the usage and misuse of pesticides. A pesticide, which is a rapid, simple, and affordable pest control option, has many uses in forestry, public health, and other areas of agriculture. In addition to their huge advantages, pesticides also pose threats to the ecosystem and unintended species when used [7]. Pesticides undergo chemical reactions including hydrolysis, photolysis, and biodegradation, which is collectively known as pesticide degradation [7]. This process results in the pesticide being converted into simpler components like water, carbon dioxide, and ammonia [8]. Both biotic transformation activities, which are mediated by microbes or plants, and abiotic transformation processes, such as chemical and photochemical reactions, are involved in the degradation of pesticides [8].

Adopting clever strategies to enhance soil quality and pest management with the introduction of new technology is urgently required. Nanotechnology has recently captured the interest and imagination of scientists and researchers due to recent advancements in the discipline [9]. With its quick development and unique applications, nanotechnology is finding new uses in the agricultural industry, particularly in the active ingredient delivery of nutrients, insecticides, and herbicides [10]. Chitin, a structural component found in multitudinous invertebrates and typically extracted from the exoskeletons of crustaceans, particularly shrimp and crabs, as well as fungi, yeast, and diatom cell walls, is converted into chitosan, a polycationic polymer [11]. It has historically been predominantly insoluble in water and several organic solvents due to its comparatively high hydrophobicity. Significant advancements have been made in the study and use of chitosan over the last 10 years. Chitosan is mostly employed in agriculture as a bio-pesticide, coating film, degradable mulch, food preservation, and plant growth regulator [12].

By alone, chitosan nanoparticles (ChNPs) have the potential to be powerful antibacterial agents and growth promoters for pathogenic fungi and bacteria [13]. Chitosan, with or without the combination of macronutrients, may function as a substituted sustained powerful biocide agent against crop pathogens such bacteria, fungi, and viruses in plant pathogen control and disease management [13]. Nanoparticle (NP)–mediated pesticide detection is a new technique that is gaining popularity in the agricultural sector. The stability, compatibility, and sensitivity of the NPs can be simply altered and range throughout a wide range. Pesticides are typically detected using NPs that are either grooved with fluorophores for surface-enhanced Raman spectroscopy or annexed with electrochemical, antibody, or enzyme-immobilized sensors [14].

This chapter elucidates the importance of chitosan in the detection of pesticides and their application for bioremediation purposes.

17.2 Pesticides and Their Chemical Nature

According to Nicolopoulou-Stamat *et al.* [15] and Hassan and Nemr [16], the term "pesticide" refers to a broad category of insecticides, fungicides, herbicides, garden chemicals, household disinfectants, and rodenticides that are used to both kill and protect against pests. The chemical and physical characteristics of these insecticides vary from one class to another. Because of this, it is admirable to classify them according to their characteristics and research their subgroups [16, 17].

Synthetic pesticides are manufactured from chemicals that do not naturally exist. Depending on their intended usage, they are divided into various groups. There are now three methods of classifying pesticides that are universally accepted [16]. These three often used methods of classifying pesticides include: (i) the chemical makeup of the pesticide; (ii) the mode of entry; and (iii) the activity of the pesticide and the species it kills [16]. According to their sources, chemical pesticides can be divided into four categories: carbamate, organophosphate, organochlorine, and pyrethroid pesticides [16]. The term "biopesticides" refers to a different type of pesticides that are naturally occurring or derived compounds, particularly from living creatures including plants, fungus, and bacteria [16, 17]. However, the classification, traits, and environmental effects of synthetic or chemical pesticides are the main emphasis of this review.

17.2.1 Classification of Synthetic Pesticides and Their Toxic Effect

The description of the active chemicals and the chemical makeup of the insecticides determine the most popular and acceptable technique for classifying them. It is a sort of classification that offers proof of the effectiveness, chemical composition, and physical characteristics of particular pesticides. Pesticides are divided into four primary groups based on their chemical makeup: organiochlorine, organic phosphorous, carbamate, pyrethrin, and pyrethroid (Figure 17.1) [16, 17].

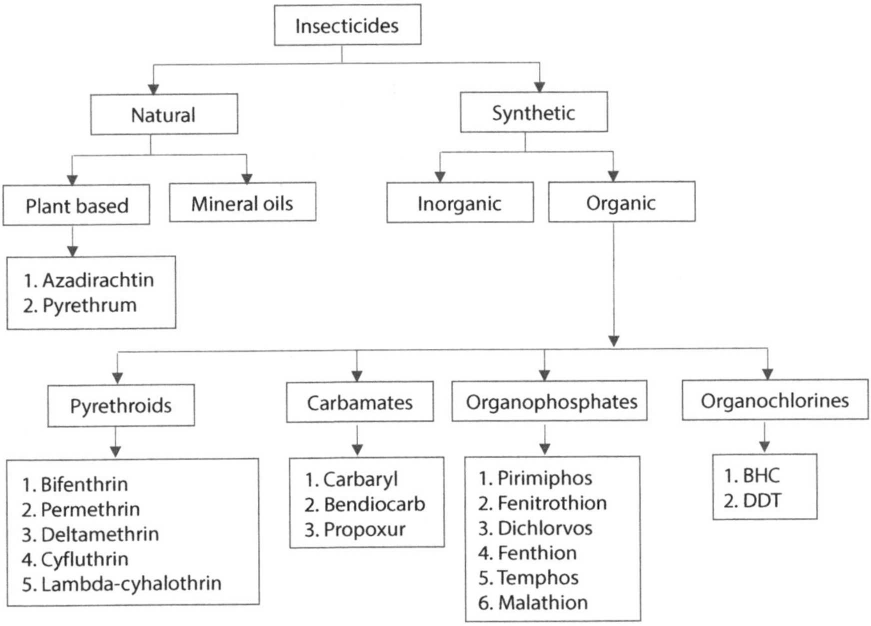

Figure 17.1 Classification of insecticides [16, 17].

17.2.1.1 Pyrethroid Pesticides

The chrysanthemum flower extracts known as pyrethrin, which are found in Kenya, are the source of pyrethroids, which are natural insecticides [16, 17]. By affecting the central nervous system, it alters the kinetics of sodium cation channels in nerve cell membranes, lengthening the time it takes for sodium channels to open. In both invertebrates and vertebrates, the sodium cation stream crosses the membrane [17, 18]. These effects may lead to a neuronal hyperexcitation [15, 16]. Scientists are now turning to the creation of synthetic pyrethroids due to the pressing need for huge quantities of these pesticides and the growing scarcity of essential oils required for the fabrication of natural organic pyrethrum [19]. The majority of pyrethroid insecticides share a number of traits, including being fast-acting, especially against chewing insects, low toxicity to birds and mammals, high toxicity to arthropods because it takes very small doses to kill insects, and highly toxic to fish if used directly on water [16, 17]. Despite the fact that many pyrethroid pesticides can be absorbed by insect pests when they walk on dry residues, they are ineffective at killing subsurface pests because of their strong adhesion to soil and organic substances. Additionally, according to

Nicolopoulou-Stamat *et al.* [15] and Hassaan and Nemr [16], they have poor water solubility. Even though there are more than 1,000 distinct pyrethroids in use today, only a handful are sold in the US. Many commercial goods, including pet shampoos, human lice treatments, pet sprays, topical mosquito repellents, and, of course, insecticide sprays for homes, businesses, and farms, contain pyrethroids as their active ingredient [15, 16].

17.2.1.2 Carbamate Pesticides

Herbicides, insecticides, nematicides, and fungicides, as well as several organic ester compounds produced from dimethyl N-methyl carbamic acid, are known as carbamates [5–18]. Cats and dogs both frequently utilize carbamates, including thiobencarb, propoxur, molinate, disulfiram (Antabuse), pyridostigmine, methiocarb, and carbaryl. The toxicity of carbamate compounds varies depending on their molecular makeup; however, they are generally less dangerous and last less time than organophosphates and organochlorines, the latter of which inhibit acetyl cholinesterase [15–18]. Acute carbamate poisoning is treated similarly to organophosphate toxicity. Given the transient nature of the carbamates, atropine administration should be done with greater caution. Organophosphate or carbamate insecticides frequently cause severe acute poisoning symptoms. The following is a list of the several organs where these poisoned symptoms can be seen: bronchial tree (dyspnea, coughing up more secretions, and muscarinic symptoms; cough, pulmonary edema, cyanosis, and bronchoconstriction), glandular stimulation (increased salivation, lacrimation, and sweating), cardiovascular effects (bradycardia, and hypotension), eye (miosis and blurred vision), bladder dysfunction (incontinence and frequency), gastrointestinal manifestations (nausea, vomiting, abdominal cramps, diarrhea, and incontinence), nicotinic receptor stimulation (including sympathetic and motor neurons), motor activity compromised (muscle twitching, depression of respiratory and circulatory, fasciculation, cramps, and weakness), sympathetic dysfunction (tachycardia, hypertension, and pallor), central nervous system effects (restlessness, emotional lability, confusion, drowsiness, Cheyne-Stokes respiration, slurred speech, tremor, ataxia, generalized weakness, coma, areflexia, convulsion, and hypothermia) [15–18].

17.2.1.3 Organophosphate Pesticides

Organophosphate insecticides are some esters made from phosphoric acid. Inhabitation of the enzyme acetylcholine, these esters form affects the human central nervous system [16–18]. Acetylcholinesterase is a neurotransmitter that disrupts nerve impulses by peacefully phosphorylating the OH group in the enzyme's active site, which is managed by this enzyme. The chemical structure of some pesticides is revealed by the intoxication symptoms of unconsciousness, nausea, headache, cramps, convulsions, and even death [16–18]. Atropine, which competes with acetylcholine at the muscarinic receptors, is the only effective treatment for organophosphate poisoning [16]. Up until the adult dose, the initial dose for adults is 2 to 5 mg IV or 0.05 mg/kg IV for children. Once the respiratory secretions have cleared and there is no bronchoconstriction, the dose is raised every 3 to 5 min if the patient does not respond to the treatment [16]. Hundreds of milligrams of atropine administered as a bolus or continuous infusion over the course of several days may be necessary in individuals with severe poisoning before they start to recover [16, 17].

17.2.1.4 Organochlorine Pesticides (OCPs)

Organochlorines are stable compounds that have the ability to build up in adipose tissue and are extremely persistent in the environment [16, 18]. These substances or their metabolites primarily affect the central nervous system in humans. This alteration of the enzymatic nerve membranes and electrophysiological properties results in altered K+ and Na+ flow kinetics through the nerve cell membrane, which may result in symptoms like acute poisoning death and seizures from breathing difficulty. Organochlorines can be divided into five groups based on their structural characteristics: (1) DDT and its analogs including DDT and dichlorodiphenyldichloroethylene (DDE) and dichlorodiphenyldichloroethane (DDD); (2) hexachlorocyclohexane (HCH), such as lindane; (3) cyclodienes including aldrin, dieldrin, endrin (sometimes referred to as "drins" in the literature), heptachlor, chlordane, and endosulfan; (4) toxaphene; and (5) mirex and chlordecone [16–18]. Some organochlorines, such as DDT, DDE, and DDD, have a 15-year field half-life, but aldrin and toxaphene have half-lives of 365 and 9 days, respectively. Death under natural conditions may be delayed and frequently viewed as a general wasting away or chronic sickness because the acute toxicity of the majority of OCPs typically occurs at doses that are greater than those thought to be environmentally realistic [16]. Most OCPs have a lipophilic, persistent character that might cause long-term storage in adipose tissue and release into the bloodstream under stressful environmental situations. Delays between the initial exposure and the onset of symptoms may result from this. According to several studies [15–18], DDT can linger in the body for up to 50 years.

The harmful effects of synthetic chemicals and their impacts on the ecosystem have made scientists seek out new technologies that are sustainable and eco-friendly in remediating the environment and alleviating the deleterious effects pesticides have on the ecosystem through the development and use of nanotechnology that utilizes chitosan in the environmental cleanup.

17.3 Properties of Chitosan Nanoparticles

Chitosan nanoparticles (ChNPs) are a form of nanomaterial that have a wide range of applications in a variety of industries, including the food, pharmaceutics, tissue engineering, agriculture, and medical fields [20].

17.3.1 Structure and Natural Origin of Chitin

Chitin is the second-most common natural polymer in the world after cellulose. Two marine crustaceans, including shrimp, crabs, lobster, and crawfish, are the main sources exploited and provide enough chitin/chitosan for the commercial sector [21]. D-glucosamine and N-acetyl-D-glucosamine are two subunits that make up the linear polymer chitosan, and they are connected by 1,4-glycosidic linkages [21]. The chitosan molecule's three rings make up its overall structure. It has three functional groups that make chemical modification simple, including amino groups, primary hydroxyl groups, and secondary hydroxyl groups [22]. These functional groups also affect the solubility and mechanical characteristics of chitosan. Additionally, chitosan has 1,4 glycosidic connections. Chitosan is more soluble than chitin in aqueous settings that are acidic. Chitosan is soluble primarily because of the protonation of -NH2 at the C-1 position of the D-glucosamine repeat unit, which, under acidic circumstances, converts the polysaccharide into a polyelectrolyte [23].

Chitosan can be used in almost all facets of human life and health, including agriculture, medicine, process engineering, and industries [24]. A number of fungi can be directly removed for their chitosan, but it can also be made by removing chitin and then deacetylating it. Chitin is a polymer that is produced by the zygomycetes, ascomycetes, basidiomycetes, deuteromycetes, and phycomycetes of the fungus kingdom more commonly than chitosan [25]. Chitosan, on the other hand, is only present in the cell walls of a small number of fungal taxa, mainly zygomycetes [25]. It is highly advantageous to extract chitosan from fungal biomass because it can be done whenever and is not affected by seasonal fluctuations. Chitin is only found in the cell walls of fungus, and it is extracted from them using the same procedure as is used to extract chitosan from crustaceans. Therefore, it is more cost-effective to use those fungi that already produce the desired output [25].

The exoskeletons of crustaceans, primarily a range of marine crustaceans including shrimp, crabs, and lobsters, are another crucial and abundant source of chitin. Currently, waste from the shrimp business serves as the primary source of chitin, with exoskeletons being used to extract both chitin and calcium. The chemical method of N-deacetylation is used to produce chitosan, and the free amino group left over after chitins' acetyl group was partially removed is what gives chitosan its cationic properties [26].

17.3.2 Chitosan Nanoparticles

Chitosan that has been altered to be smaller in size is known as nanochitosan; its size ranges from 100 nm to 400 nm [27]. It is possible to create ChNPs with the desired nanoscale properties, including a tiny size, specific surface and interface effects, and quantum size effects [27]. ChNPs have received a lot of interest recently for a variety of applications in the pharmacological, biological, and agricultural sectors [21].

17.3.2.1 Properties of Chitosan Nanoparticles

Numerous studies have demonstrated that the characteristics of chitosan NPs can range significantly depending on the processes employed for surface modification and preparation, which might result in applications in entirely different sectors [28]. For the past 10 years, chitosan has been studied as a potential material for shaping NPs. Making chitosan into NPs has improved its qualities. Chitosan NPs might exhibit typical activities because of the new properties of NP due to their small size and quantum estimation influence. They have a remarkable size and a large surface-to-volume ration and are simple, inexpensive, and easy to scale up. They are mucoadhesive and hydrophilic by nature, which increases their stability in the body and increases the security they provide to a typical tranquilizer. They fit a generic classification of drugs, tiny particles, proteins, and polynucleotides in this way. The benefits of typifying dynamic experts in a polymer framework include their regulated release and insurance from the surrounding medium or handling situations [29].

17.3.2.2 Synthesis of Chitosan Nanoparticles

Compared to the parent material, materials that have reached the nanoscale may exhibit enhanced or unexpected features. Numerous preparation methods (top-down, cross-linking, and microbiological) and potential application areas (electronics, textile, and medicine) for inorganic and organic NPs have been identified as a result of extensive research [28].

17.3.2.2.1 Emulsification and Cross-Linking Method

Chitosan NPs were originally prepared by emulsification and cross-linking using the amino group of the chitosan and the aldehyde group of a cross-linking agent. With the aid of toluene, glutaraldehyde, and Span 80 as stabilizers and cross-linkers, an emulsion composed of an aqueous chitosan solution and an oil phase is created [28]. After vigorous mixing of the phases, droplets that serve as the building blocks of the NPs are created. Centrifugation, numerous washing processes (with petroleum ether, acetone, sodium metabisulfite, and water), and vacuum- or freeze-drying are frequently used to separate NPs from the emulsion. Small-sized NPs with a restricted dispersion are attainable. However, because glutaraldehyde was found to have overt toxicity and difficulties with medication integrity, this approach is no longer in use [28].

17.3.2.2.2 Microemulsion Method

The reversed micelle (microemulsion) method is based on covalent cross-linking, just like emulsification and cross-linking, but it now uses water-in-oil reverse micelle structures to help produce ChNPs. Chitosan and glutaraldehyde from the aqueous phase are combined with an organic solvent and a lipophilic surfactant from the organic phase. Typically, n-hexane is favored as an organic solvent and cetyltrimethylammonium bromide or sodium bis (2-ethylhexyl) sulfosuccinate are employed as surfactants [28]. Chitosan NPs are created by cross-linking in a nanoreactor within the chitosan-containing micelle's core. The precipitation of surfactant with $CaCl_2$, dialysis to remove unreacted components, and freeze-drying are the three processes in the isolation of NPs. It is feasible to create ultrafine NPs with a size of less than 100 nm, which is a crucial characteristic for numerous applications where the particular surface area (loading capacity and sustained release) is crucial. Recently, the preparation process has been modified to use cross-linker and harmless solvents, which reduces the drawbacks of the traditional method based on glutaraldehyde [28].

17.3.2.2.3 Precipitation Method

Precipitation-based techniques can also be used to make ChNPs. Emulsification and precipitation are the foundations of the phase-inversion precipitation method [30]. A stabilizer (poloxamer), an organic phase (dichloromethane and acetone), and an aqueous solution of chitosan are used to create the oil-in-water emulsion [30]. To create well-dispersed emulsion droplets that are nanoscale in size, high-pressure homogenization is used. Then, acetone diffuses out of the droplets, and NPs precipitate simultaneously as methylene chloride is removed from the emulsion by evaporation at low pressure and room temperature [30]. As an alternative, there is the emulsion-droplet coalescence approach, which relies on the coalescence of two water-in-oil emulsions and causes the precipitation of NPs because NaOH acts as a precipitation agent in one of the emulsions. The two emulsions, one with chitosan and the other with NaOH, are created by first combining liquid paraffin and sorbitan sesquioleate. Using high-speed homogenization, an emulsion containing chitosan is created. When two emulsions are blended, NaOH diffuses into the ultrafine droplets, reducing the solubility of the chitosan and causing the precipitation of NPs. Three stages are required to produce NPs: centrifugation, washing with toluene, ethanol, or water, and freeze-drying [28]. Precipitation-method NPs are typically larger than 600–800 nm. The phase inversion precipitation method produces ChNPs with a high encapsulation efficiency for hydrophobic drugs like cyclosporin A, despite the fact that both methods are not particularly preferred due to the use of organic solvents and high energy requirements for homogenization applications [28].

17.3.2.2.4 Ionic Gelation Method

Ionic gelation is one of the most popular ways to create ChNPs, which have been studied for more than 20 years [31]. Its foundation is ionic cross-linking, which takes place when inversely charged groups are present. For instance, sodium tripolyphosphate (TPP) and the protonated amino groups of chitosan are negatively charged polyanion groups [32]. Aqueous TPP solution is added while being vigorously stirred after chitosan has been dissolved in an acidic aqueous solution (often an acetic acid solution). Chitosan molecules that are positively charged diffuse into one another, causing cross-linking that results in the production of NPs. Chitosan NPs can be produced using an oven method following a few centrifugations and water-washing procedures; drying, freezing, or drying [28]. It uses a simple method without solvents or hazardous cross-linkers. Additionally, the procedure may be completed at room temperature, and the size of the resulting NPs can be modified by varying the chitosan/TPP ratio, a significant characteristic that directly influences the effectiveness of drug encapsulation and delivery [28].

17.3.2.2.5 Top-Down Method

It is also possible to generate NPs via the top-down approach, which requires two steps: deacetylation (changing the acetyl group of the chitins with an amino group), and acid hydrolysis of the chitin to produce chitin NPs. Basically, a strong acid, like hydrochloric acid, is used to break glycosidic bonds, and the amorphous component is removed using a few centrifugation and washing processes. Chitin nanocrystals are isolated after several centrifugation steps. Chitosan NPs are chitin NPs that have been deacetylated to a degree greater than 60%. Unlike other techniques where drug loading occurs simultaneously with NP formation, this strategy requires a second loading step [28].

17.3.2.2.6 Green Synthesis Method

The creation of NPs utilizing microorganisms like bacteria, fungi, actinomycetes, and botanicals is known as "green synthesis." The bottom-up production of NPs is a method that integrates the fields of nanotechnology and biotechnology [33, 35]. The creation of NPs through biological means is preferred to chemical or physical means because it is a process that is safe, environmentally friendly, economically, and energetically viable; takes less time; and best utilizes the redox potential of metabolites produced by biological entities to transform macromolecules into nanoform [35]. According to Dutta *et al.* [35], the primary mechanism of biogenic production of the NP is based on redox reactions that take place when the microorganisms/biological entity binds to the metal ion and detoxifies it into the element metal through cellular enzymatic activity. Both internal and extracellular syntheses fall under this category. In intracellular synthesis, metal ions are brought inside the cell, where they undergo a reduction reaction in the cytoplasm, cell wall, and/or periplasm, leading to the formation of NPs inside the cell. But with the latter, the foreign object is trapped, and the reductase enzyme catalyzes the creation of NPs on the cell surface. Extracellular production of NPs is typically chosen since intracellular synthesis requires a laborious technique to harvest NPs from the cell matrix [35]. Due to their ability to produce NPs on a vast scale with a relative degree of control over their size and form (controlled by the surrounding environmental conditions) and a more straightforward production method, microorganisms are frequently characterized as environmentally benign, green nano-factories [35].

17.3.2.3 Characterization of Chitosan Nanoparticles

The processing environment and duration have a significant impact on NP production [35]. They are evaluated using the UV-vis spectrophotometer, electron microscope, dynamic light scattering, and Fourier transform infrared spectroscopy based on their surface plasmon resonance, morphology, particle size distribution, zeta potential, functional group analysis, etc. Additionally, atomic absorption spectroscopy is used to examine the release profile of NPs encapsulated in Ch [34, 36, 37].

17.3.2.3.1 Characterization by Dynamic Light Scattering

According to Sandhu *et al.* [38], the fundamental idea behind DLS is based on the Brownian movement of particles and molecules in a solution as a result of their interaction with randomly moving solvent particles. The photon detector measures the variation in scattered light caused by the random movement of particles after a laser beam has passed through the sample [34]. According to Karmakar *et al.* [39], DLS is used to quantify zeta potential, polydispersity index (PDI), and average particle size. The polydispersity or monodispersity of particles in an aqueous media is explained by PDI. A PDI number larger than 0.5 indicates polydispersity, and one less than 0.5 typically indicates monodispersity. Monodisperse ChNPs often have PDI values between 0.2 and 0.4 [40]. Additionally, the stability of the NPs, which is measured in the region of 30 mV, is explained by the surface charge of NPs, also known as zeta potential. Positive zeta potential measurements for the ChNPs range from 1.2–11.2 mV to 0.4–18.7 mV. Additionally, DLS's determination of the proper particle size revealed that the ChNPs are approximately spherical in form, with sizes ranging from 150 nm to 350 nm [28].

17.3.2.3.2 Characterization by UV-Vis Spectroscopy

The confirmatory investigation that establishes the creation of NPs via surface plasmon resonance is UV-vis spectroscopy. Using UV-vis absorption spectroscopy, the optical characteristics of ChNPs are investigated. ChNPs obtained from biogenic sources displayed the absorption band at 310–342 nm, in contrast to those obtained from the commercial manufacture of chitosan, which had an absorption band at 330.25 nm [34].

17.3.2.3.3 Characterization by Electron Microscopy

Transmission and scanning electron microscopes are used to examine the NPs' internal and exterior appearance [41]. According to transmission electron microscopy, ChNPs are frequently spherical in shape and amorphous in nature. Similar findings were made by other researchers using scanning electron microscopy, who discovered that the ChNPs have a nearly spherical form and a smooth exterior surface [41].

17.4 Application of Chitosan in Bioremediation of Pesticide

The agriculture industry's quick development has led to a significant increase in the amount of pesticides utilized. They must be removed since they pose a risk to others, which is why it is more important. Pesticides can be removed from contaminated water using chitosan derivatives. Following pseudo-first-order kinetics, Mostafa *et al.* [42] reported the removal of acephate, omthosate, and methyl parathion, three species of organophosphorus pesticides with adsorption capacities of 650.7 mg/g, 506.5 mg/g, and 560.8 mg/g, respectively. The pesticides adsorb to the adsorbate as monolayers, as evidenced by the data best fitted to Langmuir isotherm model. The composite demonstrated a promising activity in adsorbing the pesticides in the presence of inorganic anions such SO_4^{2}, PO_4^{3}, and NO_3, as well as the metal ions Zn^{2+}, Pb^{2+}, and Cd^{2+}. Adsorption was spontaneous, practical, and exothermic [42].

The SO_4^{2} had been reported to have a negligible impact, and other ions exhibited competitive behavior, which resulted in a sharp decline in the percentage of adsorption. For five trials, the chitosan/zeolite combination demonstrated stability in the removal of the organophosphorus insecticides [42]. It has been demonstrated that a magnetic chitosan/activated carbon bio-nanocomposite with a UiO-66 metal-organic framework is efficient at adsorbing the pesticide Imidacloprid, as well as CO^{2+} and the dye malachite green. The three contaminants mentioned above had adsorption capabilities of 25.2 mg/g, 44.5 mg/g, and 62.1 mg/g in ternary media, respectively. The data were well matched to the Langmuir isotherm model, and the pseudo-second-order kinetics of adsorption revealed that the adsorbates were adsorbed as monolayers through chemisorption. The created composite is a reusable, eco-friendly adsorbent that may be utilized to simultaneously remove several contaminants [40].

Chitosan has been used to remove pesticides from grape juice. According to Venkatachalapathy *et al.* [43], a 0.5% solution of chitosan removed 98% of chlorpyrifos and 97% of ethion in 1 h, and a 1% solution removed 96%, 95%, 94%, and 86% of phorate, fenthion, fenitrothion, and diazinon in 2 h. The effectiveness of flocculation/coagulation in removing pesticides from derivatives of chitosan that contain thymine has been reported in the past. The pesticides used are the Bordeaux mixture, Fastac 10 EC, and Karate Zeon, whose active pyrethroid constituents are alpha-Cypermethrin and lambda-Cyhalothrin, respectively; Novadim Progress, whose active organophosphoric ingredient is dimethoate; and Novadim Progress. Karate Zeon, Fastac 10EC, and Novadim Progress were all successfully removed by chitosan-thymine coordination, as well as 95% of the Bordeaux mixture. Charge neutralization caused the removal of Fastac 10 EC and Karate Zeon, whereas the combination of Novadim Progress and Bordeaux was eliminated by flocculation as a result of the combined action of charge neutralization, hydrogen bonding, and copper ion binding to the amine and thymine groups [44].

17.4.1 Nanochitosan in Pesticide Detection and Bioremediation

The significance of chitosan NPs in environmental remediation cannot be overemphasized owing to their unique properties and eco-friendliness. Below are the various ways in which nanochitosan functions in detection of pesticide in the environment.

17.4.2 Mechanism of Nanochitosan-Based Pesticide Detection

Due to its biocompatibility, biodegradability, antibacterial activity, and capacity to form NPs with different functional groups, nanochitosan has been employed as a material for the detection and degradation of pesticides. The type of sensor being utilized affects how nanochitosan-based pesticide detection works. Examples include the following:

17.4.2.1 Colorimetric Sensors

For the immobilization of enzymes or antibodies that can react with particular pesticides and cause a color change, these sensors use nanochitosan as a substrate [45]. For instance, acetylcholinesterase (AChE), an enzyme that is inhibited by pesticides containing organophosphorus and carbamates, was immobilized using nanochitosan. A spectrophotometer was able to measure a drop in the color intensity of the sensor as a result of the pesticides' inhibition of AChE [42].

17.4.2.2 Fluorescence Sensors

Fluorescent probes that can interact with pesticides and change their fluorescence properties are carried by nanochitosan in these sensors. For instance, quantum dots (QDs), which are nanoscale semiconductor crystals that generate fluorescence when stimulated by light, were enclosed in nanochitosan. Aptamers, which are short DNA or RNA sequences that may attach to particular targets, were used to functionalize the QDs. The aptamers were created to identify pesticides with an organophosphorus structure, such as chlorpyrifos and parathion-methyl. A fluorimeter was able to detect a dampening of the QDs' fluorescence as a result of the pesticides' binding to the aptamers [23].

17.4.2.3 Electrochemical Sensors

These sensors modify electrodes with nanochitosan so that the electrodes can produce electrical signals in reaction to pesticides. For example, gold electrodes that were further modified with graphene oxide (GO) and tyrosinase (TYR), an enzyme that can oxidize phenolic chemicals, were coated with nanochitosan [46]. By detecting the drop in current brought on by the herbicides' inhibition of TYR, the nanochitosan-GO-TYR electrode was able to identify phenylurea herbicides like diuron and linuron [46].

17.4.3 Bioremediation of Pesticides Using Chitosan Nanoparticles

Pesticides are being removed from the environment using a variety of ways. These include employing nanochitosan-based catalysts, adsorbents, and catalysts made of nanochitosan.

17.4.3.1 Nanochitosan-Based Adsorbent

Heavy metals, dyes, pesticides, medicines, and endocrine disruptors are just a few of the contaminants that can be removed from water using nanochitosan-based adsorbents [46]. The removal of pesticides from water can be accomplished using a variety of sustainable adsorbents, such as nanochitosan and its composites with other substances including graphene oxide, metal oxides, carbon nanotubes, and biochar [46]. In addition, the essay discusses the mechanisms of adsorption, kinetics, thermodynamics, and regeneration of the adsorbents. Singh and Saxena [47] conducted additional research to examine the possibility of nanotechnology in accelerating the bioremediation of pesticides from contaminated soil. The essay emphasizes how pesticide microbial breakdown is aided by nanomaterials like nanochitosan by improving their bioavailability, solubility, and mobility in the soil matrix [47]. An investigation on the adsorptive removal of three pesticides (atrazine, chlorpyrifos, and imidacloprid) from aqueous solutions using a chitosan/gelatin composite is presented in a research study by Attallah *et al.* [48]. The article assesses the effects of numerous factors on the adsorption process, including pH, contact time, initial concentration, and temperature. The article by Attallah *et al.* [48] also discusses the composite's adsorption isotherms, kinetics, thermodynamics, and desorption experiments.

17.4.3.2 Nanochitosan-Based Catalysts

Use of nanochitosan as a support or a modifier for other heterogeneous catalysts, such as metal oxides or metal sulfides, that may catalyze the oxidation and decomposition of pesticides under various conditions, is one strategy to improve the performance of nanochitosan for pesticide degradation. Based on the type of metal, the technique of synthesis, and the reaction mechanism, these nanochitosan-based catalysts can be divided into four groups: nanochitosan-metal NPs, nanochitosan-metal oxide composites, nanochitosan-metal sulfide composites, and nanochitosan-metal ion complexes are four examples of nanochitosan-metal materials [49].

Some examples of the application of nanochitosan in pesticide degradation using nanochitosan-based catalysts are as follows:

17.4.3.2.1 Nanochitosan-Copper Ion Complex

Nanochitosan was reacted with copper sulfate and sodium hydroxide to create this complex. Under the influence of visible light, the compound demonstrated catalytic activity for the breakdown of organophosphorus insecticides like malathion and dimethoate. The nanochitosan-copper ion complex, which could oxidize and break down the pesticides, was responsible for the degradation mechanism by producing reactive oxygen species (ROS) like hydroxyl radicals and superoxide anions [20].

17.4.3.2.2 Nanochitosan-Titanium Dioxide Composite

The breakdown of organochlorine pesticides like DDT and endrin by this composite, which was made using the sol-gel process, was accelerated when exposed to UV light [49]. The composite increased the surface area and interaction between the photocatalyst and the pollutants by enhancing the dispersion and stability of titanium dioxide NPs on nanochitosan matrix [50]. Due to the pesticides' electrostatic attraction and hydrogen bonding with nanochitosan, the composite also demonstrated strong pesticide adsorption capacity.

17.4.3.2.3 Nanochitosan-Cadmium Sulfide Composite

With the aid of a microwave, this composite was created, and it demonstrated better photocatalytic activity for the breakdown of phenylurea herbicides like diuron and linuron when exposed to visible light [51]. By doping the cadmium sulfide NPs with nanochitosan, which decreased the band gap and increased conductivity, the composite improved the optical characteristics and charge transfer efficiency of the particles. The composite also shows excellent reusability and stability [51].

17.4.3.2.4 Nanochitosan-Gold Nanoparticle Composite

With the utilization of grape juice as a reducing agent, this composite was made utilizing a green technique [52, 53]. In the aqueous solution-based degradation of carbamate insecticides like carbaryl and carbofuran, the composite exhibited excellent catalytic activity. According to Amanulla *et al.* [53] and Hashem *et al.* [54], the degradation mechanism involves the catalytic reduction of carbamate pesticides by gold NPs supported on nanochitosan, which leads to the creation of less hazardous compounds.

17.4.3.3 Nanochitosan-Based Photocatalysts

Combining nanochitosan with other photocatalysts, such as metal oxides or metal sulfides, which can capture visible light and produce ROS that can oxidize and degrade the pesticides, is one way to improve the performance of nanochitosan for pesticide degradation [55]. On the basis of the elemental composition, band gap engineering, and charge carrier migration mechanism in composite photocatalysts, these nanochitosan-based photocatalysts may be divided into four groups: solid solutions, type II heterojunction nanocomposites, pure semiconductors, and Z-scheme [56].

17.4.4 Challenges and Limitations of Nanochitosan-Based Pesticide Detection and Degradation Technologies

The monitoring and repair of pesticide pollution in the environment can be accomplished using nanochitosan-based pesticide detection and degradation technology [57]. They do, however, also experience some difficulties and constraints that must be resolved. Among these are as follows:

For the synthesis, characterization, and assessment of nanochitosan and its derivatives for pesticide detection and degradation, there are no established procedures and protocols [58]. The reliability and repeatability of nanochitosan-based materials may be impacted by different approaches' potential for diverse attributes and performances [59]. The possible negative effects of using nanochitosan and its derivatives for pesticide detection and degradation on the environment and human health as they can potentially cause oxidative stress and even toxicity. Although nanochitosan is typically thought of as a biocompatible and biodegradable substance, its interactions with other nanomaterials, pesticides, or biological systems could change how hazardous and how it behaves. To ensure the safety and sustainability of nanochitosan and its derivatives, it is important to examine their fate, transport, accumulation, and transformation in the environment [59].

There are several obstacles encountered in the use of ChNPs among them is the incomplete understanding of the mechanism and kinetics of pesticide detection and degradation using nanochitosan. According to El-Naggar *et al.* [60], multiple physical, chemical, and biological mechanisms may be involved in the interaction of nanochitosan and its derivatives with pesticides. Further research is required to determine the factors, such as the type and concentration of the pesticide, the size and shape of the nanochitosan particles, the functionalization and modification of the nanochitosan, and the environmental conditions, that may affect the effectiveness and selectivity of the nanochitosan-based pesticide detection and degradation [61].

High prices and a limited supply of nanochitosan and its derivatives for pesticide detection and degradation are among the limitations on the use of ChNPs [62]. The natural resource chitin, which may be scarce and of poor quality, is the source of nanochitosan. Nanochitosan may need specialized tools, processes, and chemicals for its synthesis, purification, functionalization, and modification, which could raise the price and complexity of nanochitosan-based materials [62]. To make pesticide detection and degradation technologies based on nanochitosan more widely used and less expensive, they must be scaled up and commercialized.

17.5 Conclusion

A substance with great potential for pesticide detection and degradation is nanochitosan. It is superior to conventional approaches in a number of ways, including high sensitivity and selectivity, affordability and ease of access, non-toxicity and biodegradability, and adaptability (may be used for a variety of pesticides). Pesticides can be found using nanochitosan in a number of ways, such as colorimetric, fluorescence, and electrochemical sensors. Through the use of nanochitosan-based catalysts, adsorbents, and catalysts composed of nanochitosan, it has also been discovered that nanochitosan is efficient in the bioremediation of pesticide-polluted environments. In general, nanochitosan is a substance that shows promise for the detection and breakdown of pesticides. It differs from standard techniques in a number of ways, and it may be used to create new and improved pesticide detection and degrading technologies.

17.6 Future Perspective

A promising area of study that promises to lessen the negative effects of agrochemicals on the environment while also enhancing soil quality and crop yield is the employment of nanochitosan in the degradation of pesticides. A biopolymer called nanochitosan is created from the natural substance chitin, which is found in the exoskeletons of insects and crustaceans. In comparison to traditional chitosan, nanochitosan has various benefits, including a larger surface area, more reactivity, greater solubility, and lower toxicity. When combined with other nanomaterials like silicon, cellulose, or zinc oxide, nanochitosan can create nanocomposites with improved photocatalytic and pesticide adsorption capabilities. Future prospects for the use of nanochitosan in the breakdown of pesticides include the creation of innovative nanocomposites based on the substance that have increased efficacy, stability, and pesticide selectivity. The mechanisms and kinetics of pesticide degradation by nanocomposites based on nanochitosan under various environmental variables, such as pH, temperature, soil type, and others, must also be investigated. In order to maximize the effectiveness of the NPs, it is crucial to assess the environmental fate, toxicity, and breakdown products of nanochitosan-based nanocomposites in soil and water systems. For acceptance and widespread use for a better environment and the ecosystem as a whole, it is also essential to optimize the synthesis, characterization, and application methods of nanochitosan-based nanocomposites for large-scale manufacturing and field application. In order to evaluate nanocomposites based on nanochitosan as an alternative to conventional pesticides and remediation methods, it is also critical to consider their economic viability and social acceptability.

References

1. Begna, T., Major challenging constraints to crop production farming system and possible breeding to overcome the constraints. *Int. J. Res. Stud. Agric. Sci. (IJRSAS)*, 6, 7, 27–46, 2020.
2. Hemathilake, D.M. and Gunathilake, D.M., Agricultural productivity and food supply to meet increased demands, in: *Future Foods*, pp. 539–553, Academic Press, Massachusetts, USA, 2022 Jan 1.
3. Lau, Y.Y., Hernandes, E., Kristanti, R.A., Wijayanti, Y., Emre, M., Exploring the Potential of Composting for Bioremediation of Pesticides in Agricultural Sector. *Ind. Domest. Waste Manage.*, 3, 1, 47–66, 2023 May 22.
4. Ali, S., Ullah, M.I., Sajjad, A., Shakeel, Q., Hussain, A., Environmental and health effects of pesticide residues. *Sustain. Agric. Rev. 48 Pestic. Occurrence Anal. Rem.*, 2, 311–36, 2021.
5. Biondo, F., Baldassarre, F., Vergaro, V., Ciccarella, G., Controlled biocide release from smart delivery systems: Materials engineering to tune release rate, biointeractions, and responsiveness, in: *Nanotechnology-Based Sustainable Alternatives for the Management of Plant Diseases*, pp. 31–147, Elsevier, Amsterdam, Netherlands, 2022 Jan 1.
6. Saroj, A., Oriyomi, O.V., Nayak, A.K., Haider, S.Z., Phytochemicals of plant-derived essential oils: A novel green approach against pests, in: *Natural remedies for pest, disease and weed control*, pp. 65–79, Academic Press, Massachusetts, USA, 2020 Jan 1.
7. Meftaul, I.M., Venkateswarlu, K., Dharmarajan, R., Annamalai, P., Megharaj, M., Pesticides in the urban environment: A potential threat that knocks at the door. *Sci. Total Environ.*, 711, 134612, 2020 Apr 1.
8. Ramalingam, G., Priya, A.K., Gnanasekaran, L., Rajendran, S., Hoang, T.K., Biomass and waste derived silica, activated carbon and ammonia-based materials for energy-related applications–A review. *Fuel*, 355, 129490, 2024 Jan 1.
9. Pansera, M., Owen, R., Meacham, D., Kuh, V., Embedding responsible innovation within synthetic biology research and innovation: insights from a UK multi-disciplinary research centre. *J. Responsible Innov.*, 7, 3, 384–409, 2020 Sep 1.
10. Acharya, A. and Pal, P.K., Agriculture nanotechnology: Translating research outcome to field applications by influencing environmental sustainability. *NanoImpact.*, 19, 100232, 2020 Jul 1.

11. Syed, N.H., Rajaratinam, H., Nurul, A.A., Chitosan from Marine Biowaste: Current and Future Applications in Tissue Engineering, in: *Sustainable Material for Biomedical Engineering Application*, pp. 87–106, Springer Nature Singapore, Singapore, 2023 Aug 2.
12. Vermoesen, E., Bodé, S., Brosens, G., Boeckx, P., Van Vlierberghe, S., Chemical strategies towards controlled release in agriculture. *Rev. Chem. Eng.*, 8, 0, 247–277, 2023 May.
13. Karamchandani, B.M., Maurya, P.A., Dalvi, S.G., Waghmode, S., Sharma, D., Rahman, P.K., Ghormade, V., Satpute, S.K., Synergistic activity of rhamnolipid biosurfactant and nanoparticles synthesized using fungal origin chitosan against phytopathogens. *Front. Bioeng. Biotechnol.*, 10, 917105, 2022 Aug 9.
14. Sur, S., Sathiavelu, M., A concise overview on pesticide detection and degradation strategies. *Environ. Pollut. Bioavailab.*, 34, 1, 112–26, 2022 Dec 31.
15. Nicolopoulou-Stamati, P., Maipas, S., Kotampasi, C., Stamatis, P., Hens, L., Chemical pesticides and human health: the urgent need for a new concept in agriculture. *Front. Public Health*, 4, 148, 2016 Jul 18.
16. Hassaan, M.A. and El Nemr, A., Pesticides pollution: Classifications, human health impact, extraction and treatment techniques. *Egypt. J. Aquat. Res.*, 46, 3, 207–20, 2020 Sep 1.
17. Kaur, R., Mavi, G.K., Raghav, S., Khan, I., Pesticides classification and its impact on environment. *Int. J. Curr. Microbiol. Appl. Sci.*, 8, 3, 1889–97, 2019 Mar 20.
18. Pandya, I.Y., Pesticides and their applications in agriculture. *Asian J. Appl. Sci. Technol.*, 2, 2, 894–900, 2018 Apr.
19. Isman, M.B., Botanical insecticides in the twenty-first century—fulfilling their promise? *Annu. Rev. Entomol.*, 65, 233–49, 2020 Jan 7.
20. Saravanan, R., Karthikeyan, S., Gupta, V.K., Sekaran, G., Narayanan, V., Stephen, A.J., Enhanced photocatalytic activity of ZnO/CuO nanocomposite for the degradation of textile dye on visible light illumination. *Mater. Sci. Engineering C*, 33, 1, 91–8, 2013 Jan 1.
21. Ingle, P.U., Shende, S.S., Shingote, P.R., Mishra, S.S., Sarda, V., Wasule, D.L., Rajput, V.D., Minkina, T., Rai, M., Sushkova, S., Mandzhieva, S., Chitosan nanoparticles (ChNPs): A versatile growth promoter in modern agricultural production. *Heliyon*, 23, 8, 11, e11893, 2022 Nov 23.
22. Sahariah, P. and Másson, M., Antimicrobial chitosan and chitosan derivatives: A review of the structure–activity relationship. *Biomacromolecules*, 18, 11, 3846–68, 2017 Nov 13.
23. Wang, W., Meng, Q., Li, Q., Liu, J., Zhou, M., Jin, Z., Zhao, K., Chitosan derivatives and their application in biomedicine. *Int. J. Mol. Sci.*, 21, 2, 487, 2020 Jan 12.
24. Morin-Crini, N., Lichtfouse, E., Torri, G., Crini, G., Applications of chitosan in food, pharmaceuticals, medicine, cosmetics, agriculture, textiles, pulp and

paper, biotechnology, and environmental chemistry. *Environ. Chem. Lett.*, 17, 4, 1667–92, 2019 Dec.
25. Abo Elsoud, M.M. and El Kady, E.M., Current trends in fungal biosynthesis of chitin and chitosan. *Bull. Natl. Res. Cent.*, 43, 1, 1–2, 2019 Dec.
26. Younes, I. and Rinaudo, M., Chitin and chitosan preparation from marine sources. Structure properties applications. *Mar. Drugs*, 13, 3, 1133–74, 2015 Mar 2.
27. Thandapani, G., Prasad, S., Sudha, P.N., Sukumaran, A., Size optimization and *in vitro* biocompatibility studies of chitosan nanoparticles. *Int. J. Biol. Macromol.*, 104, 1794–806, 2017 Nov 1.
28. Yanat, M. and Schroën, K., Preparation methods and applications of chitosan nanoparticles; with an outlook toward reinforcement of biodegradable packaging. *React. Funct. Polym.*, 161, 104849, 2021 Apr 1.
29. Aljebory, A.M. and Alsalman, T.M., Chitosan nanoparticles. *Imp. J. Interdiscip. Res.*, 3, 233–42, 2017.
30. Mansi, Gulati, S., Amar, A., Strategies for Synthesis and Chemical Modifications of Chitosan-Based Nanocomposites: A Versatile Material with Extraordinary Potential for Diverse Applications, in: *Chitosan-Based Nanocomposite Materials: Fabrication, Characterization and Biomedical Applications*, pp. 53–78, Springer Nature Singapore, Singapore, 2022 Oct 2.
31. Fan, W., Yan, W., Xu, Z., Ni, H., Formation mechanism of monodisperse, low molecular weight chitosan nanoparticles by ionic gelation technique. *Colloids Surf., B*, 90, 21–7, 2012 Feb 1.
32. Dantas MJL, F., dos Santos, B.F., A. Tavares, A., Maciel, M.A., Lucena, B.D., L. Fook, M.V., de L. Silva, S.M., The impact of the ionic cross-linking mode on the physical and *in vitro* dexamethasone release properties of chitosan/hydroxyapatite beads. *Molecules*, 24, 24, 4510, 2019 Dec 10.
33. Meena, M., Zehra, A., Swapnil, P., Harish, Marwal, A., Yadav, G., Sonigra, P., Endophytic nanotechnology: an approach to study scope and potential applications. *Front. Chem.*, 9, 613343, 2021 May 25.
34. Dutta, P., Kumari, A., Mahanta, M., *Chitosan Nanoparticle: Synthesis, Characterization, and Use as Plant Health Materials.* IntechOpen, London, 2022.
35. Patra, J.K. and Baek, K.H., Green nanobiotechnology: factors affecting synthesis and characterization techniques. *J. Nanomater.*, 2014, 219–219, 2015.
36. Ashokraja, C., Sakar, M., Balakumar, S., A perspective on the hemolytic activity of chemical and green-synthesized silver and silver oxide nanoparticles. *Mater. Res. Express*, 4, 10, 105406, 2017 Oct 24.
37. Kahdestani, S.A., Shahriari, M.H., Abdouss, M., Synthesis and characterization of chitosan nanoparticles containing teicoplanin using sol–gel. *Polym. Bull.*, 78, 1133–48, 2021 Feb.
38. Sandhu, R., Singh, N., Dhankhar, J., Kama, G., Sharma, R., Dynamic light scattering (DLS) technique, principle, theoretical considerations and applications. *Nanotechnol. Biochem. Tech. Assess. Qual. Saf. Milk Milk Prod.*, 8, 135–7, 2018.

39. Karmakar, S.A., Particle size distribution and zeta potential based on dynamic light scattering: Techniques to characterize stability and surface charge distribution of charged colloids. *Recent Trends Mater. Phys. Chem.*, 117–59, 2019.
40. Behera, S.K., Saha, D., Gadige, P., Bandyopadhyay, R., Effects of polydispersity on the glass transition dynamics of aqueous suspensions of soft spherical colloidal particles. *Phys. Rev. Mater.*, 1, 5, 055603, 2017 Oct 30.
41. Malatesta, M., Transmission electron microscopy as a powerful tool to investigate the interaction of nanoparticles with subcellular structures. *Int. J. Mol. Sci.*, 22, 23, 12789, 2021 Nov 26.
42. Mostafa, M., Jumah, M.N., Othman, S.I., Alruhaimi, R.S., Salama, Y.F., Allam, A.A., Abukhadra, M.R., Effective removal of different species of organophosphorus pesticides (acephate, omthosate, and methyl parathion) using chitosan/Zeolite-A as multifunctional adsorbent. *Environ. Technol. Innov.*, 24, 101875, 2021 Nov 1.
43. Venkatachalapathy, R., Packirisamy, A.S., Ramachandran, A.C., Udhyasooriyan, L.P., Peter, M.J., Senthilnathan, K., Basheer, V.A., Muthusamy, S., Assessing the effect of chitosan on pesticide removal in grape juice during clarification by gas chromatography with tandem mass spectrometry. *Process Biochem.*, 94, 305–12, 2020 Jul 1.
44. Ghimici, L. and Dinu, I.A., Removal of some commercial pesticides from aqueous dispersions using as flocculant a thymine-containing chitosan derivative. *Sep. Purif. Technol.*, 209, 698–706, 2019 Jan 31.
45. Singh, R.P., Handa, R., Manchanda, G., Nanoparticles in sustainable agriculture: An emerging opportunity. *J. Control. Release*, 329, 1234–48, 2021 Jan 10.
46. Ponnuchamy, M., Kapoor, A., Senthil Kumar, P., Vo, D.V., Balakrishnan, A., Mariam Jacob, M., Sivaraman, P., Sustainable adsorbents for the removal of pesticides from water: a review. *Environ. Chem. Lett.*, 19, 2425–63, 2021 Jun.
47. Singh, Y. and Saxena, M.K., Insights into the recent advances in nano-bioremediation of pesticides from the contaminated soil. *Front. Microbiol.*, 13, 982611, 2022 Oct 19.
48. Attallah, O.A., Wafa, M.M., Al-Ghobashy, M.A., Nebsen, M., Monir, H.H., Adsorptive removal of pesticides from aqueous solutions using chitosan/gelatin polymeric composite: Process monitoring and optimization. *Int. J. Environ. Sci. Technol.*, 19, 1–2, 2021.
49. Anaya-Esparza, L.M., Ruvalcaba-Gómez, J.M., Maytorena-Verdugo, C.I., González-Silva, N., Romero-Toledo, R., Aguilera-Aguirre, S., Pérez-Larios, A., Montalvo-González, E., Chitosan-TiO2: A versatile hybrid composite. *Materials*, 13, 4, 811, 2020 Feb 11.
50. Majnis, M.F., Yee, O.C., Adnan, M.A., Hamid, M.R., Shaari, K.Z., Julkapli, N.M., Photoactive of Chitosan-ZrO2/TiO2 thin film in catalytic degradation of malachite green dyes by solar light. *Opt. Mater.*, 124, 111967, 2022 Feb 1.

51. Saleviter, S., Fen, Y.W., Omar, N.A., Daniyal, W.M., Abdullah, J., Zaid, M.H., Structural and optical studies of cadmium sulfide quantum dot-graphene oxide-chitosan nanocomposite thin film as a novel SPR spectroscopy active layer. *J. Nanomater.*, 2018, 1–8, 2018 Dec 2.
52. Upadhyay, R.K., Soin, N., Bhattacharya, G., Saha, S., Barman, A., Roy, S.S., Grape extract assisted green synthesis of reduced graphene oxide for water treatment application. *Mater. Lett.*, 160, 355–8, 2015 Dec 1.
53. Amanulla, B., Palanisamy, S., Chen, S.M., Chiu, T.W., Velusamy, V., Hall, J.M., Chen, T.W., Ramaraj, S.K., Selective colorimetric detection of nitrite in water using chitosan stabilized gold nanoparticles decorated reduced graphene oxide. *Sci. Rep.*, 7, 1, 14182, 2017 Oct 27.
54. Hashem, A.H., Shehabeldine, A.M., Ali, O.M., Salem, S.S., Synthesis of chitosan-based gold nanoparticles: Antimicrobial and wound-healing activities. *Polymers*, 14, 11, 2293, 2022 Jun 5.
55. Bruckmann, F.S., Schnorr, C., Oviedo, L.R., Knani, S., Silva, L.F., Silva, W.L., Dotto, G.L., Bohn Rhoden, C.R., Adsorption and photocatalytic degradation of pesticides into nanocomposites: a review. *Molecules*, 27, 19, 6261, 2022 Sep 23.
56. Ahmed, M.A. and Mohamed, A.A., The use of chitosan-based composites for environmental remediation: A review. *Int. J. Biol. Macromol.*, 17, 124787, 2023 May.
57. Rather, M.A., Bhuyan, S., Chowdhury, R., Sarma, R., Roy, S., Neog, P.R., Nanoremediation strategies to address environmental problems. Science of The Total Environment. 2023 May 10:163998. Pandya IY. Pesticides and their applications in agriculture. *Asian J. Appl. Sci. Technol.*, 2, 2, 894–900, 2018 Apr.
58. Marques, C., Som, C., Schmutz, M., Borges, O., Borchard, G., How the lack of chitosan characterization precludes implementation of the safe-by-design concept. *Front. Bioeng. Biotechnol.*, 8, 165, 2020 Mar 10.
59. Benettayeb, A., Seihoub, F.Z., Pal, P., Ghosh, S., Usman, M., Chia, C.H., Usman, M., Sillanpää, M., Chitosan Nanoparticles as Potential Nano-Sorbent for Removal of Toxic Environmental Pollutants. *Nanomaterials*, 13, 3, 447, 2023 Jan 21.
60. El-Naggar, N.E., Shiha, A.M., Mahrous, H., Mohammed, A.A., Green synthesis of chitosan nanoparticles, optimization, characterization and antibacterial efficacy against multi drug resistant biofilm-forming Acinetobacter baumannii. *Sci. Rep.*, 12, 1, 19869, 2022 Nov 18.
61. Warsito, M.F. and Agustiani, F., A review on factors affecting chitosan nanoparticles formation, in: *IOP Conference Series: Materials Science and Engineering*, vol. 1011, IOP Publishing, p. 012027, 2021.

62. Issahaku, I., Tetteh, I.K., Tetteh, A.Y., Chitosan and chitosan derivatives: Recent advancements in production and applications in environmental remediation. *Environ. Adv.*, 6, 100351, 2023 Feb.

Index

Also of Interest

Handbook of Agricultural Biotechnology

Volume I Nanopesticides
Edited by Charles Oluwaseun Adetunji and Julius Kola Oloke
Published 2024. ISBN 978-1-119-83614-8

Volume II Nanobioherbicides
Edited by Charles Oluwaseun Adetunji and Julius Kola Oloke
Published 2024. ISBN 978-1-119-83615-5

Volume III Nanofungicides
Edited by Charles Oluwaseun Adetunji and Julius Kola Oloke
Published 2024. ISBN 978-1-119-83616-2

Volume IV Nanoinsecticides
Edited by Charles Oluwaseun Adetunji and Julius Kola Oloke
Published 2024. ISBN 978-1-119-83617-9

Volume V Nanobiofertilizers
Edited by Charles Oluwaseun Adetunji, Chukwuebuka Egbuna, Anton Ficai and Oluwatosin Ademola Ijabadeniyi
Published 2024. ISBN 978-1-394-21149-4

Next-Generation Algae, Volume I Applications in Agriculture, Food and Environment
Edited by Charles Oluwaseun Adetunji, Julius Kola Oloke, Naveen Dwivedi, Sabeela Beevi Ummalyma, Shubha Dwivedi, Juliana Bunmi Adetunji and Daniel Ingo Hefft
Published 2023. ISBN 978-1-119-85727-3

Next-Generation Algae, Volume II Applications in Medicine and the Pharmaceutical Industry
Charles Oluwaseun Adetunji, Julius Kola Oloke, Naveen Dwivedi, Sabeela Beevi Ummalyma, Shubha Dwivedi, Juliana Bunmi Adetunji and Daniel Ingo Hefft
Published 2023. ISBN 978-1-119-85728-0

www.scrivenerpublishing.com